D0935516

THE HISTORY OF BRITISH MAMMALS

The History of British Mammals

Derek Yalden

Illustrated by Priscilla Barrett and the author

First published in 1999 by T & A D Poyser Ltd
24–28 Oval Road, London NW1 7DX

This book is printed on acid-free paper

Typeset in 10pt Bembo by Phoenix Photosetting, Chatham, Kent
Printed and bound in Great Britain by University Press, Cambridge

A CIP record for this book is available
from the British Library

ISBN 0-85661-110-7

Contents

Introduction

THE history of mammals in Britain begins at the end of the Trias, about 192 million years ago, which is as early as their history begins anywhere in the world. However, the history of the mammal fauna that we have today begins only with the retreat of the last Ice Age, about 15 000 years ago, and it is primarily the story of the current fauna that I wish to tell. It falls somewhere between the interests of palaeontologists, archaeologists and zoologists. As a result, it is a story not well covered by any of them. Palaeontologists are concerned mostly with fossils millions, rather than thousands, of years old, and lose interest towards the end of the Pleistocene. Archaeologists are mostly concerned with human history, and pay attention to mammals mostly as a source of food for humans; the domestic mammals in particular are important, but wild species tend to be of only marginal interest. Zoologists have documented the changes in the fauna of the last two hundred years or so, but know less of earlier times. My own knowledge in all three areas is sketchy, and I suffer from being a 'jack of all trades' as much as a 'master of none', but I have had some good teachers and friends. I hope I can tell a good enough story to satisfy them.

As an undergraduate at University College, London, I received a good grounding in vertebrate palaeontology from Kenneth Kermack and Pamela Robinson. I also attended Royal Holloway College, where Percy Butler extended that interest. Most influentially, John Clevedon Brown contributed lectures there on the history of the British mammal fauna which initiated my abiding interest in this subject. At about that time, Gordon Corbet was publishing the results of his deliberations on the subject of small mammals reaching small islands, papers which I found, as an undergraduate, and read with considerable relief and pleasure; the tortured theorizing of the glacial relict hypothesis was suddenly unnecessary.

Subsequently, many friends in the Mammal Society and elsewhere have contributed bits of this story. I wish the late Don Bramwell could read this, for he certainly inspired some of it. My long-time friend Pat Morris has also kept me interested when I might have flagged. Most important of all, my wife Pat has been a zoological widow for most of my career, but has supported my work selflessly. I could not have studied so much, written so much, worked so much, without her tolerance. My parents also supported me staunchly during my student years, and I fervently wish my father had lived long enough to see this book.

In compiling the archaeological and place-name evidence from the literature, a number of undergraduate students have helped, willingly or unwillingly, by carrying out their own third year research projects on sections of the story. They have encouraged me, and I hope it has been mutual. They have included: Peter Langley (decline of carnivores), Charlotte Aybes (Wolf and Beaver place-names), Stephen Findlay-Wilson (archaeological Bears and Beavers), Sara Sattar (place-names in Scotland),

Sarah Beswick (deer place-names), Wayne Mason (place-names for domestic ungulates), Sneha Maroo and Nick Hand (Mesolithic faunas), Matthew Morgan (archaeological small mammals), Claire Marriage (carnivore place-names) and Will Holt (archaeological deer).

A number of archaeologists have also helped by answering detailed queries, sending reprints or letting me know of sites or records that I might have missed. Roger Jacobi shared interests in Late Palaeolithic and Mesolithic faunas, especially Pikas; Sue Stallibrass passed on information especially about small mammal faunas, and Terry O'Connor not only passed on similar data but very kindly copied large numbers of, often obscure, bone reports for me. Ros Cleal, C.A.I. French, Nancy Grace, Sheila Hamilton-Dyer, Colin Howes, Tony Legge, John Martin, Finbar McCormick, Jacqui Mulville, Graham Proudlove, Peta Sadler and Peter Woodward also told me of recent records. Needless to say, it is not their fault if I have missed some records; without their help I would have missed far more.

Dr Juliet Clutton-Brock and Professor James Fairley read whole chapters for me, Andy Currant commented critically on the tables of Pleistocene records, and Dr John Clevedon Brown read the whole of the first draft. I am very grateful for being spared a number of glaring errors. Others brought me up to date with recent information on recent mammal research. Roy Dennis told me of the fascinating studies of Wolves and Bears in the Carpathians, Ian Montgomery and Jon Russ told me about the latest bat records in Ireland. Professor Brian Huntley, Dr Henry Gree, Dr Ailsa Hall, Steve Etches and Dr Andrew Kitchener helped with their specialities.

Many authors and publishers have allowed the use of copyright figures and in some cases updated them for me: Henry Arnold, Dr K.D. Bennett, Blackwell Scientific Publications, Game Conservancy, Dr Maurice Gosling, H.M.S.O., Colin Howes, Linnean Society of London, Dr Pat Morris, Oxford University Committee for Archaeology, Dr Oliver Rackham, Dr Robert Scaife, Dr Stephen Tapper and John Wiley & Sons. Dr A.J. Morton's DMAP programme was used to create the maps, for which I am very grateful. Alexandra Erskine at the *Daily Telegraph* library kindly extracted cuttings on "Hercules" for me.

CHAPTER 1

AND THE LAND WAS COVERED IN ICE

Hippopotamus defending calf from Spotted Hyaenas

DATING THE PAST

IT is my intention to describe how the British Isles obtained the mammal fauna they have at the end of the 20th century, and only the last 15 000 years or so are relevant to this story. However, the history of mammals in these islands goes back much further, back indeed to the very beginning of mammalian history, and to the very beginning of scientific interest in this subject.

In order to discuss any history, an appropriate time scale is necessary. Unfortunately, at least three different time scales have to be applied to mammalian history. The time scale used by geologists to describe the rocks and the fossils they contain provides the long view (Fig. 1.1). Thanks to the film *Jurassic Park*, one part of this nomenclature is now well known; mammalian history in Britain began at the very end of the Triassic or at the beginning of the succeeding Jurassic, about 194 million years ago, and continued sporadically through the Jurassic and Cretaceous periods, the ages of the dinosaurs. After the dinosaurs had become extinct, at the end of the Cretaceous about 65 million years ago, the mammals became much more diverse. They are represented in Britain only in the Eocene and early Oligocene, though in the United States and elsewhere they have a much more continuous history through the Tertiary. In Britain the story restarts at the end of the Pliocene, and continues through the Pleistocene to the present. The second nomenclature for time comes from the archaeologists. They recognize a series of cultural periods, the Palaeolithic, Mesolithic, Neolithic, Bronze and Iron

GEOLOGICAL TIME PERIODS	AGE
PLEISTOCENE	2
PLIOCENE	5
MIOCENE	23
OLIGOCENE	38
EOCENE	54
PALAEOCENE	65
CRETACEOUS	135
JURASSIC	194
TRIASSIC	235

PERIOD	OXYGEN ISOTOPE STAGE	SITES
FLANDRIAN	1	Star Carr, Thatcham, to recent
LATE DEVENSIAN	2	Ballybetagh, Robin Hood's Cave
MIDDLE DEVENSIAN	3	Upton Warren, Beckford, Isleworth
EARLY DEVENSIAN	4	Wretton, Chelford
IPSWICHIAN	5	Joint Mitnor, Trafalgar Square, Kirkdale Cave
WOLSTONIAN	6	
PRE-IPSWICHIAN	7	Ilford, Aveley, Selsey, Brundon
?	8	
PRE-IPSWICHIAN	9	Gray's Thurrock, Purfleet, Cudmore Grove
?	10	
HOXNIAN	11	Hoxne, Clacton, Swanscombe
ANGLIAN	12	
?	13	Boxgrove, Westbury-sub-Mendip
CROMERIAN	?	West Runton

Figure 1.1 Geological time-scales of relevance to the history of mammals in Britain.

Ages, typified by particular sets of stone or metal tools, which grade into the historical periods, Roman, Anglo-Saxon, Norman, Medieval, Tudor, Stuart, etc. This time scale is directly relevant to the mammalian history of these islands. Then there is calendar dating, in absolute years, which is readily applicable to the historical period, but which is also applied, with increasing imprecision as one goes backwards, to archaeological and geological time. Roughly, geologists think in millions of years, archaeologists in hundreds of years, and historians hope to work in absolute years. For mammalian history, we have to slide from one time scale to another, and remember that the precision with which we date occurrences varies accordingly. A possible fourth time scale is also relevant; this is the time scale devised by pollen analysts describing the changes in the vegetation over the period of archaeological time. Pollen is remarkably resilient in many circumstances, particularly when it falls in water or on wet ground; because tree pollen (in particular) is produced so prolifically, it can provide a continuous record in the right places of biological change over about 12 000 years. As a practical time scale for discussing the recent history of the mammal fauna of Britain, this is invaluable.

Both pollen zones in peat and mammal bones themselves can be dated absolutely by radiocarbon dating, and this has revolutionized our understanding of events over the last 30 000 years or so. As plants photosynthesize, they incorporate carbon, taken in as carbon dioxide, into their tissues, and animals which eat them similarly incorporate that carbon. However carbon exists in three forms, chemically identical but separable physically because their molecules are different weights. The commonest, about 99% of atmospheric carbon, is ^{12}C; this is a stable isotope, like ^{13}C which con-

14C DATES (years b.p.)	POLLEN ZONES	ZONE NAMES	ARCHAEOLOGICAL PERIODS	MAMMALS
1000	VIII	SUB-ATLANTIC	NORMAN ANGLO-SAXON ROMAN	*Oryctolagus, Dama*
2000				*Rattus*
3000	VIIB	SUB-BOREAL	IRON AGE	*Mus*
4000			BRONZE AGE	
5000			NEOLITHIC	*Ovis, Capra*
6000	VIIA	ATLANTIC	MESOLITHIC	
7000				
8000	VI	BOREAL		
9000	V			
10000	IV	PRE-BOREAL		*Alces, Bos, Cervus*
11000	III	YOUNGER DRYAS	LATE PALAEOLITHIC	*Ochotona*
12000	II	WINDERMERE		*Megaloceros*
13000	I	OLDER DRYAS		
14000				

tributes about 1%. The rarest is ^{14}C, contributing only $1\times10^{-10}\%$, but this is a radioactive isotope with a half-life of 5568 years (that is, half of the amount present at any time has decayed after 5568 years). Thus if the tiny amount in an archaeological specimen can be estimated, and compared with the amount that should have been present in fresh plant material, it gives an age for the specimen. This gives years that are always quoted as 'years b.p.', before present (and present is conventionally taken as 1952). However, when the technique is used to date the rings in ancient trees, which correspond to actual calendar years and can be counted back, it turns out that ^{14}C dates are increasingly rather younger than actual dates as one goes back in time. Back to about 2000 b.p. there is no great discrepancy, but by 10 000 radio-carbon years ago, there is a discrepancy of about 1000 years, with the radio-carbon years being too young. When it is possible to correct the radio-carbon dates to absolute years (so-called calibrated dates), they are quoted as years B.P. (or indeed as years B.C. and A.D.). For the most part, years b.p. are used in this account, because calibrated dates have not been available for the earlier part (Late and early Post Glacial) of this story, the part that is of most interest for the history of our mammal fauna.

Mesozoic Mammals

Just north of Oxford lies the village of Stonesfield, where in a now abandoned and buried quarry, an 'ancient stonemason' collected in 1812 or 1814 two small jaws that were taken to Oxford University (Simpson 1928). One is still there, and among the most famous fossils that the university museum possesses, while the other is in the Natural History Museum in South Kensington (Fig. 1.2). Although recognized as mammalian jaws, they were not announced to the world until 1824 and not formally described, as *Didelphis prevostii* and *Didelphys bucklandi,* until 1832 and 1828 respectively. The reason for the caution was that these bones came from Lower Jurassic slates, well back in the Mesozoic, the age of reptiles, and long before mammals were supposed to have existed. The description of these two first specimens was therefore accompanied by much debate, first as to whether they really did come from Mesozoic deposits, and then when that point was conceded, whether they were really mammals rather than reptiles or even fish. The use of the name *Didelphis* (or *Didelphys*) signalled the belief of the original describers that these were indeed mammalian (the North American Opossum is *Didelphis virginiana*). Both their identity and their provenance was finally settled by Richard Owen in 1838, and Mesozoic mammals are now a well studied but still infuriatingly rare group. However, they belong to groups that are quite distinct from any of the present-day mammalian orders, certainly not to *Didelphis,* and the two original specimens are now in different genera, and indeed orders, being known as *Amphitherium prevostii* and *Phascolotherium bucklandi.*

The earliest mammals in a geological (rather than historical) sense are now known to come from the very end of the Trias or beginning of the Jurassic, about 194 million years ago. Again, British mammals figure largely, and play an important part in

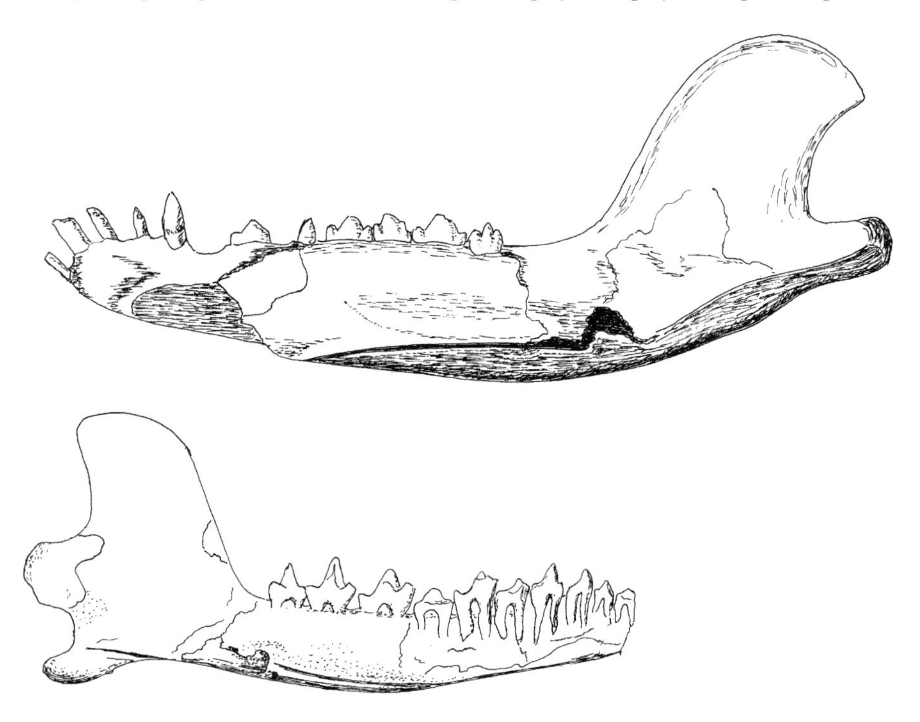

Figure 1.2 The jaws of *Amphitherium prevostii* (below) and *Phascolotherium bucklandi* (above), the first Mesozoic mammals to be described (based on photographs in Simpson 1928 and Goodrich 1894).

discussions about how and when mammals evolved. The British specimens come from fillings of reddish Triassic soil washed into fissures in the Carboniferous limestones of the Mendips and Glamorgan; at that time, these hills were dry tropical islands. The remains tend to be fragmentary, isolated teeth and broken bones, but much of the anatomy of one species, *Morganucodon watsoni*, has been established (aided by more complete specimens of close relatives from China and South Africa), and teeth of several other forms (*Eozostrodon*, *Kuhneotherium*, *Haramiya*, *Thomasia*) are also described from those early deposits. *Morganucodon* has a recognizably mammalian dentition with four incisors, a strong canine, five (upper) or four (lower) conical premolar teeth and four multicusped molar teeth in each jaw (Kermack *et al.* 1973, 1981). The lower teeth are set in a jaw that looks mammalian, with an articular condyle at the back and both upper (coronoid) and lower (angular) processes (Fig. 1.3). Similarly the upper teeth are carried by a skull that has a single bony nostril at the front (not two nostrils like birds or reptiles), a secondary palate that separates the airway from the mouth, and two condyles at the rear where the skull articulates with the first vertebra (not a single condyle as have birds and reptiles). The post-cranial skeleton also has the look of a mammal about it; the rib cage is short, so that a chest can be distinguished from a trunk (in reptiles, the ribs extend all the way down the body), and its limb bones are slender with well-developed articular surfaces. However, it also had some features that are decidedly primitive by the standards of most modern mammals. For

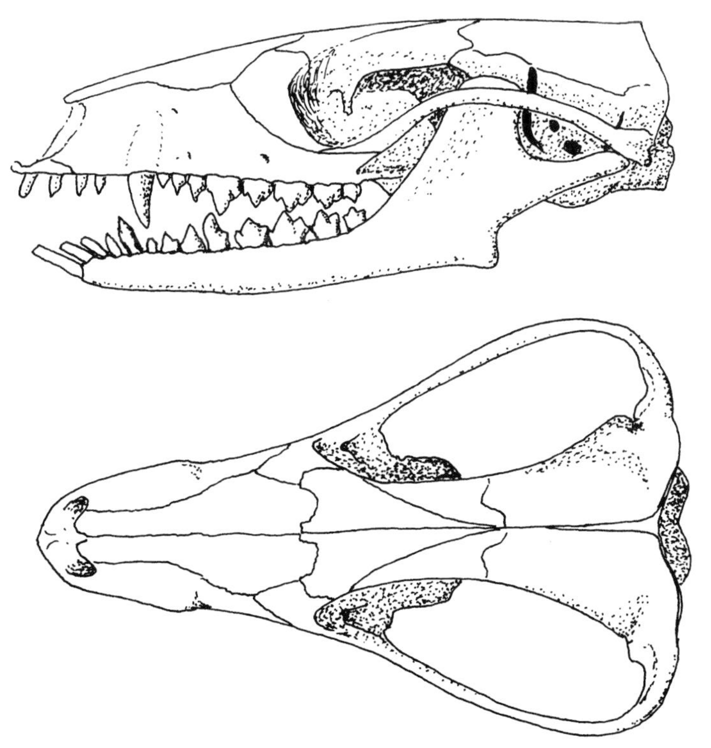

Figure 1.3 The skull of *Morganucodon watsoni*, the earliest British mammal, from the Triassic/Jurassic boundary fissure-fillings of South Wales (after Kermack *et al.* 1973, 1981).

Morganucodon

instance, it did not have the chain of three ear ossicles by which modern mammals conduct sound from the ear drum to the inner ear; the equivalent bones are all there, but two of them, the articular/malleus and quadrate/incus, formed a jaw joint, as they do in reptiles and birds (they are called the articular and quadrate in reptiles or birds, but the malleus and incus in mammals). As a consequence, *Morganucodon* must have had its ear drum tucked in behind its lower jaw rather than on the side of its head, and it probably did not have a pinna (external ear flap) like modern mammals. Its shoulder girdle and hip bones were more like those of the Australian monotremes (the echidna and platypus) in their anatomy, and like them it probably had a rather broad stance, with its elbows and knees sticking out sideways rather than tucked in underneath its body. *Morganucodon* was small, as indeed were all the Mesozoic mammals; with a skull about 26 mm long, it was slightly larger than a Water Shrew (Kermack *et al.* 1981). While it was nothing like as specialized as modern shrews in its anatomy, it probably did feed on insects and arachnids, may well have been nocturnal (because its skull has a good snout but smallish orbits) and possibly scampered around in bushes as well as on the ground.

The next glimpse of Mesozoic mammals in Britain is the Stonesfield slate fauna already mentioned. Eventually four jaws of *Amphitherium prevostii,* three jaws of *Phascolotherium bucklandi* and three jaws of a third species, *Amphilestes broderipii,* were recovered from the mines before they closed in 1909. They remain the most important glimpse of mammals of this age anywhere in the world. Only a few isolated teeth are known from a few sites elsewhere (in Oxford and on Skye) of comparable age (Simpson 1928, Savage 1989).

A further glimpse of Mesozoic mammals in Britain comes from the end of the Jurassic, about 135 million years ago. The sea cliffs of Durlston Bay, in the Isle of Purbeck just south of Swanage in Dorset, have yielded a comparatively rich fauna, with about 18 species, referred to five orders (Fig. 1.4). Mostly the specimens are only jaws,

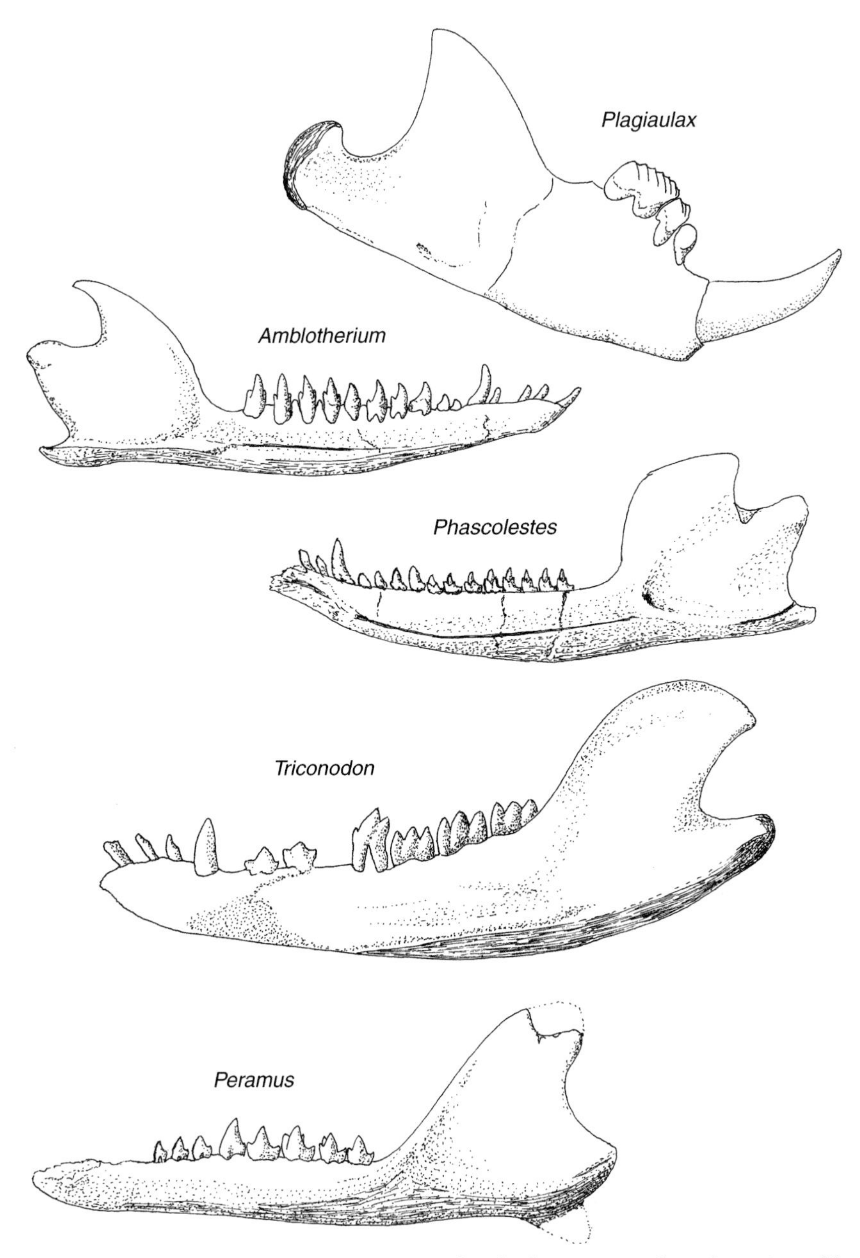

Figure 1.4 Mammals from the Upper Jurassic of Purbeck, Dorset, to show the variety of forms then living in Britain (after Simpson 1928). *Plagiaulax* was a seed-eating multituberculate, the others were insectivores or small carnivores.

like the Stonesfield mammals, but there are parts of the skulls of two forms. The mammals are still only small, shrew-sized, animals, and four of the orders were insectivores or small carnivores. The fifth group, the Multituberculata, is of particular interest because it comprises small rodent-like mammals that were clearly herbivorous, probably seed- or fruit-eaters and the first such mammals to evolve. They had a pair of strong incisors at the front of the jaws, then a small gap before three or four cutting premolar teeth, like those seen in certain small kangaroos. Behind them were two molars, each with two rows of three cusps. These teeth obviously chewed with a backwards-and-forwards action (rather than a side-to-side action like ourselves), a specialized action also seen in many rodents. However, true rodents do not appear in the fossil record until 75 million years later, after which they appear to displace the multituberculates and the latter became extinct. The beds which contain the Purbeck mammals appear to be high up near the top of the cliffs, and the mammals seem to have come from a small pocket of fossil-rich deposits which has been worked out. Blocks that have fallen onto the shore below after winter storms might yet yield new specimens, but none had been found since the original collections in the 1850s and 1860s, despite several serious attempts. Then, during 1986, excavations in a quarry about 5 km inland from Durlston Bay uncovered a remarkable trackway of a sauropod dinosaur in the limestone, and the underlying clay was investigated over the next three years for smaller vertebrates. A rich collection of salamanders, frogs, lizards, dinosaurs, crocodiles and many more mammals has been obtained, though these are mostly isolated teeth and small bones (Ensom 1987, Kielan-Jaworowska and Ensom 1992). Most have not yet been described, but they seem to be mostly the species already known from the earlier collections; two new species of multituberculate have been described already, however, and other novelties may yet appear. However, faunas of comparable age, and with very similar species, are now known elsewhere in the world, notably from Portugal and from the famous dinosaur locality of Como Bluff in Wyoming, USA, so Upper Jurassic mammals are considered to be relatively well known.

For the whole of the succeeding Cretaceous period, there are few mammal specimens from Britain, and those few are only isolated teeth. These come from the deposits in the Weald of south-east England, where the famous first teeth of the dinosaur *Iguanodon* were discovered. The situation is little better in the rest of the world until near the end of the Cretaceous, when, in both Mongolia and the United States, the last survivors of the Mesozoic groups occur along with rather more modern mammals. These and the earlier Mesozoic mammals are fully reviewed by Lillegraven *et al.* (1979), who describe their anatomy, geological and geographical distribution and evolutionary significance; a short review of the British faunas is given by Savage (1989).

TERTIARY MAMMALS

When the dinosaurs became extinct at the end of the Cretaceous, about 65 million years ago, the mammals which survived were still small, shrew- to marmot-sized, but they included both marsupials and placentals, the two major groups of mammals in the present fauna. Both in North America and in Europe, there were animals very like the present-day opossums, but most of the mammals were not readily recognizable as examples of the modern families. Unfortunately, this transitional period is not represented in Britain; the earliest faunas we have are confined to the very end of the Palaeocene and early Eocene about 54 to 47 million years ago (Ma) and to the later Eocene and early Oligocene, about 42 to 35 Ma (Hooker 1989).

In the London Clays of the Thames Basin, which were laid down under fresh water, there is an important angiosperm (flowering plant) flora, and some scant remains of the mammals that fed on them (Collinson and Hooker 1987). These include the famous early horse *Hyracotherium*, formerly known as *Eohippus* ('dawn horse'), an animal the size of a fox terrier with five front and four back toes on each foot rather than the single effective toe of modern horses. It was a browser, with low-crowned cheek teeth, unlike its modern descendants which are grazers with high-crowned teeth. Hooker (1980) discusses the sequence of species present in Britain; a smaller species, *H. cuniculus*, in the earlier deposits is replaced by the larger *H. vulpiceps* and *H. leporinum* in later levels. Other members of the fauna include the last record (*Charlesmooria*) of the multituberculates in Britain, surviving from the Mesozoic, early marsupials (*Didelphodus*), primitive primates (*Phenacolemur, Teilardina, Pelycodus*), rodents (*Paramys, Microparamys*), perhaps displacing the multituberculates, as well as primitive ungulates (*Hyopsodus, Coryphodon*) and carnivores (*Oxyaena*). The faunas are dominated by smaller mammals, particularly the seed- and fruit-eating rodents. Fruits of the palm *Nipa* and other waterside and aquatic vegetation indicate that appropriate foods were readily available, and suggest a tropical environment like present-day Malaya (Collinson and Hooker 1987).

The later Eocene to Oligocene faunas are located mostly in the Hampshire Basin, particularly on the Isle of Wight at both its western end, on Headon Down, and at the eastern end at Whitecliff Bay near Bembridge. These deposits cover some 7 million years, and include over 100 species of mammals, belonging to nine orders. Didelphid marsupials are still present, though they died out in Europe in the Miocene so survive now only in North and South America. Among the rodents, early dormice (Gliridae) as well as extinct families of early rodents (Paramyidae, Pseudosciuridae, Theridomyidae) are present in the Eocene levels, and the earliest hamsters (Cricetidae) and beavers (Castoridae) appear in the Oligocene. Among the insectivores, both primitive hedgehogs (Erinaceidae) and moles (Talpidae) are present. Bats are sparsely represented – they are always scarce fossils – but a few teeth indicate the

Hyracotherium

presence of both horseshoe (Rhinolophidae) and primitive (Archaeonycteridae) bats in the Eocene, and a vespertilionid *Stehlinia* related to the woolly bats *Kerivoula* of Africa and Asia appears in the Oligocene. Primates, essentially a tropical group today, are present in the form of lemur-like members of the families Omomyidae and Adapidae, as well as the rather rodent- or cobego-like Plesiadapidae. The carnivores are scarce, and not closely related to any of the modern forms, but the ungulates are well represented. The Artiodactyla include early relatives of the ruminants (Amphimerycidae) and camels (Xiphodontidae) as well as numerous primitive forms in the Eocene: in the Oligocene, distant relatives of the hippopotamuses (Anthracotheriidae) and pigs (Entelodontidae) appear. The Perissodactyla include a distinctive European family the Palaeotheriidae, related to the horses (true horses, Equidae, were evolving in North America at this time), as well as a primitive rhinoceros *Hyrachyus* and the tapir-like Lophiodontidae (Hooker 1989).

As a whole, these later Eocene and especially the Oligocene mammal faunas have a distinctly more modern look to them than the early Eocene faunas from the London Clay. They change during the 7 million years, due both to evolutionary changes and to immigration of new forms. There is a major break at the end of the Eocene, when five of the 13 existing species become extinct and 13 new immigrants appear. At any one level, these faunas contain between 13 and 45 species (Hooker 1989), the latter being reasonably comparable with the number of mammals found in Britain now (especially when so many of the modern fauna are bats, which are so under-represented in the fossil community). At that time, however, Europe had a tropical climate, fauna and flora, though somewhat cooler than in the earlier Eocene; the plant and other fossils at Headon Hill suggest an open lagoon, with the Water Soldier *Stratiotes* and sedges (Hooker *et al.* 1995), yet the mammals imply that there were forests nearby. Large browsing species were more numerous than in the Lower Eocene, particularly the tapir-like *Palaeotherium*. The habitat was clearly rather more open than the forests of the Lower Eocene, and Hooker (1994) suggests an environment analogous to the Florida Everglades.

PLEISTOCENE MAMMALS

For most of the Tertiary, there are no appropriate rocks and therefore no mammal faunas from Britain, though the story continues elsewhere, especially in North America but also in Europe and Africa. So, when the record restarts in the Pleistocene, the species which appear in Britain are well known from elsewhere, and their source in both a geographical and evolutionary sense is clear.

The Pleistocene is a complex period in northern latitudes, characterized by a series of severe glacial periods interrupted by warmer intervals, the interglacials, like the one in which we are living. These alternating colder and warmer periods were first recognized by geologists working in the Alps in southern Europe, who named the successive glacial periods the Günz, Mindel, Riss and Würm glaciations after river valleys in which the evidence of glacial activity was clear. The pattern was then recognized in northern Germany, where equivalent glacials were named Elster, Holstein, Saale and Weichsel, and in North America (Nebraskan, Kansan, Illinoan and Wisconsinan). In Britain, there is clear evidence of only three glaciations, the Anglian, Wolstonian and Devensian. There is no doubt that the last of these (Würm, Weichsel, Wisconsinan and Devensian) are contemporaneous, but this becomes less certain with the earlier periods, which is why the diferent names are retained. The interglacials are named Hoxnian, Ipswichian and Flandrian (= Post Glacial, = Holocene) in Britain and, being warmer periods, they have much better mammal faunas. However, there remain problems of correlation between what are usually small pockets of fossiliferous deposits, a situation exacerbated by the fact that succeeding glaciations often obliterated the evidence of earlier periods. Much the best record of what happened to temperatures, and to marine protozoans, is recorded in the deep-sea oozes, and cores of these deposits provide a sound basis on which to evaluate events on land.

The best known of these records for the later Pleistocene was compiled by Shackleton (1977), and the record has been extended back to the earlier Pleistocene by Shackleton *et al.* (1991) based on a deeper core from the eastern Pacific Ocean in the Panama Basin (Fig. 1.5). This shows the first signs of cooling as early as 2.5 Ma, and signs of cooling are taken to define the start of the Pleistocene. More precisely, the appearance of an arctic mollusc, *Artica islandica*, in the Mediterranean at about 2.4 Ma has been taken to be the conventional start (West 1977). Current convention starts the Pleistocene slightly later, at 1.8 Ma (Funnell 1995), at which time *Cytheropteron testudo*, an ostracod crustacean, appears at the type site for this period, in southern Italy (Jones and Keen 1993). The most significant feature of the deep sea temperature record is the very much more complicated pattern of change that it reveals, and the relatively short duration of either extremely cold or really warm periods. Instead of four glacial periods, at least seven or eight might be recognized; and instead of three interglacials, there might be as many as 11 (Fig. 1.5).

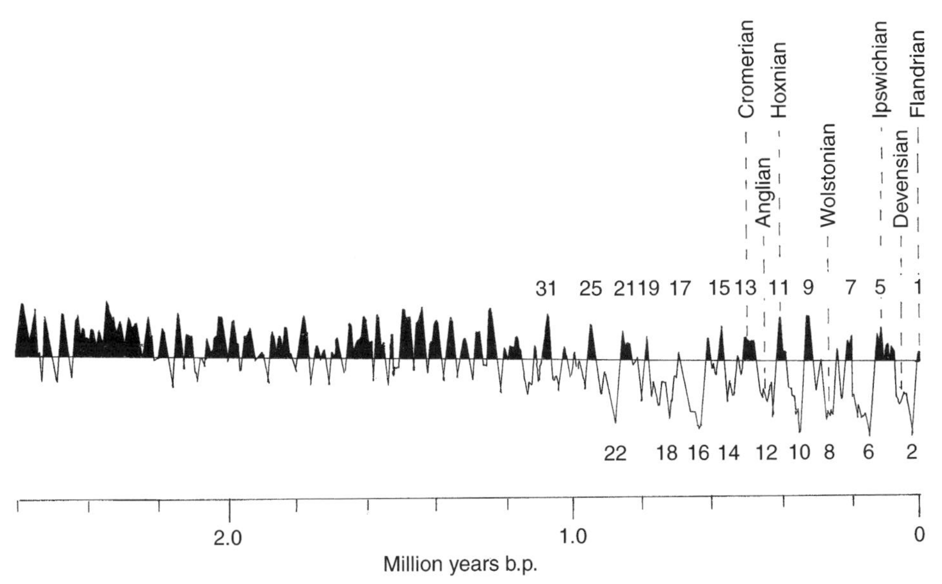

Figure 1.5 The curve of palaeotemperatures over the last 2.6 million years (Ma), derived from analysis of deep-sea cores (after Shackleton 1977, Shackleton *et al.* 1991). The warmer interludes (black) have become shorter and less marked during the Pleistocene.

The record of terrestrial pollen floras, and faunas, indicates minor periods of warming known as interstadials, not sufficient for the growth of a full forest cover (which defines an interglacial), within the glacial periods, and the deep sea record confirms this more complicated pattern. However, the difficulty of correlating particular terrestrial fossil faunas and floras with each other, and with the marine temperature curve, remains formidable, and certainly not resolved (Jones and Keen 1993, Stuart 1995). It is therefore easiest to follow the Pleistocene history of mammals in Britain by looking at a few well-known fossil faunas, whose ages relative to each other are reasonably clear. The classical divisions of the Pleistocene will be used as headings, but they imply a much more simple pattern of change than what really happened. Fortunately, Tony Stuart has devoted a great deal of research effort to studying the mammals of this period, and has summarized existing knowledge in a book (Stuart 1982) which provides the basis for this account.

Pliocene and Earliest Pleistocene

In East Anglia, thick deposits of marine clays, silts and shelly gravels, the Red Crag and Norwich Crag, include a pollen record of the vegetation of the nearby land, and occasional specimens of mammal bones as well. The forest was largely conifer, with the Hemlock *Tsuga*, now extinct in Europe but surviving in North America, pine *Pinus* and spruce *Picea* prominent. Birch, alder, elm, oak and hornbeam also contributed in the warmer phases, along with the now Asian Wing-nut *Pterocarya*. In colder periods an oceanic heath vegetation developed. The mammals from these periods include, among the rodents, beavers *Castor fiber* and *Trogontherium minus*, and several relatives of the Water Vole, *Mimomys pliocaenicus*, *M. blanci*, *M. newtoni*, *M. reidi* and *M. rex*, of different sizes. Among the ungulates, there were large deer *Eucladoceros* related to the later Pleistocene *Megaloceros*, a gazelle *Gazella anglica*, mastodons *Anancus arvernensis* and *Zygolophodon borsoni* as well as a true elephant *Archidiskodon meridionalis*, both three-toed horse *Hipparion* and true horse *Equus stenonis*, a tapir *Tapirus arvernensis* and a rhinoceros *Dicerorhinus megarhinus* related to the modern Sumatran Rhinoceros. The carnivores included a panda *Parailurus anglicus*, the large bear-like *Agriotherium*, a clawless otter *Aonyx reevei* and a hyaena *Pachycrocuta perrieri*. Only one site of this age is outside East Anglia; a cave in the Carboniferous limestone at Doveholes, near Buxton in Derbyshire, yielded a small fauna which included one of the very few British specimens of a sabre-toothed cat *Homotherium latidens*, as well as *Pachycrocuta* sp., *Anancus arvernensis*, *Archidiskodon meridionalis* and *Equus* cf. *bressanus*. This is obviously broadly contemporary with the East Anglian faunas; unfortunately, the cave was quarried away, and the site has been filled with a municipal rubbish dump.

These are faunas broadly comparable with the better known faunas from southern Europe known as the Villafranchian faunas, and dated to around 2.4 to 1.7 Ma, that is from the later Pliocene just into the Pleistocene, though they were previously regarded (see above) as earliest Pleistocene. They serve to emphasize, first, that Britain was not at that stage an island, but shared the continental fauna. Secondly, they show that the warm-climate fauna of the Pliocene had not yet been extinguished by the deteriorating climate. Thirdly, they emphasize that the species present were not much like those now found in Europe; only *Castor fiber* is a familiar species.

Cromerian

On the north coast of Norfolk, the famous Freshwater Bed exposed in the soft cliff at West Runton has long yielded mammalian fossils in some abundance, and this is the type site for the Cromerian Interglacial. The deposits represent a broad shallow channel, about 400 m across, cut through sands and gravels of a previous glacial period. They start with marls and muds at the bottom, passing upwards into riverine shelly sands and organic muds. The pollen record shows that the base corresponds to the beginning of the interglacial, with pine and birch dominant, but these were displaced by spruce and alder, then elm, oak, lime, hornbeam and hazel as a full temperate forest developed. The fauna has a much more familiar look about it, and several modern species appear in Britain for the first time. Among the insectivores were a hedgehog, perhaps *Erinaceus europaeus*, and two moles *Talpa europaea* and the smaller *T. minor*, the Russian Desman *Desmana moschata* and several shrews *Sorex runtonensis*, *S. savini* and *Neomys newtoni*, though not the modern ones. The rodents and ungulates similarly contain a mixture of familiar and extinct species; the commonest mammal is a water vole *Mimomys savini*, along with European Beaver *Castor fiber* and the extinct *Trogontherium cuvieri*, the Common Hamster *Cricetus cricetus*, Wood Mouse *Apodemus sylvaticus*, Bank Vole *Clethrionomys glareolus*, Root Vole *Microtus oeconomus* and extinct Pine Voles *Pitymys arvaloides* and *P. gregaloides*. The ungulates include Red, Roe and Fallow Deer and Wild Boar, as well as the extinct large deer *Megaloceros verticornis* and *Alces latifrons*, the rhinoceros *Dicerorhinus etruscus* and a horse. There was also an extinct bison *Bison schoetensacki*. The Spotted Hyaena *Crocuta crocuta* makes its appearance, along with Wolf, Weasel and a cave bear *Ursus deningeri*; a marten and an otter are also present, as well as two poorly diagnosed cats, one small and one much larger.

Just above this layer is an estuarine or marine gravel which contains few mammal specimens, but is notable for the occurrence of a macaque, close to or the same as the Barbary Ape *Macaca sylvana*, and a squirrel *Sciurus whitei*, as well as Wood Mouse and *Mimomys savini*. From the fauna as well as the pollen, Norfolk at this time was a wooded landscape, with a meandering broad river valley, but the large deer, horse and bison suggest that it was open woodland rather than closed forest.

Other sites are needed to complete the record for the later part of the Cromerian Interglacial. At Sugworth near Oxford, a smaller fauna included many of the same species, notably *Mimomys savini*, *Sorex savini* and *Dicerorhinus etruscus*, but added perhaps the first record of Pigmy Shrew *Sorex* cf. *minutus* and a large extinct shrew *Beremendia* and vole *Pliomys episcopalis*. At this site, believed to be coincident with the maximum forest cover in the Interglacial on the basis of its pollen record, Wood Mouse and Bank Vole were the dominant species. The latest Cromerian site, at Ostend on the Norfolk coast, contains *Arvicola cantiana*, the direct ancestor of the modern Water Vole, rather than *Mimomys savini*. *Arvicola* differs from *Mimomys* only in that its molar teeth do not develop roots, but remain growing throughout life. This is comparable to the familiar difference between *Microtus* and *Clethrionomys*, and represents an evolutionary change within the Cromerian, allowing the voles to cope with a more abrasive diet.

There are two other very important mammal sites that may belong at the end of the Cromerian Interglacial, and be contemporaneous with the Sugworth fauna, but alternatively may represent another interglacial, later than the Cromerian but earlier than the Anglian Glaciation. These are the fossil cave complex at Westbury-sub-Mendip and the site at Boxgrove in Sussex. Full accounts of these sites have not yet been published, but they are important sites for their archaeology as well as their fauna. Claims that Westbury contained early but crude stone tools were disputed, but have been fully vindicated by the discovery at the contemporaneous Boxgrove site of (a few) human bones (Roberts *et al.* 1994). The dating of these sites depends especially on their mammal faunas. As a cave site, Westbury has a fauna dominated by carnivores, with the 'European Jaguar' *Panthera gombaszoegensis* common and the European Hunting Dog *Xenocyon lycaonoides* making its only known British appearance. The Lion *Panthera leo* appears in Britain for the first time, spreading out from Africa, and Spotted Hyaena *Crocuta crocuta* is present. The smaller ancestors of the Wolf and Cave Bear, *Canis mosbachensis* and *Ursus deningeri*, are also present (Turner 1995). The small mammal fauna has been described by Bishop (1982) and analysed further by Andrews (1990). It derives from a variety of predators, owls and mustelids, which complicates further the fact that an extensive period of time, covering changes in habitat, is represented. The common Water Vole at both Boxgrove and Westbury is *Arvicola cantiana*, indicating a time later than the typical Cromerian. Other species include some definite teeth of the Field Vole *Microtus agrestis*, as well as many uncertain *M. agrestis/arvalis* specimens, Root Vole, Bank Vole and Wood Mouse. The extinct voles *Pliomys episcopalis*, *Pitymys gregaloides* and *P. arvaloides/subterraneus* are present, and the commonest species present are the *Pitymys* and *Microtus*. In the earlier part of the deposits, the dormice *Muscardinus avellanarius* and *Eliomys quercinus* are present, suggesting warm temperate woodland conditions, but towards the end the lemmings *Lemmus lemmus* and *Dicrostonyx torquatus* appear, suggesting a decline to cooler conditions. The Hedgehog *Erinaceus europaeus*, both moles *T. europaea* and *T. minor*, the Desman, and the shrews *Sorex minutus*, *S. savini* and *S.* cf. *runtonensis* are also present, along with Stoat, Weasel and Pine Marten. At Boxgrove, the small mammals include *Sorex savini* and a smaller shrew, *Lemmus lemmus*, *Clethrionomys glareolus*, *Pitymys arvaloides*, *P. gregaloides*, *Microtus arvalinus*, *M. oeconomus* and *Oryctolagus cuniculus*. Larger mammals include *Canis mosbachensis*, *Ursus* sp., Mink *Mustela lutreola*, *Meles meles*, *Megaloceros* sp., *Bos primigenius* and *Capreolus capreolus* (Currant, in Roberts 1986). This is very like the Westbury fauna, and looks as though it belongs similarly between the Cromerian Interglacial and Anglian Glacial in age.

Anglian Glaciation

Glacial periods produce a sparse fossil record, and few mammals are known from the Anglian cold stage. A ground squirrel *Spermophilus* sp. from Mundesley in Norfolk, and a small fauna including Norway Lemming *Lemmus lemmus*, the voles *Arvicola cantiana* and *Microtus gregalis*, Horse *Equus ferus* and Reindeer *Rangifer tarandus* from near Benson in Oxfordshire certainly belong to this period. There is also a good record of Red Deer from Hoxne, Suffolk, dated to the end of this glaciation.

Hoxnian Interglacial

By contrast, there are several sites of Hoxnian Age, and the faunas are much richer. Hoxne itself, the type site for this interglacial, is in Suffolk. It yields a mammal fauna including Common Shrew *Sorex araneus* for the first time. The Desman *Desmana moschata*, beavers *Castor fiber* and *Trogontherium cuvieri*, Field Vole *Microtus agrestis* and the ancestral water vole *Arvicola cantiana* were again present, along with Red, Roe and Fallow Deer, and *Macaca* appears again in Britain. The 'Irish Elk' *Megaloceros giganteus* appears in this interglacial, both at Hoxne and at Swanscombe in Kent, where a richer fauna than at Hoxne includes also Lion *Panthera leo*, Pine Marten *Martes martes*, Wild Cat *Felis silvestris*, Wild Boar *Sus scrofa*, Wolf *Canis lupus*, Cave Bear *Ursus spelaeus*, Straight-tusked Elephant *Palaeoloxodon antiquus*, and Rabbit *Oryctolagus cuniculus*, along with the famous human skull known as Swanscombe Man. Towards the end of the Interglacial, at Hoxne, the fauna indicates the cooling of the climate and opening out of the vegetation by the presence of *Lemmus lemmus*.

Wolstonian Glaciation

The type site for this glaciation in Britain, Wolston in Warwickshire, yields a large mammal fauna including Spotted Hyaena, Woolly Mammoth *Mammuthus primigenius*, Woolly Rhinoceros *Coelodonta antiquitatis*, Horse *Equus ferus*, Reindeer *Rangifer tarandus* and either *Bos* or *Bison*, which seem to comprise an appropriate steppe fauna for a glacial, but the additional report of Straight-tusked Elephant *Palaeoloxodon antiquus*, a southern species, is out of place.

At Tornewton Cave, at Torbryan in Devon, there is a complex sequence of faunas representing at least two glacial and two interglacial periods. The Glutton Stratum, the oldest layer, seems to belong in the Wolstonian. Additional to the Glutton *Gulo gulo* which gives it its name are Wolf, Lion, Badger *Meles meles*, Red Fox *Vulpes vulpes*, a clawless otter *Aonyx antiqua* and a bear. Among the ungulates are Horse and Reindeer, along with rhinoceros (*Coelodonta*?) and *Bos/Bison*. Mountain Hare *Lepus timidus* appears for perhaps the first time (but hare species are very difficult to distinguish). The small mammals include Common Hamster *Cricetus cricetus* and a smaller extinct hamster *Allocricetus bursae*, lemmings *Dicrostonyx torquatus*, *Lemmus lemmus* and *Lagurus lagurus* and Root Vole *Microtus oeconomus*. This sounds like a steppe fauna, though the presence of some woodland species (like the Badger) may indicate either an admixture of faunas, or slightly specialized conditions not quite like any modern analogue.

Probably contemporaneous deposits are present on Jersey at the archaeological cave site of La Cotte de St Brelade. Woolly Mammoths and Woolly Rhinoceroses were the main prey of the humans there, but Horses, Red Deer, Reindeer and *Bos/Bison* were also taken. Most intriguingly, a few specimens of Chamois *Rupicapra rupicapra* were also recorded, the furthest north-west, and the nearest to Britain, that they seem ever to have reached. Carnivores included Wolf, Arctic Fox and a bear, perhaps Cave Bear. The rodents were mostly Arctic Lemming *Dicrostonyx torquatus* and Narrow-headed Vole *Microtus gregalis*, but *M. arvalis* and *M. nivalis (malei)* were also reported, along with a single tooth of birch mouse *Sicista* sp., the nearest that has been recorded to Britain (Scott 1986b, Chaline and Brochet 1986).

Ipswichian Interglacial

Conventional reckoning had only one interglacial between the Hoxnian and Flandrian Interglacials, with its type site at Bobbitshole, Ipswich, Suffolk; this site yields a good pollen record but a poor fauna, though Beaver *Castor fiber* is present. Typically *Hippopotamus amphibius* is present in Britain in faunas of this age, classically at Trafalgar Square in London. However, there has been increasing unease among those studying fossil mammals about the correctness of this view, and there is now a consensus that at least two interglacials, which might be loosely termed the 'Pre-Ipswichian' and 'Ipswichian Interglacials', should be recognized (Currant 1989, Stuart 1995). (The oxgen-isotope temperature curves from deep sea sediments actually suggest three interglacials, and a few mammal faunas can be assigned to the different Oxygen Isotope Stages, but too many records are at present unassignable; the tables of mammal records give the information that is available.)

The sites that represent the earlier of these two include the brickfields around Gray's Thurrock, the lower part of the sequence at Aveley, and the Cudmore Grove Channel, all in Essex, and a site at Itteringham in Norfolk. These sites do not have *Hippopotamus*, but they do have white-toothed shrews *Crocidura* (Currant 1989). Some of the fauna from Tornewton Cave, the Otter Stratum, must also belong here (Sutcliffe 1995). Among the species present in Britain were *Macaca sylvana*, for the last time, along with Brown Bear (for the first time), Wolf, Otter, Common, Pigmy and Water Shrews, Mole, Beaver, the three common rodents (Field Vole, Bank Vole and Wood Mouse), Red Squirrel, Wild Boar, Red and Roe Deer and Aurochs. The intermediate nature of the water voles is an important feature; they are transitional between *Arvicola cantiana* and its descendant *A. terrestris*, the modern Water Vole. A less familiar look to the fauna is provided by the presence of Narrow-nosed Rhinoceros (*Dicerorhinus hemitoechus*, now *Stephanorhinus hemitoechus*) and Straight-tusked Elephant. The otter of Tornewton Cave's Otter Stratum was a clawless otter *Cyrnaonyx*, its only British record (Sutcliffe 1995).

If two interglacials are correctly recognized (and the deep-sea cores certainly show at least two warmer intervals that correspond to them), then evidence for an extra glaciation has also been confused. There is at present insufficient evidence to disentangle the records that are presumably included in the supposed Wolstonian sites already discussed.

Sites that belong to the conventional Ipswichian Interglacial include Trafalgar Square, Peckham and Brentford in London; Barrington, Cambridge; and cave sites including Victoria Cave in Yorkshire, and Joint Mitnor Cave and the Hyaena Stratum of Tornewton Cave, both in Devon. The Hippopotamus at Barrington was accompanied by Lion and Spotted Hyaena, as well as Brown Bear, Badger and Wolf. Among the other ungulates were Red and Fallow Deer as well as Irish Elk, Aurochs, extinct bison (*Bison priscus*), Straight-tusked Elephant and Narrow-nosed Rhinoceros. At Trafalgar Square, the Hippopotamus was also accompanied by Lion, Red and Fallow Deer, Irish Elk, Aurochs, and Straight-tusked Elephant. The northern-most record for Hippopotamus in Britain is apparently Victoria Cave, near Settle, where Straight-tusked Elephant and Narrow-nosed Rhinoceros are also present. Cave sites are often hard to correlate confidently with open sites, but dating of the flowstones in Victoria Cave to between 135 000 and 114 000 b.p. places it firmly in the Ipswichian (Gascoyne *et al.* 1981). Horse is a surprising absence from these middle Ipswichian faunas, for it was present at the end of the previous glacial, and the other ungulates indicate that conditions must have included some open glades, in the woodlands, that should have suited it (Stuart 1995). However, there is also a surprising absence of evidence for Human presence during this period, and Stuart (1995) speculates that Britain was an island, cut off from Europe by the high sea-levels, for at least some of this period.

Towards the end of the Ipswichian Interglacial, pollen analysis shows an early decline in the extent of deciduous woodland and an expansion in Pine, Birch and herbaceous pollens, indicating the deteriorating climate towards the impending Devensian Glaciation. At Aveley, in Essex, the skeleton of a Straight-tusked Elephant, with an appropriate Oak/Pine/Hazel woodland pollen flora for the middle of the interglacial, lay physically very close below the skeleton of a Woolly Mammoth, with a pollen flora indicating more open conditions, with Hornbeam, Pine and Oak contributing about 30% of the pollen rain, but herbs contributing 65% (West 1969). Horse was also present, as it was at similar sites at Brundon and Sotton, Suffolk. These late Ipswichian faunas include a mixture of open ground (e.g. Root Vole, Mammoth, Horse, Irish Elk) and forest (e.g. Red Deer, Straight-tusked Elephant, Aurochs) species. At the very end of the interglacial, sites such as Crayford, Kent, and Stoke Tunnel, Suffolk, produce souslik *Spermophilus*, Musk Ox *Ovibos moschatus*, Woolly Rhinoceros and lemmings, clearly indicating colder conditions.

Devensian Glaciation

If evidence of earlier glaciations has been obscured by the subsequent ones, then at least the most recent, the Devensian which began about 70 000 years ago, is clearly represented. There are numerous sites in the river-gravels of this age and a few, much better preserved and therefore more informative, lacustrine sites. There is an abundance of sites that yield evidence about the climate, from geological, pollen and other lines of evidence, but fewer that contain mammals. Many cave sites, with abundant remains of mammals, also belong to this period, but unfortunately many of them were excavated last century, when the complex history of the Pleistocene was not realized, and the techniques to attempt correlations were not available; most are poorly stratified and poorly dated. Some of these older sites are being re-examined, and radiocarbon dates on bone specimens are sometimes able to 'rescue' sites from a state of being useless to being informative (e.g. Currant 1986, on Gough's Cave; Woodman *et al.* 1997 on Irish cave faunas).

The earliest Devensian sites include Wretton and Coston in Norfolk. At both, *Bison priscus* is the dominant animal, but Wolf, Arctic Fox, (Brown?) Bear, Reindeer, Red Deer, Horse, Woolly Mammoth and Woolly Rhinoceros are also present. The deposits in which these occur include a succession of vegetation zones, from their pollen flora, indicating both three open rather marshy grassland and two more wooded phases. It is reasonable to speculate that the Red Deer, at least, was a resident during the latter. There is a well-recognized wooded phase with Birch, Pine and Spruce, the Chelford Interstadial (Chelford is in Cheshire), with a date of 60 000 b.p. or earlier, which probably corresponds to the later wooded period at Wretton. It implies a short milder episode in an otherwise fairly Arctic climate, and the beetle faunas of this period certainly attest to its severity: a temperature in July of about 10°C (Coope 1977), and severe winters.

The climate seems to have ameliorated again about 42 000 b.p., at a time represented at Upton Warren between Droitwich and Bromsgrove in Worcestershire and therefore termed the Upton Warren interstadial. The flora is an open herb-rich community, one with some interesting salt-tolerant species like *Glaux maritima*, *Triglochin maritima* and *Plantago maritima,* as well as Dwarf Willow *Salix herbacea*, *Thalictrum alpinum* and *Draba incana*. These meadows were grazed by Woolly Mammoths, Woolly Rhinoceros, Reindeer and *Bison priscus*. A very mixed beetle fauna, containing both northern and southern species as well as a few continental ones, suggests a July temperature of about 17°C to allow for the southern forms, but the open conditions and perhaps the very rapid change in climate facilitated the survival of the northern species. A number of other sites contain mammal faunas of comparable age, including Tattershall Castle in Lincolnshire (Reindeer, Bison), Isleworth, Middlesex (Bear, Bison, Reindeer, *Microtus gregalis*, *M. oeconomus*), Fladbury, Worcestershire (Horse, Woolly Mammoth, Woolly Rhinoceros) and Oxbow near Leeds (Woolly Mammoth). These sites are all just within the limits of radio-carbon dating, and have dates in the range 43–38 000 b.p. At Beckford, Worcestershire, a rather later site with a date of 27 650 b.p., mammals include Horse, Woolly Mammoth, Woolly Rhinoceros, Reindeer, Bison (*B. priscus)* and one of the rare occurrences of Musk Ox *Ovibos moschatus* in Britain. Clearly this is the fauna of an Arctic tundra, and the suggestion that the isolated vole teeth might be the southern *Microtus arvalis* is incongruous; probably they in fact indicate the Narrow-headed Vole *Microtus gregalis*, whose teeth are indistinguishable, though its skull is very different.

Cave sites that might belong to this period include some of the most famous examples of Hyaena dens, though their dating is often uncertain unless recent radio-carbon dates have been obtained. At Tornewton Cave, The Elk Stratum (but the Elk is actually a large Red Deer, Lister 1984a) and the Reindeer Stratum belong in this period. The main accumulator of bones was apparently the Spotted Hyaena; its prey included Horse, Reindeer, Woolly Rhinoceros and *Bos/Bison* (hard to distinguish on broken or partial specimens). Other species present include Red Fox, Wolf, Arctic Lemming *Dicrostonyx torquatus* and the voles *Clethrionomys glareolus*, *Arvicola terrestris*, *Microtus agrestis*, *M. oeconomus* and *M. gregalis*. The Bank Vole looks out of place in this assemblage, but the other species range well into the Arctic, except most obviously for the hyaena. It seems clear that the Spotted Hyaena is not now restricted to Africa by climate, but that some combination of restricted prey availability and the viscissitudes of the later Devensian climate pushed it out of Europe and it failed to return. Charles and Jacobi (1994) summarize the available ^{14}C dates for British specimens of Spotted Hyaena, and demonstrate that they coincide with those for Woolly Rhinoceros (Table 1.1). The tundra conditions indicated by the vegetation were perhaps more productive than present-day tundra, because of the longer growing season at lower latitudes. Picken's Hole near Compton Bishop, Somerset, is another hyaena den, with a ^{14}C date of 34 000 b.p. obtained on bone specimens. Souslik, Arctic Fox, Reindeer, Horse, Woolly Rhinoceros and Woolly Mammoth accompany the Hyaena in this layer. Kent's Cavern, Devon, a famous hyaena den, has given ^{14}C dates of 38 270 b.p. on Horse, 28 720 b.p. on Brown Bear, 28 160 b.p. on Woolly Rhinoceros and 27 730 b.p. on *Bison priscus* (Table 8.3 of Stuart, 1982). Very large Red Deer at this site are also of interest, having been made into a different species (*Stongyloceros spelaeus*) at one time (Lister 1987). Both Irish Elk and Red Deer are present, and this has caused

Table 1.1 Available ^{14}C dates for British specimens of Woolly Rhinoceros *Coelodonta antiquitatis* (★ and ★★: repeat dates on the same specimen) and Spotted Hyaena *Crocuta crocuta* (from Charles and Jacobi 1994).

Coelodonta antiquitatis		
Pin Hole Cave, Derbyshire	SK5377	>41 400 b.p.
Kennington, Kent★	TR0245	35 000 b.p.
Kennington, Kent★	TR0245	34 000 b.p.
Kennington, Kent★	TR0245	33 600 b.p.
Leadenhall Street, London	TQ3381	29 450 b.p.
Kent's Cavern, Devon	SX9264	28 160 b.p.
Bishopbriggs, Lanarkshire	NS6070	27 550 b.p.
Earls Barton, Northamptonshire★★	SP8563	28 800 b.p.
Earls Barton, Northamptonshire★★	SP8563	28 000 b.p.
Earls Barton, Northamptonshire★★	SP8563	26 300 b.p.
Earls Barton, Northamptonshire★★	SP8563	25 810 b.p.
Earls Barton, Northamptonshire★★	SP8563	23 750 b.p.
Pin Hole Cave, Derbyshire	SK5377	22 500 b.p.
Ogof-yr-Ychen, Dyfed	SS1497	22 350 b.p.
Crocuta crocuta		
Pin Hole Cave, Derbyshire	SK5377	42 200 b.p.
Sandford Hill, Avon	ST4259	36 000 b.p.
Bench Tunnel Cavern, Devon	SX9356	34 500 b.p.
Pin Hole Cave, Derbyshire	SK5377	34 500 b.p.
Pin Hole Cave, Derbyshire	SK5377	32 200 b.p.

some confusion, especially since the Red Deer were nearly as big as the Irish Elk. Lister says that they had antlers with basal circumferences of 25–30 cm, compared with 27–32 cm for Irish Elk at the same site; modern Red Deer rarely exceed 20 cm. Modern Scottish stags weigh around 90–120 kg, but the full glacial animals were the size of Canadian Wapiti, around 300–450 kg. The radio-carbon dates indicate that at least a large part of the Kent's Cavern fauna belongs in mid-Devensian times, but the site is a complex one with later faunas as well, and needs further study.

One other important cave site that is of this age deserves discussion. Castlepook Cave, Co. Cork, was thought to have the only Irish mammal fauna that is earlier than the Late Devensian. Excavated last century, the fauna is a bit confused, with domestic, post-glacial, mammals and Wood Mouse as well as Mammoth, Wolf, Red and Arctic Fox, Spotted Hyaena, Brown Bear, Irish Elk, Reindeer and both lemmings (*Dicrostonyx torquatus* and *Lemmus lemmus*). The occurrence of the hyaena strongly suggests a mid-Devensian date, and this has fortunately been confirmed by a series of ^{14}C dates directly on the bones. The first to be done was on Mammoth, at 33 400 b.p.; recently, additional dates of 32 060 b.p. on Irish Elk, 27 930 b.p. on *Lemmus*, 20 300 b.p. on *Dicrostonyx* and 19 950 b.p. on Arctic Fox have been obtained. The same dating programme has also shown that Reindeer (28 000 b.p.) and Brown Bear (26 340 b.p.) from Foley Cave, Co. Cork and Mammoth (27 150 b.p.) and Red Deer (26 090 b.p.) from Shandon Cave, Co. Waterford, belong to this period (Woodman and Monaghan 1993, Woodman *et al.* 1997; see also Chapter 8). This indicates that

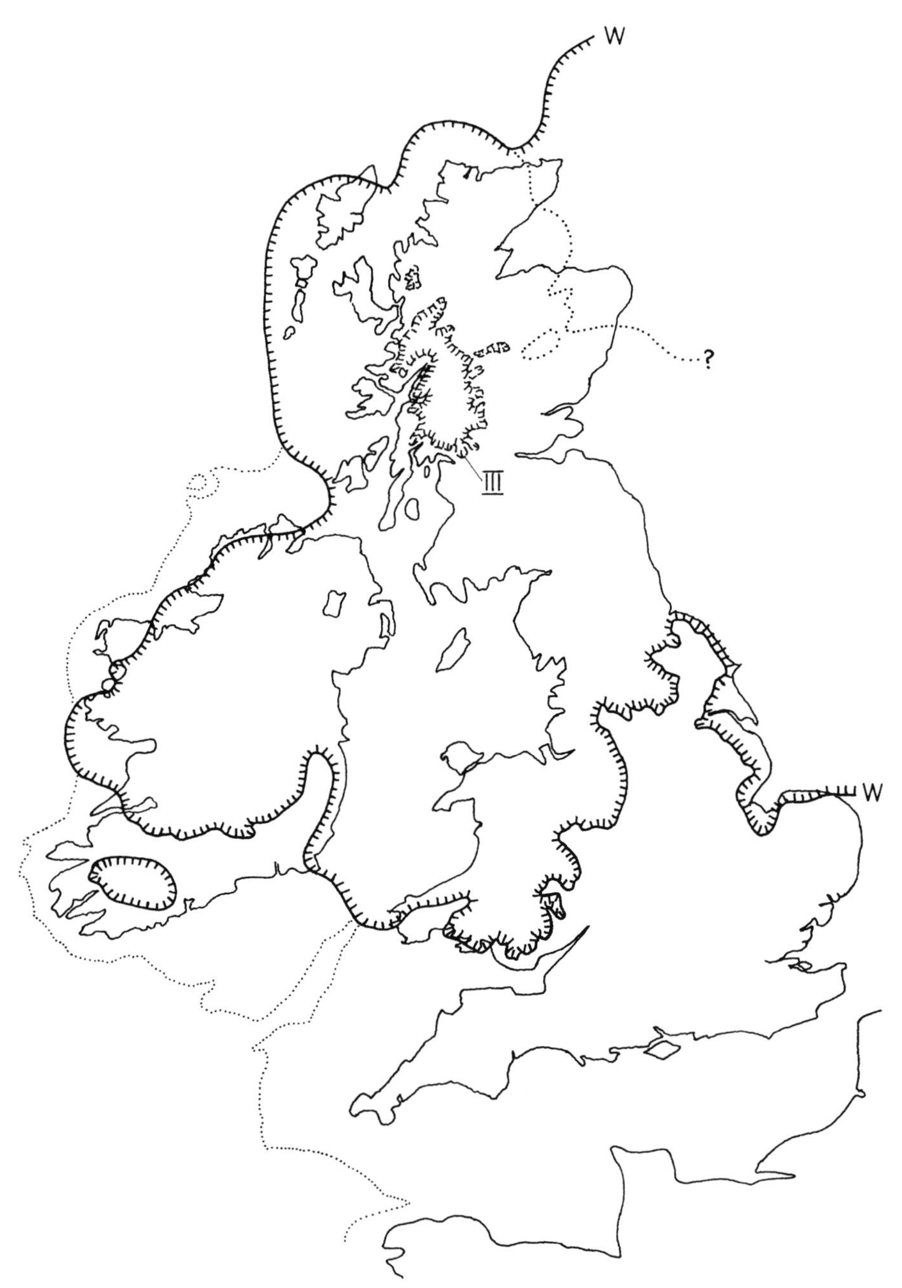

Figure 1.6 The maximum extent of the ice sheet covering Britain at the height of the Last (Devensian) Glaciation, about 20 000 years ago (W), and in the Loch Lomond Advance/Younger Dryas Period (III). (From Yalden 1982 in *Mammal Review* **12**, after various sources, by permission of Blackwell Science.)

the lowered sea level of the early Devensian was reduced sufficiently to allow some mammal fauna to reach the island.

However, the maximum of the Devensian Glaciation was reached in the period, 25 000 to 15 000 b.p., that is represented by the dates for Arctic Fox and Arctic Lemming at Castlepook. In England, the ice sheet extended down from the Scottish mountains to cover the whole of northern Britain down to north Yorkshire, with a tongue of ice extending down the North Sea coast to touch the Norfolk coast in the east, and down over Wales to reach the Gower coast in the west. Virtually all of Ireland, except a narrow strip across the south, was also covered by ice, with an additional ice-cap over the mountains of Cork and Kerry (Fig. 1.6). South of the ice sheet was a very bare tundra, with sparse vegetation and probably few animals. The Arctic Fox and Arctic Lemming would be expected in such a landscape, and one might expect Stoats to join the Arctic Foxes in hunting lemmings; Mountain Hares might be another prey, but there are few sites in Britain to document this period. Barnwell Station near Cambridge has a ^{14}C date on peat of 19 500 b.p., and a mammal fauna of Woolly Mammoth, Reindeer, Woolly Rhinoceros and Horse. A single, but undated, record of Polar Bear (*Ursus maritimus*) from the gravels of the River Thames at Kew and a record of Saiga *Saiga tartarica* from a similar site at nearby Teddington may date from this period. The brickearth at Fisherton, Salisbury, Wiltshire has Mammoth, Woolly Rhinoceros, Reindeer, Musk Ox and a souslik, and might also belong. These species are all characteristic of tundra or steppe conditions, and tolerant of extreme

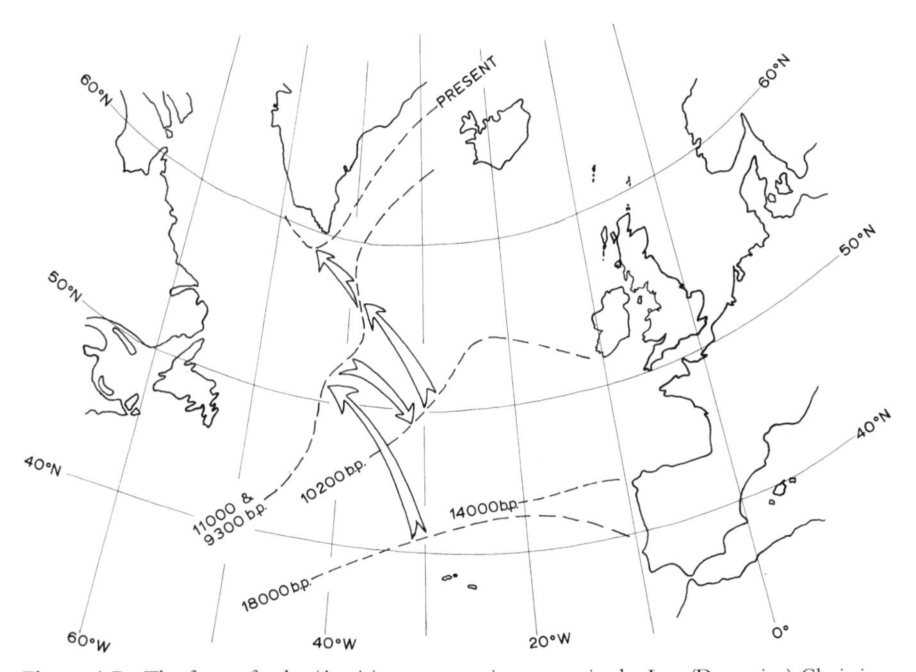

Figure 1.7 The front of polar (Arctic) water at various stages in the Last (Devensian) Glaciation, based on Ruddiman *et al.* (1977). In the Devensian Glaciation, the polar front was well south of the British Isles, and even the west coast of Ireland must have experienced a Greenland-like climate. (From Yalden 1982 in *Mammal Review* **12**, by permission of Blackwell Science.)

Figure 1.8 The reappearance of temperate land-mammal faunas in Britain at successive Interglacial Periods, showing the gradual appearance of faunas more like the present (Recent) one (after Currant 1989).

INTERGLACIAL/ OXYGEN ISOTOPE STAGE	POST-CROMERIAN	HOXNIAN	PRE-IPSWICHIAN	PRE-IPSWICHIAN	IPSWICHIAN	RECENT
	13	11	9	7	5	1
Erinaceus europhaeus Hedgehog	*	*				*
Sorex araneus Common Shrew			*	*	*	*
Sorex minutus Pigmy Shrew	?	?	*	*	*	*
Neomys newtoni A water shrew	*	*				
Neomys fodiens Water Shrew			*	*		*
Crocidura sp. A white-toothed shrew				*		
Talpa europaea Mole	*	*		*		*
Macaca sylvanus Macaque	*		*	*		
Homo sapiens Man		*	*			*
Lepus timidus Mountain Hare	*	*	*		*	*
Sciurus whytei A squirrel	*	*				
Sciurus vulgaris Red Squirrel				*		*
Trogontherium cuvieri Giant Beaver	*	*	*			
Castor fiber Beaver	*	*	*	*		*
Muscardinus avellanarius Dormouse		*				*
Clethrionomys glareolus Bank Vole	*	*	*	*	*	*
Pliomys episcopalis An extinct vole	*	*				
Mimomys savini A water vole	*					
Arvicola cantiana A water vole		*	*	*	* ↘	
Arvicola terrestris Water Vole					↘ *	*
Microtus agrestis Field Vole	?	(*)	*	*	*	*
Pitymys subterraneus Pine Vole	*	*	*			
Apodemus sylvaticus Wood Mouse	*	*	*	*	*	*
Canis mosbachensis A wolf	*	* ↘				
Canis lupus Wolf			↘ *	*	*	*
Vulpes vulpes Red Fox					*	*
Ursus deningeri/spelaeus Cave Bear	*	*	*			
Ursus arctos Brown Bear				*	*	*
Martes martes Pine Marten	*	*	*			*

Figure 1.8—*contd*

INTERGLACIAL/ OXYGEN ISOTOPE STAGE	POST-CROMERIAN	HOXNIAN	PRE-IPSWICHIAN	PRE-IPSWICHIAN	IPSWICHIAN	RECENT
	13	11	9	7	5	1
Mustela nivalis Weasel	*	*				*
Mustela erminia Stoat		*			*	*
Meles meles Badger		*			*	*
Lutra lutra Otter	*		*	*		*
Felis silvestris Wild Cat	*		*		*	*
Equus ferus Wild Horse	*	*	*	*		
Dicerorhinus etruscus Etruscan Rhino	*	*				
Dicerorhinus kirchbergensis Merck's Rhino			*	*		
Dicerorhinus hemitoechus Narrow-nosed Rhino			*	*	*	
Hippopotamus amphibius					*	
Sus scrofa Wild Boar	*		*	*	*	*
Cervus elaphus Red Deer	*	*	*	*	*	*
Capreolus capreolus Roe Deer	*	*	*	*	*	*
Bos primigenius Aurochs			*	*	*	*
Temperature +						
Curve −						

winter cold. The crucial point for the history of the present fauna is that very few of the extant species (perhaps only Mountain Hare, Stoat and Weasel, if there were lemmings to hunt) could have survived such glacial conditions, even in the extreme south of England or Ireland. In the past, some have thought to argue, particularly with respect to Ireland, that the Gulf Stream Drift might have ameliorated the climate in the extreme west. Even this is not feasible. Analysis of the foraminifera in the deep-sea cores, and direct calculation of the temperatures from them, has shown that the edge of the polar waters lay far to the south, about at the latitude of Lisbon, during the glacial maximum (Fig. 1.7).

CONCLUSIONS

This brief survey of the longer-term history of mammals in these islands provides a suitable background for the main story, the history of the present fauna. At the beginning of the story, at the end of the Trias, mammals had only just evolved from reptiles, though they were some 65 million years ahead of birds in this respect. The brief glimpses afforded by the fossil record of Jurassic and Cretaceous times confirms that there were still small, but primitive, mammals in these islands. The early Eocene record shows that the modern groups, marsupials and placentals, had almost completely displaced the Mesozoic groups, and by the later Eocene/early Oligocene some of the modern orders had evolved, and one or two familiar families had also appeared.

The history of Pleistocene mammals shows a similar story at a finer level of detail.

The earliest faunas have mostly modern genera, but few of the modern species. At each succeeding interglacial, a temperate fauna (and flora) reassembled, a little more like the present fauna at each stage (Fig. 1.8; Currant 1989). Many of the large mammals were already in existence by the Cromerian, though unfamiliar additional species like various elephants and rhinoceroses also appear. Other large mammals look exotic, because they have long been extinct in Europe, but survive in Africa. For the smaller mammals, the record shows continuing evolution through the period, and also adaptation to the increasing cold of successive glaciations.

The evolution of the voles and lemmings is particularly striking, and several lineages

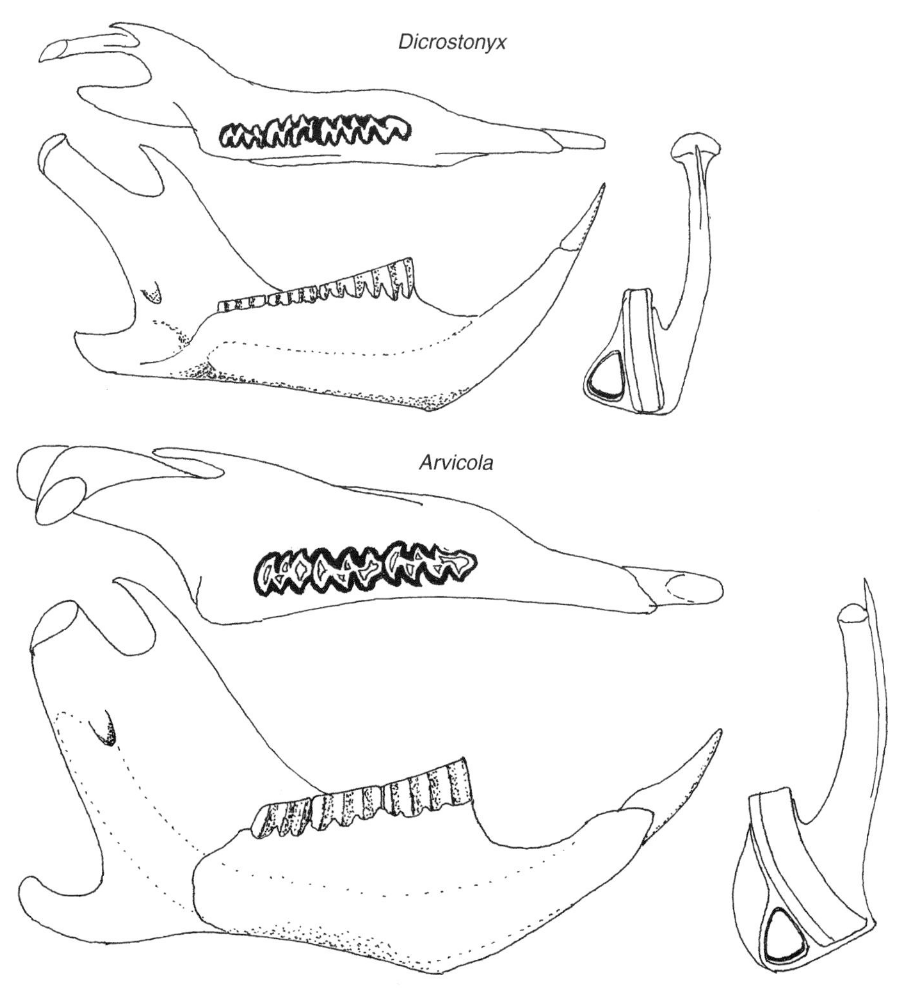

Figure 1.9 The teeth and jaws of lemmings and voles to show the different patterns of molar teeth and the position of the lower incisors. In lemmings (e.g. *Dicrostonyx*), the lower incisor runs inside the roots of the molars, and ends at the base of the vertical (coronoid) process; in voles (e.g. *Arvicola*), it crosses between the roots of the second and third molars, so can extend further up the coronoid process. This gives voles a better chance of grinding grasses, which are very abrasive but available food.

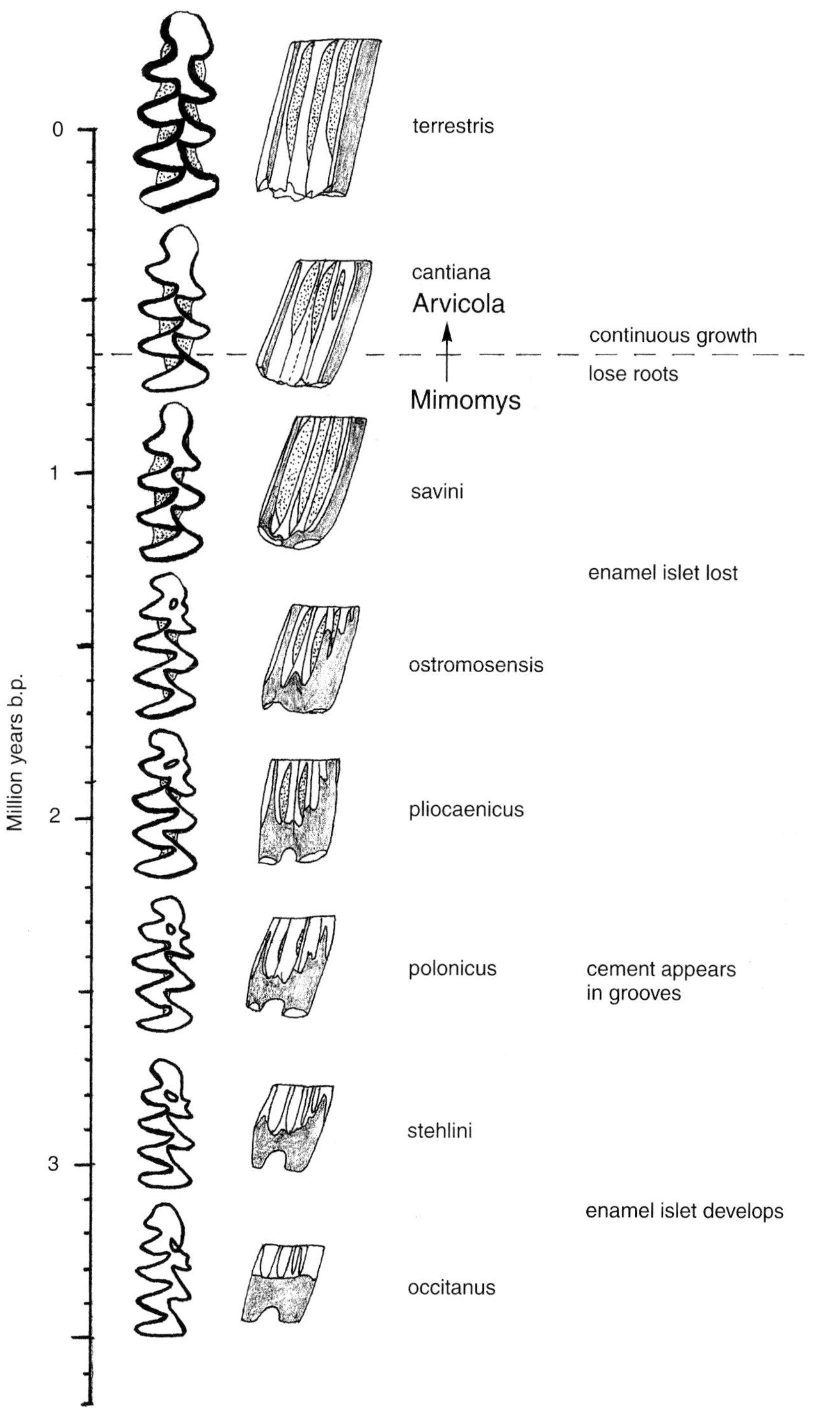

Figure 1.10 The evolution of the *Mimomys–Arvicola* lineage, as shown by the form of the first lower molar tooth (after Chaline 1987). The crown gets taller and the roots relatively shorter; when they no longer form, the molars grow throughout life, and this feature is used to distinguish *Arvicola* from its ancestor *Mimomys*. Cement (stipple) appears in the angles between the prisms of the teeth, and dentine tracts spread up the front and back of the crowns (grey), while the animals, and therefore their teeth, get larger. These changes collectively occur over more than 3 million years.

show the evolution in parallel of open-rooted molars to cope with the increasingly available, but abrasive, diet of grasses, sedges and rushes. The two genera of lemmings belong to lineages that have been distinct since well back in the Pliocene. The ancestral form for *Dicrostonyx*, *Praedicrostonyx hopkinsi*, appears in the earliest Pleistocene glacial period in North America, and in the mid-Pleistocene in Asia. Early forms of *Dicrostonyx*, *D. antiquitatis* and *D. simplicior* appear in Europe for the first time in the earlier (Gunz and Mindel) glacial periods. They have simpler molars, very like the form found now in Labrador, *D. hudsonicus* (Kowalski 1980). The latest form, with more complex molars, known loosely as *D. torquatus* but perhaps including several closely related species (Jarrell and Fredga 1993), appears in the Wolstonian and Devensian Glaciations. The other lineage has a longer and more diverse record, beginning in the Pliocene of Asia with an early *Synaptomys*, a genus that survives only in North America. *Lemmus* evolved from it in Europe by the early Pleistocene, and reappeared in the fossil record of western Europe at each glacial period. However, it prefers wetter tundra than *Dicrostonyx*, and feeds more on sedges, grasses and mosses; *Dicrostonyx* prefers dwarf shrubs, particularly *Salix* and *Dryas* (Batzli 1993). This may be reflected in its absence from drier steppe periods during glacial maxima. The lemmings were the earliest of the voles to evolve open-rooted molars, by the early Pleistocene, which they did by shortening their incisors (Fig 1.9). In voles such as *Arvicola* and *Microtus*, the incisors are long, but lie between the roots of the molars, not beneath them as in ancestral rodents.

The best recorded lineage of voles in Europe is that of *Mimomys–Arvicola* already mentioned in discussing the sequence of faunas in Britain. From a late Pliocene ancestor, *M. occitanus*, through *M. stehlini*, *M. polonicus*, *M. pliocaenicus*, *M. ostramosensis* to *M. savini*, the crowns of the molar teeth get higher, the first lower molars lose the islet of enamel from the front loop, and the cement appears on the outside of the teeth and becomes progressively taller (Fig. 1.10; Chaline 1987). Most of this sequence is missing from the British record because the early Pleistocene is largely absent. The change from *Mimomys (savini)* to *Arvicola (cantiana)*, marking the loss of roots to the molars, has already been noted as an indication of the difference in time between Cromerian and later intergacials. Modern *Arvicola terrestris* and, in south-western Europe, *A. sapidus* are differentiated from *A. cantiana* by the uneven thickness of the enamel on the cutting surfaces of their molars, and only evolve in the last glaciation.

The genus *Mimomys*, in the form of another species *M. burgondiae*, gave rise also to *Clethrionomys* in the early Pleistocene (Chaline 1987), but *Microtus* and the closely-related *Pitymys* had a separate origin in the early Pleistocene, possibly from yet another *Mimomys*, *M. newtoni*, in Asia. Their immediate ancestor evolved open-rooted molars at about 1.9 Ma, becoming *Allophiomys deucalion*.

Chapter 2

The Beginning of History

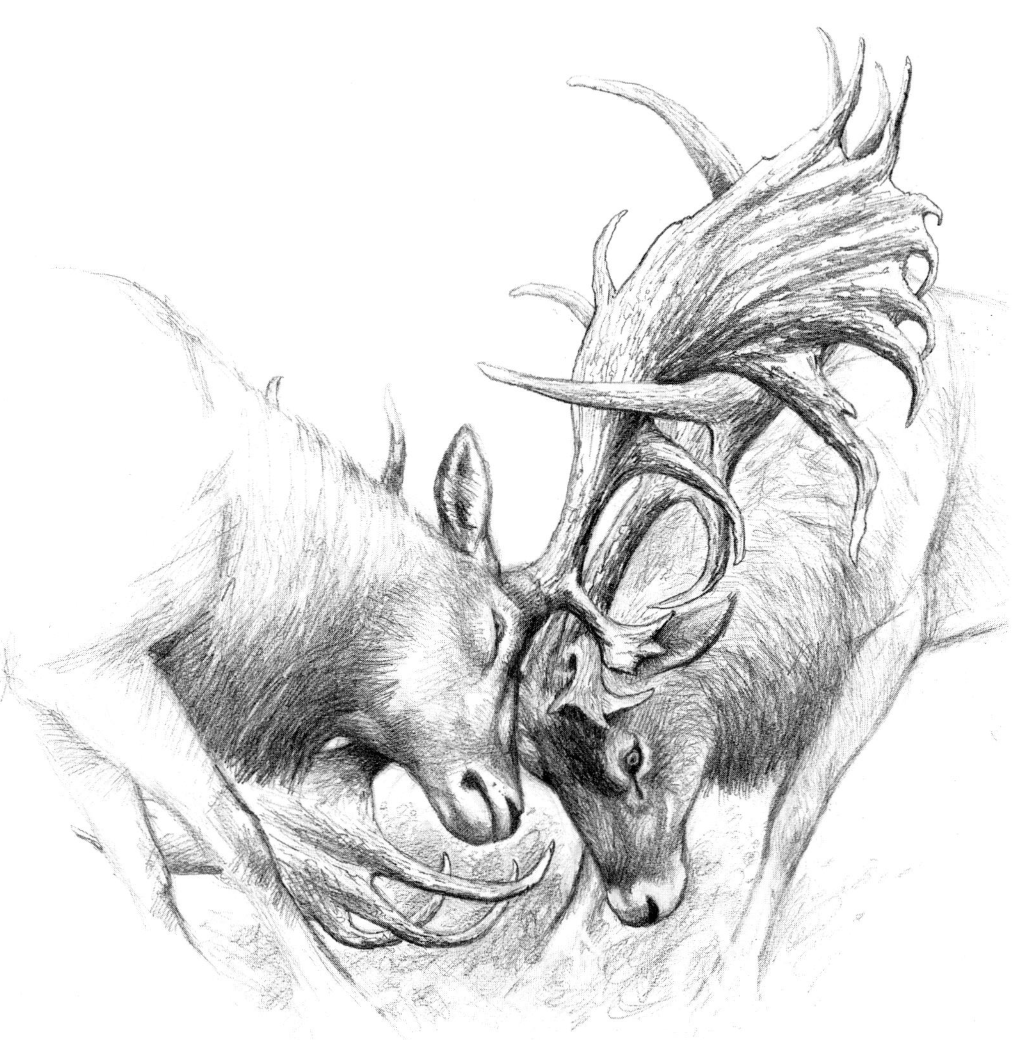

Irish Elk Stags fighting

THE end of the last (Devensian, in Britain) glaciation was a climatically complicated event. This period, always known as **the** Late Glacial (though all the other glaciations must have had late periods too!), is never-the-less an important one for the history of mammals in Britain and indeed elsewhere in Europe. It is no less important for the history of Man. Some discussion of the climatic events is therefore essential. The period covers, roughly, 15 000 to 10 000 b.p. (perhaps 16 000 to 11 000 b.p.), and marks the transition in Britain from a bleak tundra, with little biological activity, to a readable biological record.

The initial interpretation of the sequence of events came from archaeological sites in Denmark, which showed a sequence of cold, then warmer, then back to cold, environments. These were termed the Older Dryas, Allerød Interstadial and Younger Dryas, respectively. The Mountain Avens, *Dryas octopetalla*, is a conspicuous plant as a macrofossil in the two colder periods, while the warmer interlude was first recognized at the archaeological site of Allerød.

With the more complete record now available from pollen analysis, and the additional evidence from other sources such as fossil beetles and molluscs, the story is somewhat more detailed and complicated, but also confirmed in essentials. In Danish archaeological sites, an additional fluctuation to warmer conditions and back again can be recognized, so the sequence Oldest Dryas – Bølling Interstadial – Older Dryas – Allerød Interstadial – Younger Dryas may be used. In Britain, perhaps with a more oceanic climate during the warmer periods, the alternations from warm to cooler and back to warmer in the middle of this sequence are less evident, but analysis of both pollen and beetle remains from a site in Lake Windermere at Low Wray Bay provides evidence of a warmer interlude, going gradually cooler and then suddenly much colder. The convention in Britain is therefore to recognize a Windermere Interstadial, the whole of the intermediate period and covering both the Bølling Interstadial and the Allerød Interstadial (Coope and Pennington 1977). The sharp cooling of the Younger Dryas is well recognized everywhere in northern-western Europe (though less evident further south and east), and has a crucial impact on how mammalian history is interpreted.

Late Glacial Habitats

What, in more detail, does the evidence show about the habitats available for mammals in this period? At the beginning of the interlude, the ice sheet was melting rapidly, exposing much bare ground. Even in the south of England, not actually covered by ice, the vegetation suggested by the pollen record is a sparse tundra. The bare ground was more extensive than the plant cover, and Mugwort *Artemisia* pollen was dominant. This was replaced by grassland with *Rumex* (dock or sorrel, probably Sheep Sorrel), and this in turn by Crowberry *Empetrum* heath with Juniper *Juniperus communis* scrub. By about 12 000 b.p., birch *Betula* woodland was well established over most of England, Wales and southern Scotland, though in northern and western Scotland Crowberry heath remained predominant (Tipping 1991). This presumably reflects the wetter and more exposed conditions there. Conditions in the rest of north-western Europe were essentially similar (Kolstrup 1991). This birch woodland flourished to about 11 000 b.p., when the pollen record shows a reversion to the open tundra vegetation characteristic of 2000 years earlier. Birch and Juniper disappeared, at least in northern and western Britain, and while Crowberry remained in some places, mostly grasses dominated the pollen rain, with such indicators of open conditions as Thrift *Armeria maritima*, Dwarf Birch *Betula nana*, and, of course, *Dryas octopetalla* indicating the tundra-like conditions. The rather dry nature of the vegetation is indicated by the presence of *Artemisia* again, and at these low latitudes, with a long growing season, the vegetation was probably rather unlike the present high-arctic tundra, more a hybrid between the tundra and the steppe environments of present-day Europe.

The colder conditions indicated by the Younger Dryas period were recognized long ago by geologists, who had noted that glaciers reformed on the mountains of western Scotland, and spread far enough to reach Loch Lomond, whence they had named the period the Loch Lomond Readvance. Though much too far south to experience direct effects of this small ice-cap, conditions in southern England were sufficiently cold to register periglacial effects such as ice-wedges and soil marks. Ice-wedges result from alternate freezing and thawing of the ground, during the annual cycle, in which cracks fill with water during summer and then widen as the water freezes in winter. Similar action causes stones to gather in polygons on flat ground, and to creep down sloping ground in stripes; these events only occur in present-day tundra conditions, with annual temperatures around -6°C, so it is reasonable to assume that conditions were similarly severe. On the downlands of Kent and Sussex, soil creep created such stripes. Birch woodland characteristic of the Windermere Interstadial was replaced by open vegetation with Crowberry, Dwarf Birch, and especially sedges and grasses even at such southern sites as Hawks Tor on Bodmin Moor (Brown 1977), on the Isle of Wight (Scaife 1982) and in the New Forest. In lowland Britain, pollen of *Rumex* and *Artemisia* was abundant; Pennington (1977) remarks that this abundance of *Artemisia* implies rather dry conditions, and a vegetation type that is no longer present in Europe. A computer-generated analysis of pollen communities of a wide range of sites from across Europe and across the 13 000 years at 1000 year intervals by Huntley (1990) reached the same conclusion; many of the communities recognized by the analysis have no modern equivalents.

The most sensitive record of the climate during the Late Glacial probably comes from the beetles. Coope (1977) has argued convincingly that the changes in the vegetation indicated by the development of birch woodlands in the Windermere Interstadial were somewhat delayed, relative to the actual change in temperature, by the slow migration and development of woodlands. Beetles, he argued, can migrate faster, and respond more quickly to improving or deteriorating climatic changes. On this basis, the climatic improvement probably reached its maximum about 12 000 b.p., and the climate was actually deteriorating during the period to 11 000 b.p. when the birch woodland was expanding to its maximum. The evidence for the severity of the

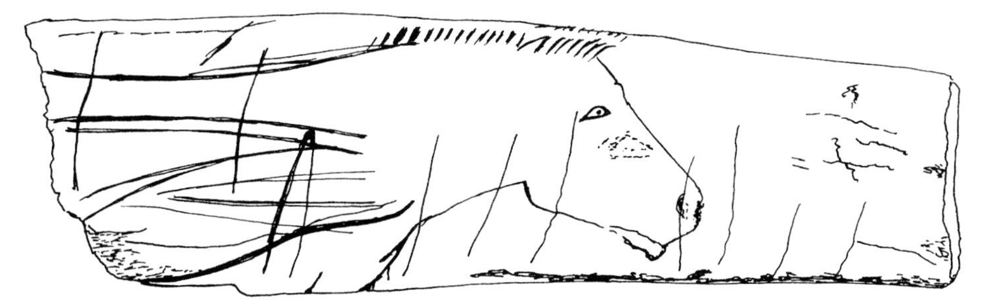

Horse head, engraved on rib, Creswell Crags Upper Palaeolithic

cold conditions during the Younger Dryas is decidedly confirmed by the beetle evidence; for instance, the high-arctic *Boreaphilus hennigianus*, *Olophrum boreale* and *Acidota quadrata* occur at Hawks Tor on Bodmin Moor (Coope 1977). Coope also remarks on a number of beetles that occurred in southern Britain at this time but which now survive only in eastern Siberia, such as *Pterostichus magnus* (at Farmoor, Oxford), *Helophorus jacutus* (at Comberton near Ludlow) and *Tachinus jacuticus* (on the Isle of Man). These clearly indicate a tundra-type open vegetation and a continental climate.

If the beetle fauna reacted more quickly to the change of temperature than the vegetation, it is obvious that mammals too would have reacted quickly; unfortunately, because they are warm-blooded, they have wide climatic tolerances and equally wide ranges, in many cases, so it is harder to predict what the presence of a particular mammal at a site indicates about the ecology or climate of that site, and conversely harder to predict what mammals should be present given some particular climate. There are some inconsistencies in the fossil communities which suggest that the mammal faunas present in Late Glacial times have no exact analogues at the present.

Late Glacial Mammals

So what were the mammal faunas like? This is the last period in which Woolly Mammoths roamed Britain. It was thought that they did not return to Britain after the maximum of the last glaciation, but the discovery of four, an adult and three young ones, trapped in a kettle-hole at Condover in Shropshire, has changed our perceptions (Coope and Lister 1987). The adult was about 30–32 years old, and probably a male; the juveniles were two smaller ones, 3–4 years old, and an older one 5–6 years old. They can be aged because teeth of elephants erupt in sequence, wear out and drop out of the front of the jaw, being replaced from behind by the next teeth, and the sequence is sufficiently similar in both modern species to suggest that the similar-sized Woolly Mammoth had much the same pattern of tooth replacement. The remains have been ^{14}C dated to 12 800 b.p., well into the Windermere Interstadial. The kettle-hole contained a sticky grey sandy clay, which was severely churned; it seems likely that it acted as a natural pit-fall trap from which the animals were unable to escape. Blowflies *Phormia terraenovae* and dung beetles *Aphodius prodromus* were also present in the deposits as evidence of the fate of the animals.

Woolly Mammoths of Windermere Interstadial age are confirmed now from at least four further sites, thanks to radio-carbon dating; Gough's Cave, Somerset; Pin Hole Cave and Robin Hood Cave, Derbyshire; and Kent's Cavern, Devon (Lister 1991). These other specimens are only small bones or fragments of ivory, so do not yield more information about the biology of the animals, but this is well known from studies elsewhere; numerous cave paintings and engravings show the small ears, high domed forehead and shaggy fur, and the frozen specimens preserved in Siberia confirm these features. The stomach contents of the Beresovka mammoth contained mainly sedges, grasses and mosses, while the Shandrin mammoth contained cotton-grass, sedges and mosses, as well as willow, alder and dwarf birch twigs (Vereshchagin and Baryshnikov 1982).

The most numerous large mammal during this period was probably the Reindeer, which Campbell (1977) lists from 30 out of 33 Late Upper Palaeolithic sites, that is, roughly of Windermere Interstadial age. However, certainly dated specimens are few, and it is possible that most records belong to the colder periods either at the beginning of the Windermere Interstadial, or at the end, in the Younger Dryas. Specimens from Aveline's Hole, Somerset at 12 480 b.p. and Pin Hole Cave at 13 050 b.p. do belong in this period. At Gough's Cave, Somerset, a well-dated fauna (12 dates, all between 12 800 and 11 900 b.p.) is dominated by Tarpan (Wild Horse); other dated herbivores include Aurochs, Saiga and Red Deer. The Mountain Hare was also common, and these five species were certainly among the prey of humans, for their bones bear cut marks (Currant 1986). Other species recorded from the cave, but less certainly of the same date, include both Red and Arctic Fox, Wolf and Brown Bear, Pika, Beaver, Norway Lemming and Water Vole, and Reindeer. Of these, Pika and Reindeer, at least, probably date to the later, Younger Dryas, cold period. The Saiga is perhaps the most interesting of these species. It is a small antelope, about the size of a sheep, albeit a long-legged sheep, currently found in the semi-deserts and on the steppes, the dry grasslands, that stretch from the borders of Romania across the southern Ukraine and Khazakstan to Mongolia. Its most notable feature is a greatly inflated nose, believed to be a structure that prevents dust from entering the lungs, but perhaps at least as important for conserving heat and moisture during the cold dry winters that characterize the steppes. Only the males carry horns, but these are much valued, as indeed is the meat. Last century, it numbered in the millions, but overhunting reduced its numbers to a few hundred. Under protection after 1920, it increased to between 2 and 6 million, and carefully regulated commercial hunting was resumed in the 1960s (Bannikov *et al.* 1961). More recently, agricultural development and overhunting have again threatened the species (Teer *et al.* 1996, Bekenov *et al.* 1998). During the last glaciation, Saiga spread out to reach into Alaska, across the then dry Bering Straits in the east, and across western Europe as far as England in the west (Fig. 2.1). However, there are only a handful of British records, and their dating is possibly spread across several millenia (Table 2.1). The Twickenham specimen is believed to date to before the maximum of the Last Glaciation, perhaps to the Upton Warren Interstadial (see Chapter 1). The four specimens from the Mendip area are believed all to date from the Late Glacial, with the specific dates of 12 380 b.p. and 12 100 b.p. on two of them (Currant 1986, 1987).

One of the rarest, but most significant archaeologically, of Late Glacial mammals is the Elk *Alces alces*. There are two well dated specimens, from Neasham, County Durham, and from High Furlong near Blackpool, Lancashire. The Neasham specimen came from lake muds that have been dated to 11 561–11 011 b.p., well in the Windermere Interstadial. A male with attached antlers, it could have died anytime from late summer through the winter to early spring. The High Furlong specimen

Table 2.1 Records of Saiga *Saiga tatarica* from Britain (from Currant 1986, 1987).

Orleans Road, Twickenham, Middlesex	TQ170735	Devensian	Stuart 1982
Sun Hole, Cheddar Gorge, Somerset	ST467540	Late Glacial	Collcutt *et al.* 1981
Gough's Cave, Cheddar Gorge, Somerset	ST467539	Late Glacial, 12 380 b.p.	Currant 1986
Soldier's Hole, Somerset	ST468540	Late Glacial, 12 100 b.p.	Currant 1987
Wolf Den, Wavering Down, Somerset	ST359563	Late Glacial	Currant 1987

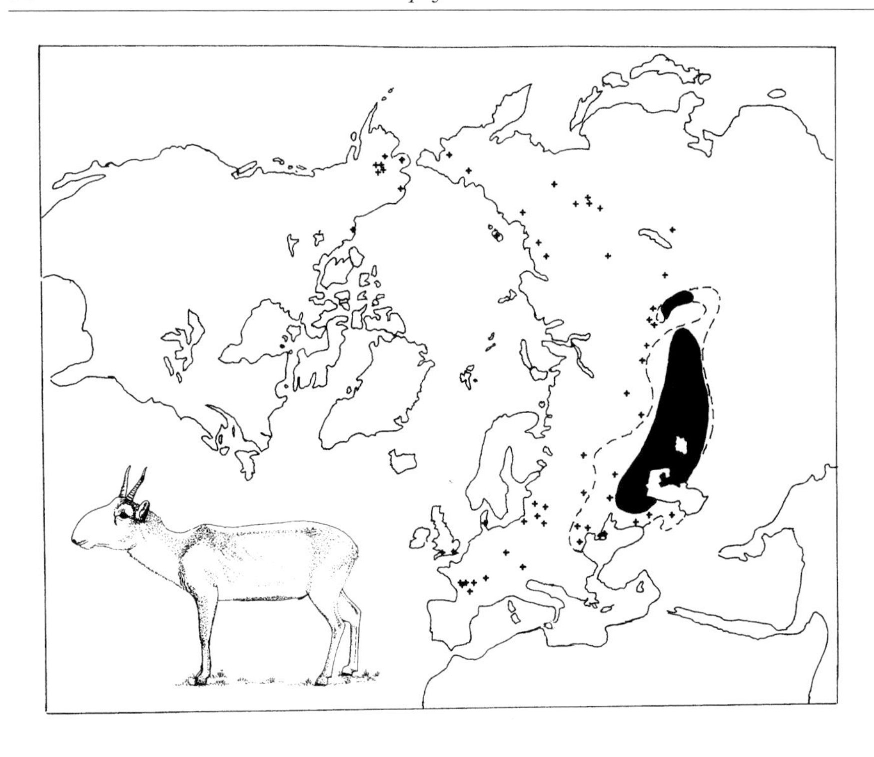

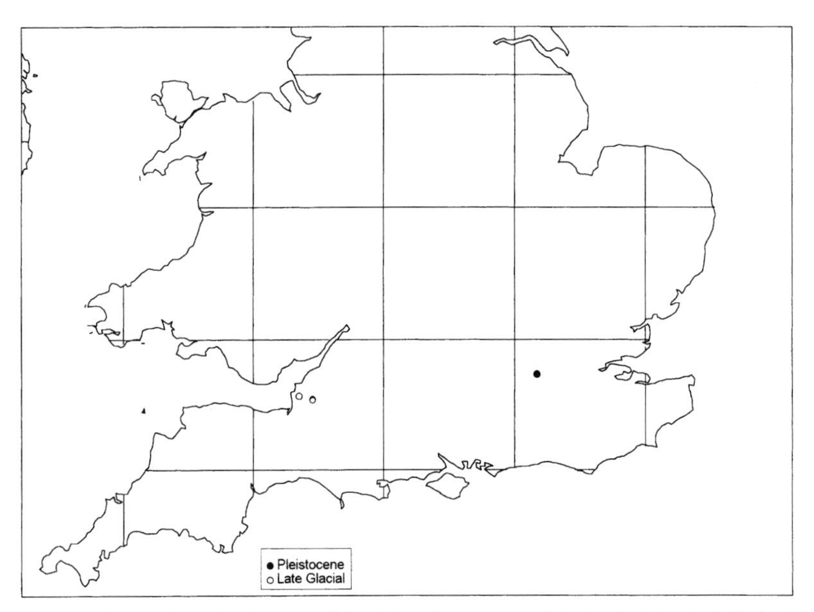

Figure 2.1 The modern distribution of the Saiga *Saiga tatarica* (black), its historical (dashed line) and fossil occurrences (crosses), and its fossil occurrences in Late Glacial England.

was directly dated on its bone to 12 400 b.p. (Housley 1991); also a male, it was in the process of shedding its antlers, so it died in late winter. The special interest of this specimen derives from the two barbed projectile (arrow or spear) points, made of bone, found with the skeleton, one in the rib cage and one in the left hind foot. There were another 15 or so other scars, probably caused by flint tips, on the ribs, scapula and limb bones which indicate that the animal had been ambushed by human hunters on previous occasions. Evidently it had escaped from them, presumably by swimming into the lake where it had then drowned (Hallam *et al.* 1973). The pollen analysis of the site shows a classical Late Glacial sequence from open herbaceous (Older Dryas) through birch woodland (Windermere Interstadial) and back to more open conditions (Younger Dryas), and the Elk skeleton was clearly in the birch woodland stage; willows were also present, and some herbs, as well as crowberry and juniper, just the sort of open woodland that Elk favour in modern Scandinavia.

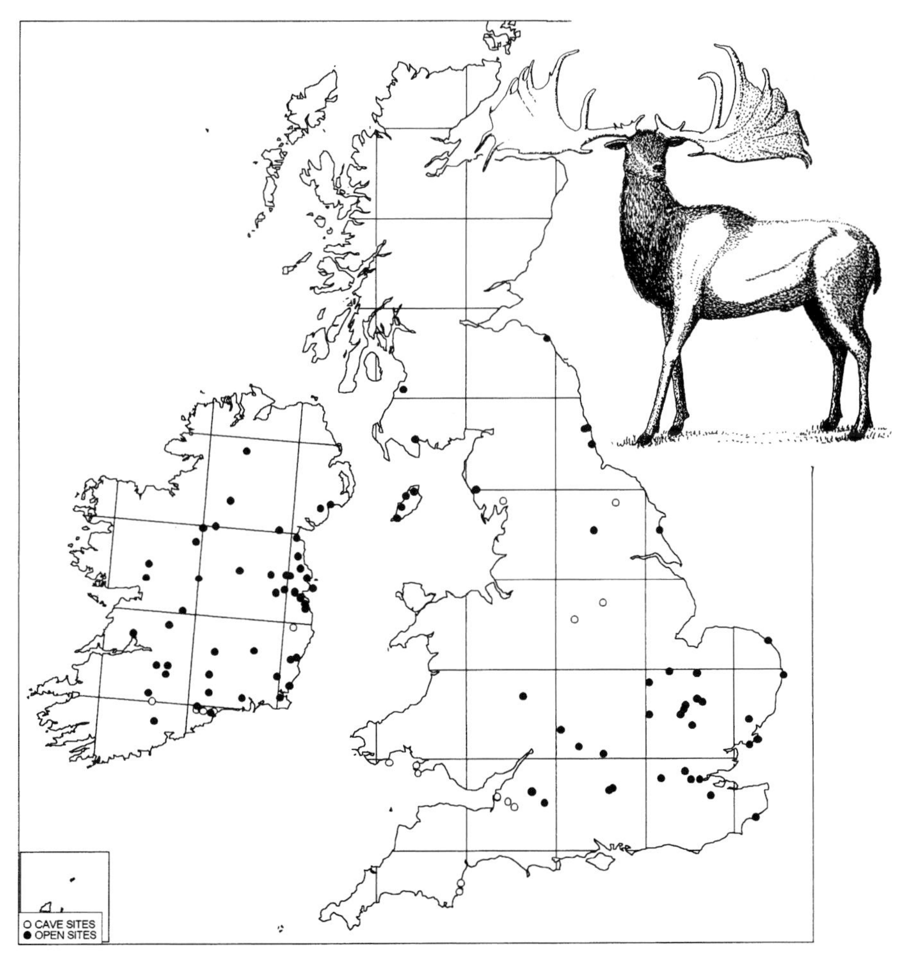

Figure 2.2 The distribution of the Irish Elk *Megaloceros giganteus* in the British Isles (after Reynolds 1929).

Not closely related, except in name, the Irish Elk *Megaloceros giganteus* was in fact a large relative of the Red Deer lineage, rather than a true Elk, but like the Elk had palmate antlers. This was among the largest deer known, standing up to 1.8 m at the shoulder and weighing perhaps 500 kg (the largest extant deer is the Elk, which attains a similar shoulder height and weight, but has much smaller antlers; Gould 1974). Though extremely abundant in Ireland, where its remains are invariably associated with the blue clays that lie below the peat, it was certainly not confined there, but had a wide range right across Europe, Russia and northern Asia, and into northern Africa. In Britain, Reynolds (1929) records it from Ballaugh and four other sites on the Isle of Man, and from 57 sites elsewhere, compared with 54 sites in Ireland (Fig. 2.2, Table 2.2). Among directly dated specimens are examples from Newhall Cave, Co. Clare (11 750 b.p.), Brandesburton, Yorkshire (12 850 b.p.) and Kent's Cavern, Devon (12 180 b.p.) (Campbell 1977, Woodman and Monaghan 1993). The most famous site is undoubtedly Ballybetagh Bog, just south of Dublin, where the skulls of at least 100 animals have been recovered. In an interesting analysis of this population, Barnosky (1985, 1986) showed that they were all males (even though collectors were looking for females to complete museum displays) but they were generally smaller or poorer males than those generally preserved in museums from other sites. No skulls with scars of recently shed antlers were found, though isolated shed antlers were discovered; thus the stags were living at the site in spring (when they shed their antlers), but not dying then. It looks as though this was a valley bottom site where weakened stags were likely to die of starvation in mid-winter, much as Red Deer do (cf. Lowe 1968, Putman *et al.* 1996), and the smaller stags were more likely to die than the bigger 'prime' stags. One jaw has been dated to 10 600 b.p., right at the end of the Windermere Interstadial (strictly, since in Ireland, the Woodgrange Interstadial). The Irish Elk was obviously polygamous, like the other herd-living cervids, and probably the sexes separated to different feeding grounds in winter, as do Red Deer; stags, being larger, need to eat more than hinds, but can cope by eating larger quantities of roughage. However, they tend to enter winter in poorer condition than the hinds, because of the energy demands of the autumn rut. The enormous antlers, with their 4 m span and a combined weight of up to 35 kg, have attracted a lot of popular and academic attention. Gould (1974) showed that the sizes of antlers of deer in general, and of Irish Elk in particular, were allometrically correlated with body size. That is, larger deer have proportionately bigger antlers than smaller deer of different species, and the Irish Elk falls on the upper extension of the line for modern deer (Fig. 2.3). Further, individual Irish Elk have proportionately bigger antlers if they are bigger (that is, if their skulls are longer). Gould went on to argue that the antlers were too large to have been used as fighting weapons, but that they were ideally shaped to have been used in display, either to impress hinds or intimidate rivals. However, a biomechanical analysis of the structure of the antlers by Kitchener (1987) indicates that they were strengthened in all the right places (like the tips of the tines) that one would expect in fighting weapons, and would have locked with the antlers of rivals in just the way they do in other deer. Clearly the stags engaged in shoving matches just like their modern Red Deer relatives.

The Red Deer is perhaps the most surprising of the species present at Gough's Cave, for conditions in the Late Glacial were always thought to have been too severe

Table 2.2 Records of Irish Elk *Megaloceros giganteus* from Britain (from Reynolds 1929 and Godwin 1975).

Aldermaston, Berkshire	SU5965	Reynolds 1929
Audley End, Essex	TL5237?	Reynolds 1929
Aylesford, Kent	TQ7359	Reynolds 1929
Barnwell, Cambridgeshire	TL0385	Reynolds 1929
Barrington, Cambridgeshire	TL3949	Reynolds 1929
Bedford	TL0449	Reynolds 1929
Bleadon Cave, Somerset	ST4357	Reynolds 1929
Brentford, Middlesex	TQ1778	Reynolds 1929
Brixham, Devon	SX9255	Reynolds 1929
Burwell Fen, Cambridgeshire	TL5767	Reynolds 1929
Caldey Island, Pembroke	SS1496	Reynolds 1929
Chesterton, Cambridgeshire	TL4560	Reynolds 1929
Clacton, Essex	TM1715	Reynolds 1929
Cowthorpe, Yorkshire	SE4252	Reynolds 1929
Coldingham, Berwickshire	NT9065	Reynolds 1929
Cresswell Crags, Derbyshire	SK5374	Reynolds 1929
Dogger Bank	–	Reynolds 1929
Erith, Kent	TQ5177	Reynolds 1929
Folkestone, Kent	TR2336	Reynolds 1929
Fulbrook, Warwickshire	SP2513	Reynolds 1929
Grantchester, Cambridgeshire	TL4355	Reynolds 1929
Grays, Essex	TQ6177	Reynolds 1929
Gough's Cave, Somerset	ST4653	Reynolds 1929
Happisburgh, Norfolk	TG3731	Reynolds 1929
Hilgay, Norfolk	TL5795	Reynolds 1929
Hoe Grange Cave, Derbyshire	SK2155	Reynolds 1929
Hutton Cave, Somerset	ST3458	Reynolds 1929
Ilford, Essex	TQ4486	Reynolds 1929
Ipswich, Suffolk	TM1644	Reynolds 1929
Kent's Cavern, Devon	SX9364	Reynolds 1929
Kirkdale Cave, Yorkshire	SE6785	Reynolds 1929
Jarrow Dock	c. NZ3265	Reynolds 1929
Knighton-on-Teme, Worcestershire	SO6370	Reynolds 1929
Long Hole, Glamorgan	SS4684	Reynolds 1929
London	?	Reynolds 1929
Loxbrook, Bath, Somerset	c. ST7464	Reynolds 1929
Lulham, Peterborough	c. TL1998	Reynolds 1929
Maybole, Ayrshire	NS3009	Reynolds 1929
Newmarket	TL6463	Reynolds 1929
off Lowestoft	c. TM5693	Reynolds 1929
Oxford	SP5305	Reynolds 1929
Ravensbarrow Hole, Lancashire	SD4187?	Reynolds 1929
R. Rye, Yorkshire	?	Reynolds 1929
Seaton Snook, Durham	c. NZ4049	Reynolds 1929
Skipsea, Yorkshire	TA1655	Reynolds 1929
South Shields	NZ3667	Reynolds 1929

Table 2.2 *continued*

Spritsail Tor, Glamorgan	SS4493		Reynolds 1929
Stanway, Gloucestershire	SP0532		Reynolds 1929
Twerton, Bath, Somerset	ST7263		Reynolds 1929
Ufton, Berkshire	SU6367		Reynolds 1929
Walton, Essex	TM2521		Reynolds 1929
Wastwater	NX10		Reynolds 1929
Westbury, Wiltshire	ST8751		Reynolds 1929
Whittlesey Mere, Cambridgeshire	TL2797		Reynolds 1929
Wigtown Bay	c. NX4355		Reynolds 1929
Wookey Hole, Somerset	ST5347		Reynolds 1929
Isle of Man			
Ballalheaney, Andreas	SC4298		Reynolds 1929
Ballaterson, Ballaugh	SC3393		Reynolds 1929
Close-y-Garey, St. John's	c. SC2281		Reynolds 1929
Kentraugh	SC2269		Reynolds 1929
Strandhall	SC2368		Reynolds 1929
Ireland			
Ask Bog, Gorey, Wexford	T1762		Reynolds 1929
Ballinderg, Athlone	N04		Reynolds 1929
Ballybetagh, Dublin	O2122	15 170 b.p.	Woodman *et al.* 1997
Ballybetagh, Dublin	O2122	10 610 b.p.	Woodman *et al.* 1997
Ballymackward, Fermanagh			Reynolds 1929
Ballynamore, Londonderry			Reynolds 1929
Belfast	J3373		Reynolds 1929
Ballynamintra Cave, Waterford	X108955	11 110 b.p.	Woodman *et al.* 1997
Buttevant, Cork	R5409		Reynolds 1929
Cappagh, Waterford	X1696		Reynolds 1929
Cappoquin, Waterford	X102997		Woodman *et al.* 1997
Cappagh Cave, Waterford	X1696		Woodman *et al.* 1997
Castlebellingham, Louth	O0695		Reynolds 1929
Castlepook Cave, Doneraile, Cork	R603009	37 200 b.p.	Woodman *et al.* 1997
Castlepook Cave, Doneraile, Cork	R603009	32 200 b.p.	Woodman *et al.* 1997
Castlepook Cave, Doneraile, Cork	R603009	32 060 b.p.	Woodman *et al.* 1997
Chapelizod, Dublin	O0934		Reynolds 1929
Cloggy Bridge, Cavan			Reynolds 1929
Cloone, Leitrim	N1399		Reynolds 1929
Cool Bog, Limerick	R7441	Late Glacial	Godwin 1975
Dangan River, Meath	N8151		Reynolds 1929
Dee River, Louth	O0-9-		Reynolds 1929
Drogheda, Louth	O0975		Reynolds 1929
Drumsna Bridge, Leitrim	M9997		Reynolds 1929
Dundrum, Dublin	O1628		Reynolds 1929
Dundrum, Down	J4036		Reynolds 1929
Dungarven, Waterford	X2693		Reynolds 1929
Dunmow, Galway (Dunmore?)	M5603?		Reynolds 1929

Table 2.2 *continued*

Dunshaughlin, Meath	N9652		Reynolds 1929
Edenvale Caves, Clare	R322747	11 750 b.p.	Woodman *et al.* 1997
Enniscorthy, Wexford	S9739		Reynolds 1929
Enniskerry, Wicklow	O2217		Reynolds 1929
Garransdarragh Bog	W645785	11 820 b.p.	Woodman *et al.* 1997
Griffinrath Td, Kildare	N9735	Late Glacial	Godwin 1975
Howth, Dublin	O2939		Reynolds 1929
Johnstown Castle, Wexford	T0216		Reynolds 1929
Kilgreany Cave, Waterford	X172944	10 960 b.p.	Woodman *et al.* 1997
Killowen Cave, Wexford	T1159?		Reynolds 1929
Killuragh Cave	R782488	11 510 b.p.	Woodman *et al.* 1997
Kilskeer, Meath	N6671		Reynolds 1929
Kiltoghert, Leitrim (Kiltyclogher?)	G9845?		Reynolds 1929
Knocklong, Limerick	R7231		Reynolds 1929
Knocknacran, Monaghan	H2830	11 310 b.p.	Godwin 1975
Legan Bog, Longford			Reynolds 1929
Leigh Td, Tipperary	S2215	Late Glacial	Godwin 1975
Leighlin, Carlow	S6965		Reynolds 1929
Longford Pass, Tipperary	S2561	Late Glacial	Godwin 1975
Mountshannon, Lough Derg, Clare	R7186		Reynolds 1929
Lough Gur, Limerick	R6140		Reynolds 1929
Lough Naglack, Carrickmacross, Mon.	H85		Reynolds 1929
Mobarnan, Fethard, Tipperary	c. S2135		Reynolds 1929
Mullingar, West Meath	N4352		Reynolds 1929
Naul Bog, Dublin	O1361		Reynolds 1929
Newtownbabe Td, Louth	J3030	Late Glacial	Godwin 1975
Newtonstewart, Tyrone	H3985		Reynolds 1929
Portumna, Galway	M8503		Reynolds 1929
Rathcannon Bog, Limerick			Reynolds 1929
Rathcoffey South Td, Kildare	N8931	Late Glacial	Godwin 1975
Ratoath, Meath	O0252	Late Glacial	Godwin 1975
Schiale, Limerick			Reynolds 1929
Shandon Cave, Waterford	X292950		Woodman *et al.* 1997
Shortalstown	T1130	12 160 b.p.	Godwin 1975
Strokestown, Roscommon	M9380		Reynolds 1929
Tuam, Galway	M4352		Reynolds 1929
Turloughmore, Galway	M4136		Reynolds 1929
Turvey, Dublin	O2151		Reynolds 1929
Waterford	S6012		Reynolds 1929

for this, an essentially woodland species. It is, indeed, quite rare at this time, but the climatic record indicates that parts of the Windermere Interstadial were quite warm enough, and the birch woodlands presumably provided suitable forage and winter shelter. Other dated specimens from this period include examples from Misbourne (at 12 530 b.p.), King Arthur's Cave (12 210 and 12 120 b.p.) and Aveline's Hole (12 100 b.p.); the Gough's Cave specimen has a date of 12 800 b.p. (Housley 1991),

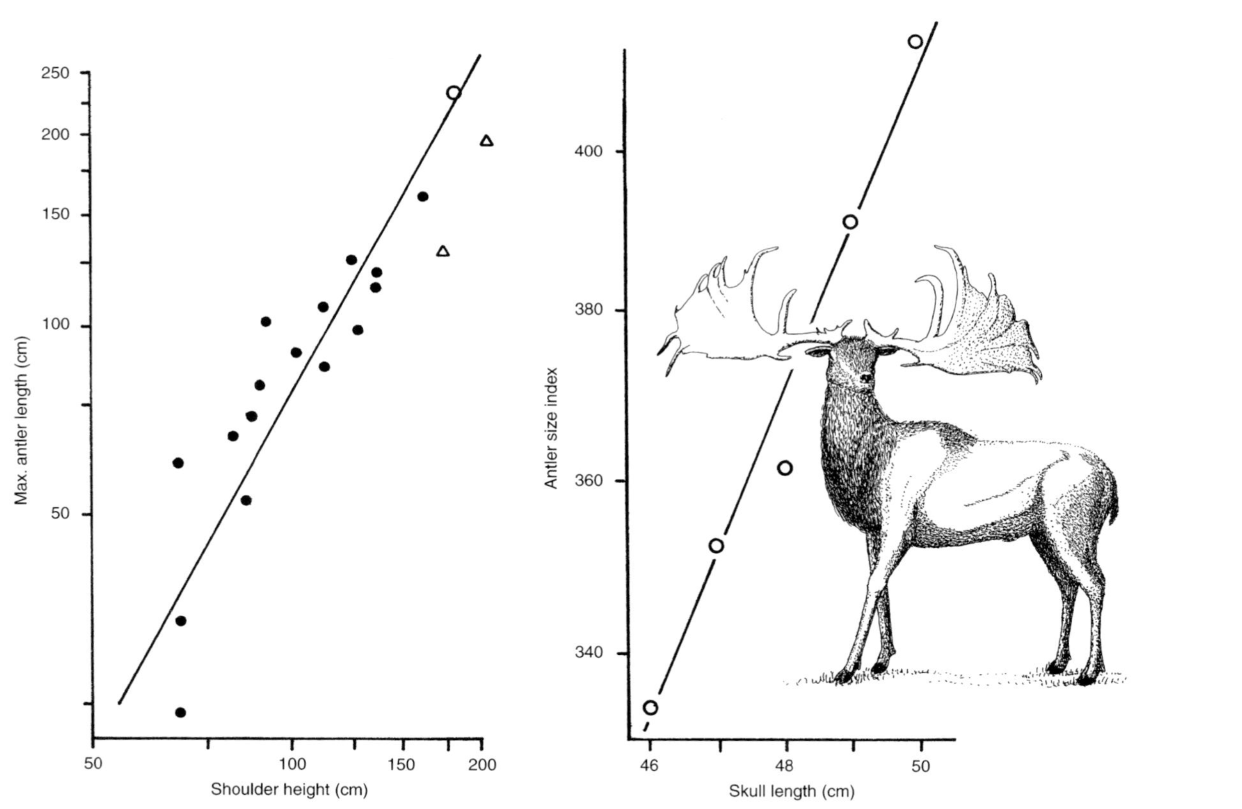

Figure 2.3 The allometric relationships between body size and antler size in deer in general and in Irish Elk *Megaloceros giganteus* (after Gould 1974). Deer of the subfamily Cervinae, to which *Megaloceros* belongs, are shown as solid dots (left); the regression line is extended to *Megaloceros* (circle). The triangles mark the positions of European Elk and American Moose, which are about as large as *Megaloceros*, but have smaller antlers. Within *Megaloceros*, deer with larger skulls have larger antlers (right). Thus antlers are proportionately bigger both in larger deer in general, and in Irish Elk in particular.

so these animals do indeed all date from a short period in the warmest part of the Windermere Interstadial. The Aurochs (Wild Ox) was also a woodland animal, which often occurs with Red Deer in postglacial sites, and which might therefore have been unexpected in Late Glacial ones. It too is found at Gough's Cave (at 11 900 b.p.) and a few sites of similar age (Kent's Cavern, 11 880 b.p.; Pin Hole, 12 480, 12 400 and 10 979 b.p.; Housley 1991).

The Wild Horse is the most abundant ungulate at Gough's Cave, with seven dates all clustered between 12 530 and 12 260 b.p. (Housley 1991). Campbell (1977) regards it as the second most frequent prey of Late Upper Palaeolithic hunters, after Reindeer, reporting it from 21 out of 33 sites. Another site with Tarpan dated to the Windermere Interstadial is Three Holes Cave (11 970 b.p.; Housley 1991).

At Robin Hood Cave, Derbyshire, the main prey of the hunters was not ungulates but Mountain Hares *Lepus timidus*; at least 42, and possibly another 20, of the hare bones have cut marks created by stone tools (Charles and Jacobi 1994). Several of these bones have been directly dated to 12 600–12 290 b.p.; similar dates have also been obtained on cut specimens from Pin Hole (12 510 and 12 350 b.p.) and on unmodified bone from Church Hole (12 240 b.p.). It seems that specialized hunting of Mountain Hares, perhaps for their white winter fur, was being carried out in the Creswell Crags area, but the species was also hunted at Gough's Cave in Somerset, and Currant (1986) remarks that the species was common in many Late Glacial sites. It was rather larger than present-day Mountain Hares, and is often listed as *Lepus anglicus* in the older literature.

The presence of other ungulates in this period is doubtful. Campbell (1977) includes the Woolly Rhinoceros, Ibex and Bison among the species present in the Late Upper Palaeolithic, i.e. broadly the Windermere Interstadial (his Table 31). However, there is no convincing evidence that either the Rhinoceros or Bison returned to Britain after the maximum of the Devensian Glaciation, and it is suspected that both are actually contaminants in cave deposits from earlier in the Devensian. The status of the Ibex as a British mammal at any time is doubtful, and the record from Robin Hood Cave discussed by Campbell (1977, pp. 130–131) is felt by Charles and Jacobi (1994) to be more probably a Post Glacial contaminant sheep or goat; the record only comprises one molar tooth, indistinguishable from modern goats, and a piece of horn core which they could not relocate. They also reconsidered the status of Woolly Rhinoceros from Robin Hood Cave; they point out that all the dated specimens of the species in Britain are between 41 400 and 22 350 b.p., and moreover quote a date for a specimen from that cave of 42 900 b.p. All the rhino teeth are in poor condition, and seem to belong to an older fauna than the certain prey of the hunters, like the Mountain Hares discussed above. Elsewhere, the latest occurrences of Woolly Rhinoceros in Siberia are also between 33 000 and 27 300 b.p. (Stuart 1991), but less certain dates from western Europe from Late Glacial times suggest the possibility that, like Woolly Mammoth, it may yet be found in the Late Glacial in Britain. It was clearly much rarer than the Mammoth, however, throughout its range (Stuart 1991).

Ungulates were certainly not the only mammals present in Britain at this time. Carnivores included Lynx, Brown Bear, Wolf, Red and Arctic Fox and perhaps Glutton. However, there is no evidence that two very distinctive Devensian carnivores, the Spotted Hyaena and Lion, returned to Britain after the glacial maximum

Steppe Pika gathering hay, with Saiga behind

(although the Lion at least did occur elsewhere in Europe) and their presence in a fauna may be assumed to indicate that it is not of Windermere date. At Gough's Cave, the Arctic Fox has been directly dated to 12 400 b.p., and the Wolf and Lynx are confidently assigned to the same period by Currant (1991). The Arctic Fox is one of the less common carnivores in Late Glacial Britain, though known from at least seven sites, including Cathole, Robin Hood Cave, Wetton Cave, Soldier's Hole, Gough's Cave, Long Hole and Ightham Fissure (Reynolds 1909, Campbell 1977, Currant 1991, David 1991). Some of these specimens may belong to the full Devensian (e.g. Long Hole) and some may belong to the Younger Dryas, but at least the Gough's Cave specimen belongs here. The Brown Bear was present at many cave sites, and directly dated examples come from Kent's Cavern (14 275 b.p.), Sun Hole (12 378 b.p.) and, in Ireland, Keshcorran and Red Cellar Caves (11 920 and 10 650 b.p., respectively) (Campbell 1977, Woodman and Monaghan 1993). The Wolf was also dated at Keshcorran Caves, to 11 150 b.p. The Glutton, or Wolverine, is one of the rarest mammals in the Late Glacial of Britain, known from only a handful of sites (Table 2.3) and none of them well dated. The example from Plas Heaton Cave, Clwyd (Fig. 2.4), a lower jaw, has an associated fauna of Hyaena, Bear, Badger, Wolf, Fox, Reindeer, Bison and Horse; the presence of Hyaena and Bison suggests a full Devensian Glacial date. One from Wetton Cave, Staffordshire is associated with a fauna including Arctic Fox and Reindeer, but also a mixture of Mesolithic elements; it probably does belong in the Windermere Interstadial or the succeeding Younger Dryas (Bramwell 1976), while one from Chelme's Combe Rockshelter, Somerset, may belong with the Younger Dryas Age Reindeer fauna there which is now well dated (Currant 1991). In modern-day Scandinavia, the Glutton frequently preys on Reindeer and has been persecuted by the Lapplanders as a consequence. Of 3445 compensation claims paid to Norwegian farmers in 1985 for losses due to Gluttons, 370 were for adult Reindeer and 254 were for calves; the rest were for sheep (Kvam *et al.* 1988).

Table 2.3 Records of Wolverine *Gulo gulo* from Britain. The Cromerian record may belong to *Gulo schlosseri*, the smaller ancestor of the modern Wolverine.

Mundesley, Norfolk	TG3830	Cromerian	Stuart 1982
Tornewton Cave, Devon	SX8167	Devensian, 22 160 b.p.	Kurtén 1973
Pin Hole Cave, Creswell Crags	SK533742	Devensian?	Reynolds 1912
Ash Tree Cave, Whitwell, Notts.	SK515376	Devensian	Jacobi pers. comm.
Port Eynon Point Cave, Glamorgan	SS4684	Devensian	Jacobi pers. comm.
Plas Heaton Cave, Denbigh	SJ031691	Devensian	Reynolds 1912
Kent's Cavern, Devon	SX9264	Devensian	Jackson 1953
Yealm Bridge, Devon	SX5951	?	Reynolds 1912
Whatley Bone Fissure, Somerset	ST730482	Devensian	Savage 1969
Banwell Cave, Somerset	ST382588	Devensian?	Reynolds 1912
Wetton Mill Rock Shelter, Staffs.	SK096563	Late Glacial	Bramwell 1976
Chelm's Combe Rockshelter, Somerset	ST463545	Late Glacial, Y. Dryas	Bramwell 1976

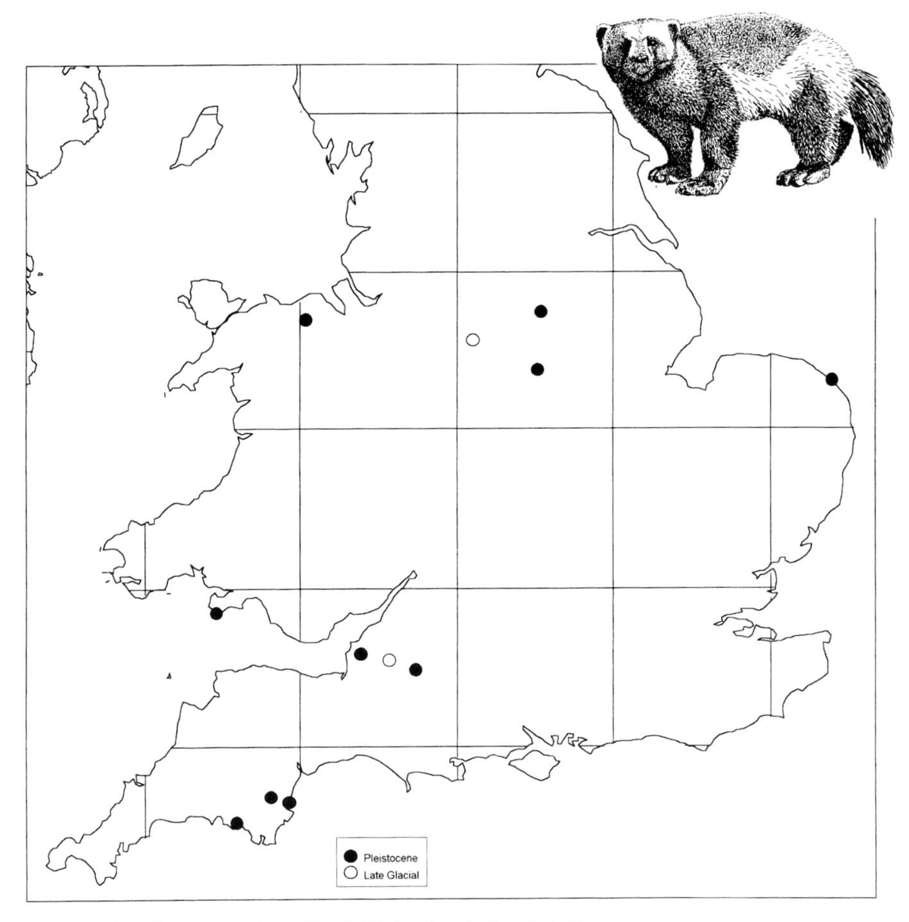

Figure 2.4 The distribution of fossil Wolverine *Gulo gulo* in Britain.

One additional but unexpected carnivore deserves a mention; the sabre-toothed cat *Homotherium latidens* has been reported from the Late Glacial of both Kent's Cavern and Robin Hood Cave. Since this is a species that occurred much earlier in the Pleistocene, in both Britain and elsewhere in Europe (see Chapter 1), but not at any later period, these records have caused much discussion (Campbell 1977, Jenkinson 1984, Charles and Jacobi 1994). The possibility that *Homotherium* survived in Britain long after it became extinct elsewhere has even been considered (Kurtén 1968). The specimen from Robin Hood Cave, a single canine, was dug out on the 3rd July 1876, a date when the senior archaeologist Tom Heath had been absent for 4 days, and it has been widely argued that a hoax, along the lines of the better known Piltdown forgery, was perpetrated on the other excavators (Jenkinson 1984, p. 37). However, there appears to be a hole bored in the canine, and a more likely explanation is that Upper Palaeolithic cave dwellers had collected the specimen elsewhere as an ornament (Charles and Jacobi 1994). The known site near Buxton is not too far away, as the

Reindeer-hunter might have travelled, but the specimen apparently has different chemical properties from all *Homotherium* known from other sites, a factor which also argues against it being a forgery (Oakley 1980). The Kent's Cavern specimens are two canines, but they are in a cave that has a depth and variety of deposits covering a long period of time, including Cromerian faunas, and careless excavation could easily have mixed them up.

The characteristic rodents of this period are the lemmings *Lemmus lemmus* and *Dicrostonyx torquatus,* and the Narrow-headed Vole *Microtus gregalis.* These are listed from, respectively, 24, 40 and 19 sites in the British Isles by Sutcliffe and Kowalski (1976), but these records extend across a wide time-span. Unfortunately, such rodents are too small to be of interest to archaeologists as human prey, and until the advent of the more refined radio-carbon dating techniques they were too small to be dated directly themselves. Most of the useful records therefore are dated by association with other specimens, with artefacts, or with pollen or other plant remains. This is a pity, for the more detailed records for these species from elsewhere in Europe indicate that they could tell an interesting story of the subtle changes in climate and habitat (cf. Chaline 1972 and Marquet 1993 for France, or Cordy 1991 for Belgium). Some of the records listed by Sutcliffe and Kowalski (1976) certainly belong to earlier glaciations, and others to the maximum of the Devensian Glaciation (see Chapter 1). Records that seem genuinely to belong in the Windermere Interstadial include both *Lemmus* and *Dicrostonyx* at Sun Hole (dated by Late Upper Palaeolithic artifacts and the larger mammals; Campbell 1977) and at Robin Hood Cave (dated on associated fauna, artefacts, and the ^{14}C dates on Mountain Hare bones; Campbell 1977, Charles and Jacobi 1994). At present, both lemmings have a circumpolar distribution, but only *Lemmus* occurs in Scandinavia (Fig. 2.5), while only *Dicrostonyx* occurs in Greenland (Fig. 2.6). They do however differ quite markedly in their ecology. *Dicrostonyx* tends to be more common furthest north; it turns white in winter, like the Mountain and Arctic Hares, and it also develops forked claws in winter to assist in digging into snow. It feeds much more on dwarf shrubs, willow and birch, and tends to live in drier areas. *Lemmus* does not turn white, nor does it have the complex winter claws. It prefers wetter areas, and feeds particularly on grasses, sedges and mosses, foods which are less nutritious but also less well equipped with anti-digestive compounds such as alkaloids (Batzli 1993). The current distribution of the Narrow-headed Vole *Microtus gregalis* is interesting in the light of the differences between the two lemmings; it has a split range, on the tundra of northern Russia and on some of the mountains in Siberia (Fig. 2.7). The vole occurs in a variety of grasslands, sub-alpine meadows and the sandy banks of streams in the tundra covered in willow scrub and horsetail stands. It seems, unlike *Dicrostonyx*, to favour wetter areas, and feeds on a wide variety of lilies, herbs such as Ox-eye Daisy and Wormwood, and the grasses, willows and horsetails of the tundra (Ognev 1964), but the two are usually associated in fossil communities (Cordy 1991). The converse association is between *Lemmus* and *Microtus oeconomus*, the Root Vole. This has a wider and more southern distribution than the other three, with a relict population in Holland, and from northern Germany and Poland northwards through Scandinavia and Russia (Fig. 2.8). It prefers wetter areas than other *Microtus*, as demonstrated in Holland by De Jonge and Dienske (1979).

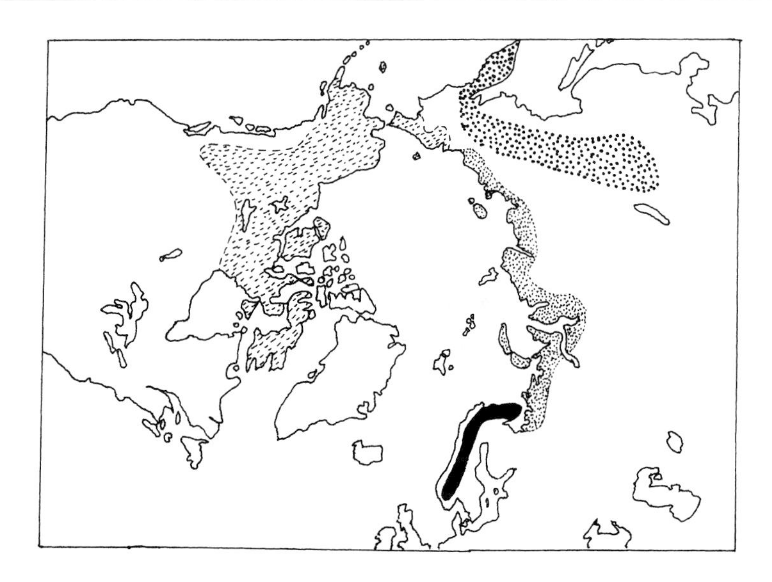

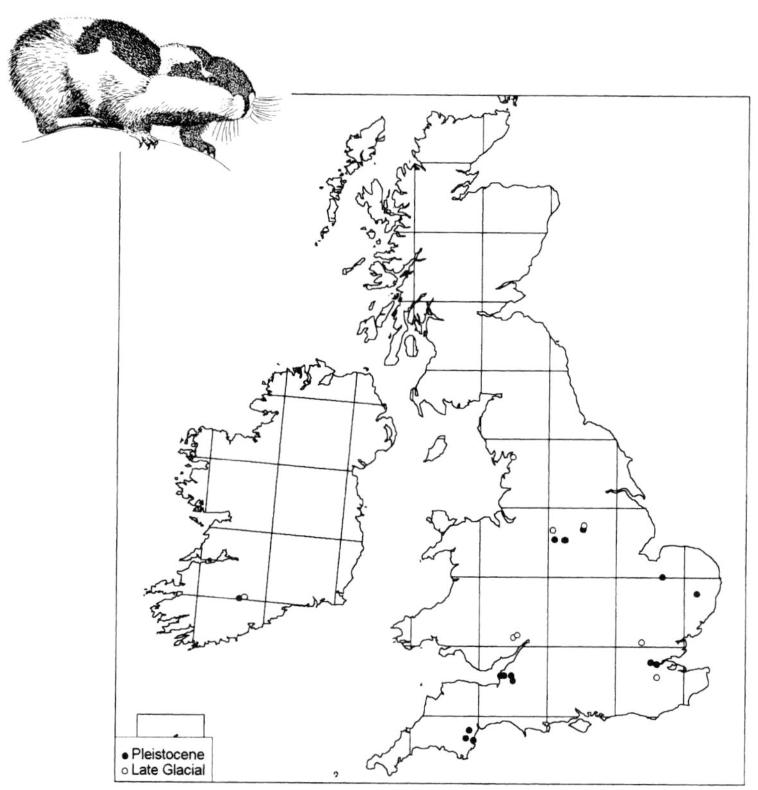

Figure 2.5 The modern distribution of the Norway Lemming *Lemmus lemmus* (black) and its relatives, and its fossil distribution in the British Isles.

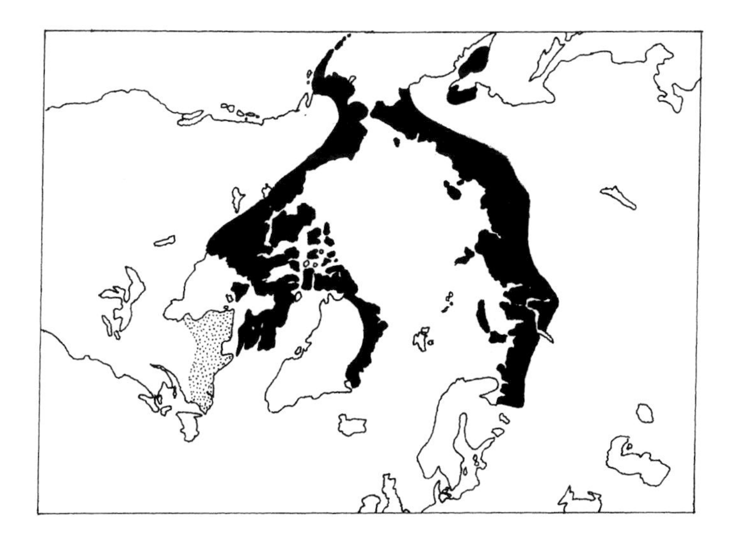

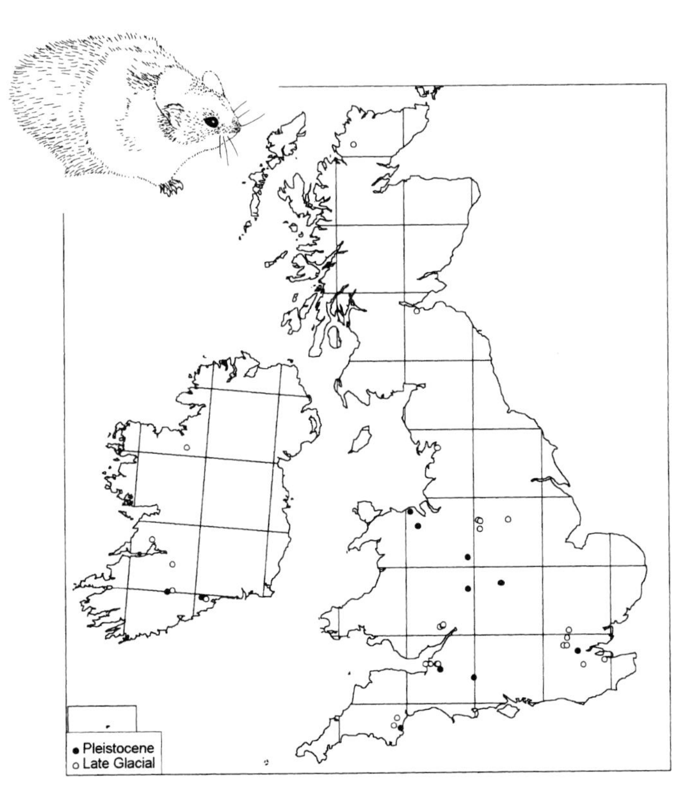

Figure 2.6 The modern distribution of the Arctic Lemming *Dicrostonyx torquatus*, and its fossil distribution in the British Isles. The genus is more restricted to the Arctic than *Lemmus*.

Figure 2.7 The modern distribution of the Narrow-headed Vole *Microtus gregalis*, and its fossil distribution in the British Isles. It occurs both in the Arctic and on the steppes of Russia.

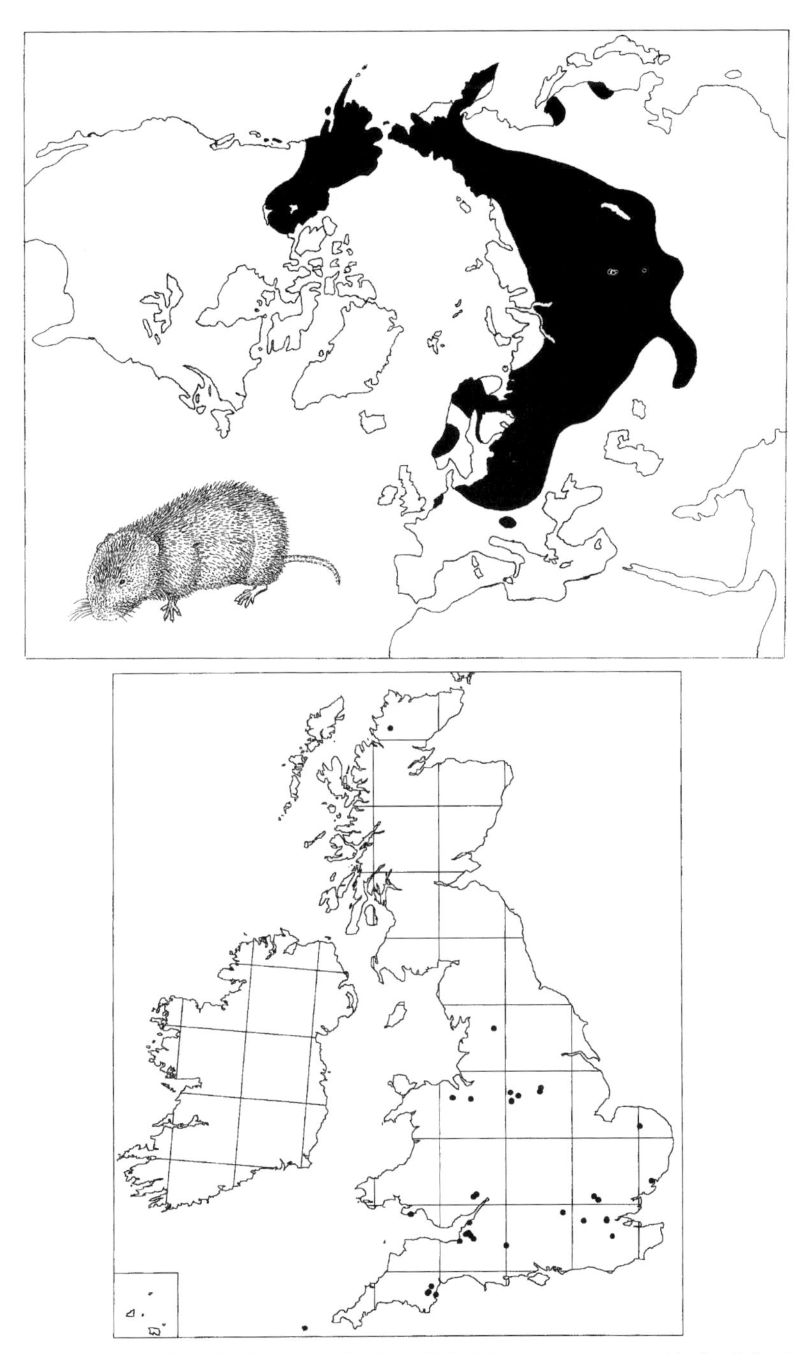

Figure 2.8 The modern distribution of the Root Vole *Microtus oeconomus*, and its fossil distribution in the British Isles. It is more widespread than *M. gregalis* at present, and has been found at more British sites.

The Younger Dryas

The reversion to colder conditions which is well reflected in the pollen, beetle and geological evidence is also hinted at by the mammal faunas, though their dating is rarely good enough to provide a convincing story on their own. Perhaps the most convincing is the fauna at Chelm's Combe Cave (Currant 1991), which includes Reindeer (with five ^{14}C dates between 10 910 and 10 190 b.p.) and Tarpan (10 370 b.p.). An interesting archaeological example, an antler modified into a so-called 'Lyngby' axe, from Earls Barton in Northamptonshire, has a date of 10 320 b.p. (Cook and Jacobi 1994). At Ossom's Cave in the Manifold Valley, Staffordshire, Reindeer was the predominant species (at least five individuals), and two young animals were dated to 10 780 and 10 600 b.p. (Scott 1986a). Reindeer is also well-dated from Gough's Cave (10 450 and 9920 b.p.; Currant 1991) and Tarpan from Robin Hood Cave (10 590 and 10 390 b.p.; Charles and Jacobi 1994). Grigson (1978) speculates that these were the only two ungulates in Britain at this time. One site with a good record of rodents from this period is at Nazeing, where *Dicrostonyx*, *Microtus gregalis*, *M. oeconomus* and *Arvicola terrestris* were recorded in a plant bed with Pollen Zone III plants (Hinton 1952). However, there are many other sites, particularly caves, that must also date from this period. In particular, it seems that the Steppe Pika *Ochotona pusillus* only occurred in Britain for the limited period of this time, and other associated species may therefore be dated to this period by proxy. Two Pika specimens have been specifically dated, from Great Doward Cave at 10 020 b.p. (Housley 1991) and from Broken Cavern, Torbryan at 10 180 b.p. (Price 1996). A third specimen, from Bridged Pot, also has a ^{14}C date associated with it, of 9090 b.p., but there are suspicions that this date is too late; perhaps the large mammal bone on which it was obtained is younger than the rest of the fauna with which it is associated (Burleigh *et al.* 1984, Bowman *et al.* 1990). Other records are undated (Table 2.4) but appear to be confined to southern Britain. One possible exception is the specimen listed from Robin Hood Cave, Creswell by Campbell (1977), but Charles and Jacobi (1994) were unable to confirm its presence there: however, the specimen has since been relocated and the identity confirmed (R. Jacobi, pers. comm.). It is striking that none of the many caves that have been excavated in the Peak District has yielded remains of *Ochotona*. The Pika is an interesting mammal, a small relative of the hares and rabbits which is best known for its habit of gathering hay in the autumn as a food store to see it through the winter. There are about 18 species (Erbayeva 1988), most of them confined to eastern Asia but with two species in North America (Fig. 2.9).

So what other small mammals do occur with Steppe Pika at these sites? The two lemmings, Root Vole and Narrow-headed Vole are frequent, and a specimen of Root Vole from Broken Cavern, Torbryan, has been ^{14}C dated to 10 370 b.p., confirming its presence in the Younger Dryas period (Price 1996). There are a few records of Sousliks (Ground Squirrels) *Spermophilus superciliosus* (in Sutcliffe and Kowalski 1976, but almost certainly the same as the extant European *S. citellus*), from Pin Hole Cave, Picken's Hole, Ightham Fissures, Fisherton, Langwith Cave and various Mendip sites, including Bleadon. One intriguing possible associate of the Sousliks is a large form of Polecat, originally regarded as a full species *Mustela robusta*, but probably simply a large Late Pleistocene form of the more familiar *M. putorius*. This was originally recognized in Germany, and described from Britain for the first time from Ightham Fissures in

Table 2.4 Records of Steppe Pika *Ochotona pusilla* from Britain (from various sources and Jacobi pers. comm.).

Westbury-sub-Mendip, Somerset	ST506504	Anglian Glacial?	Andrews 1990
Pontnewydd, Denbigh	SJ006714	Wolstonian Glacial?	Stuart 1983
Kent's Cavern, Devon	SX9264	Devensian?	Jackson 1953
Windmill Hill Cave, Brixham, Devon	SX931565	Devensian?	Jacobi pers. comm.
Great Doward Cave, Monmouthshire	SO5012	Late Glacial, 10 020 b.p.	Housley 1991
King Arthur's Cave, Hereford	SO546154	Late Glacial	Jackson 1953
Hutton Cave, Somerset	ST361582	?	Jackson 1953
Nazeing, Essex	TL3806	Late Glacial, P.Z.III	Hinton 1952
Broken Cavern, Torbryan	SX815674	Late Glacial, 10 180 b.p.	Price 1996
Chelme's Combe, Somerset	ST463545	Late Glacial, Y. Dryas	Jackson 1953
Rowberrow Cavern, Somerset	ST459580	Late Glacial	Jackson 1953
Wolf Den, Wavering Down, Somerset	ST359563	Late Glacial	Balch 1948
Aveline's Hole, Somerset	ST475588	Late Glacial	Jackson 1953
Tom Tivey's Hole, Somerset	ST705445	Late Glacial	Barrett 1966
Bridged Pot, Ebbor Gorge, Somerset	ST526488	Late Glacial, Y. Dryas	Balch 1928
Cow Cave, Chudleigh, Devon	SX864787	Late Glacial?	Jacobi pers. comm.
Robin Hood's Cave, Derbyshire	SK534742	Late Glacial	Campbell 1977
Cavall's Cave, R. Wye		Late Glacial	Jacobi pers. comm.
Badger Hole, Wookey Hole, Somerset	ST532481	?	Balch?
Gough's Cave, Somerset	ST468540	Younger Dryas?	Currant 1986
Sun Hole, Cheddar, Somerset	ST467541	Younger Dryas	Collcutt *et al.* 1981
Soldier's Hole, Cheddar, Somerset	ST468540	c. 10 300 b.p.	Campbell 1977
Ightham Fissures, Kent	TQ6056	Late Glacial?	Newton 1894
Happaway Cave, Torquay, Devon	SX9164?	Late Glacial	Jacobi pers. comm.

Kent (Newton 1894, 1899). It has been largely ignored ever since, but some specimens from Robin Hood Cave, Creswell, shown to me by Roger Jacobi, are evidently the same. Their skulls are about 12% bigger than modern Polecats. Were they adapted to feed on larger prey, perhaps Ground Squirrels, or were they another example of the larger size, perhaps in response to climate, seen in other Late Glacial mammals, such as Red Deer, Wolverine and Spotted Hyaena?

Comparisons with Elsewhere in Europe

During the Late Glacial, Britain was still a part of continental Europe, since sea-level was at about −100 m relative to present Ordnance Datum. The faunas of Belgium, France and Spain to the south of us are therefore highly relevant to our understanding of the British fauna of this period. In Belgium (Cordy 1991), a sequence of small mammal faunas provides a detailed story of faunal change which is what we ought to find in Britain, if only so many important sites had not been so hastily and (by modern standards) badly excavated. A full sequence is represented there. The Oldest Dryas fauna is predominantly *Dicrostonyx* (46%) and *Microtus gregalis* (26%). The Bølling

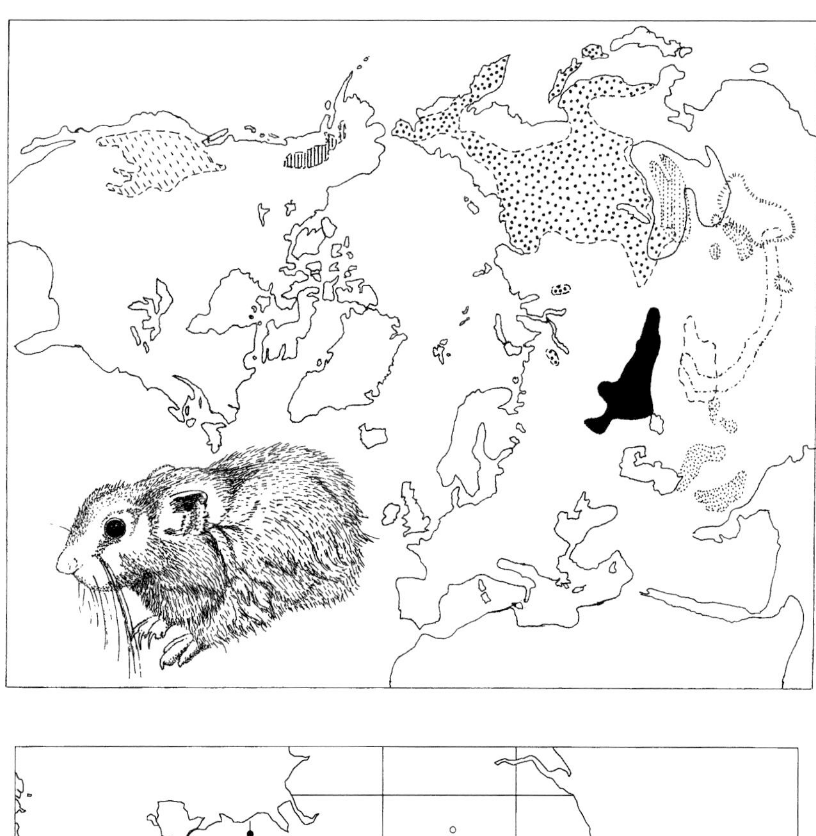

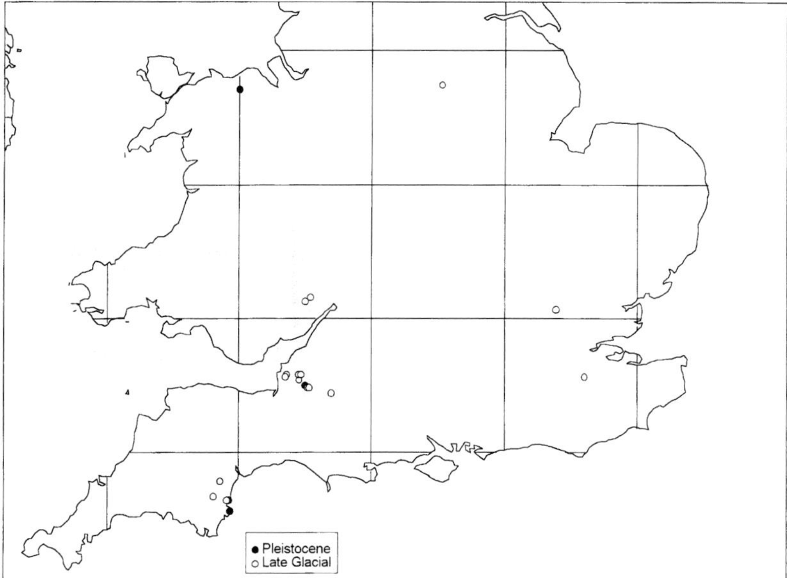

Figure 2.9 The modern distribution of the Steppe Pika *Ochotona pusilla* (black) and other species of *Ochotona*, together with its Late Glacial distribution in Britain.

Interstadial is recognizable, with *Microtus oeconomus* dominant (up to 39%) and *Ochotona* important (13%). In the Older Dryas, *Dicrostonyx* (up to 39%) and *M. gregalis* (up to 62%) become numerous again. The Allerød Interstadial has *M. agrestis* as the dominant species (up to 78% of the small mammal fauna), though with *M. gregalis* or *M. oeconomus* as its associates. In the Younger Dryas, *M. agrestis* remains numerous, but *Dicrostonyx* appears again (up to 20%), and *Ochotona* also makes a last appearance. Two interesting rarer rodents appear during the Late Glacial in Belgium, and also in France (Chaline 1972), the Birch Mouse *Sicista betulina* and the Common Hamster *Cricetus cricetus*. The Birch Mouse is at present a rather northern species, with a patchy distribution across Scandinavia and northern parts of Germany and Poland, but typically as its name implies in open birch woods. Although it looks like a mouse, with a long tail and large ears, it is more nearly related to the Jerboas (Dipodidae), and like hamsters has only two rows of cusps down its upper cheek teeth, rather than three rows as in mice. Moreover, it has an extra upper cheek tooth, a small premolar (Fig. 2.10). In any site it is rare; at Vaufrey in the Dordogne, for example, there were only seven individuals in some 9400 small mammals dating from the Riss Glaciation (Marquet, 1993) and it is recorded at only two Late Glacial sites by Chaline (1972), La Garenne and Poron des Cueches. Similarly in Belgium, in a sample of 3100 small mammals from several Late Glacial faunas, there was only one Birch Mouse (Cordy 1991). The species has never been recorded from Britain, but given its rarity it may yet be discovered (there is one specimen of Riss age from Jersey, see Chapter 8). The Common Hamster (Fig. 2.11) is a larger, rat-sized, relative of the familiar household pet (*Mesocricetus auratus*, a Middle Eastern species), weighing 200–500 g (Niethammer 1982). It is an inhabitant of open steppe grassland, but has become an agricultural pest.

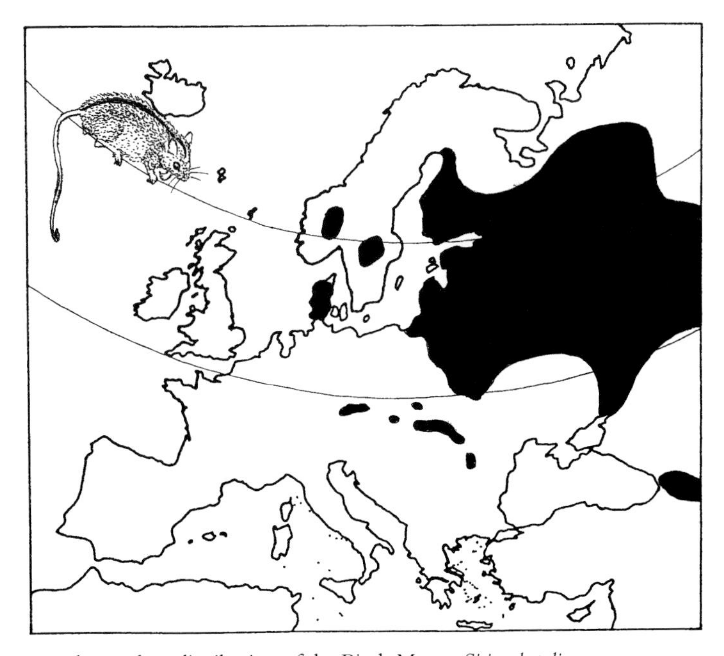

Figure 2.10 The modern distribution of the Birch Mouse *Sicista betulina*.

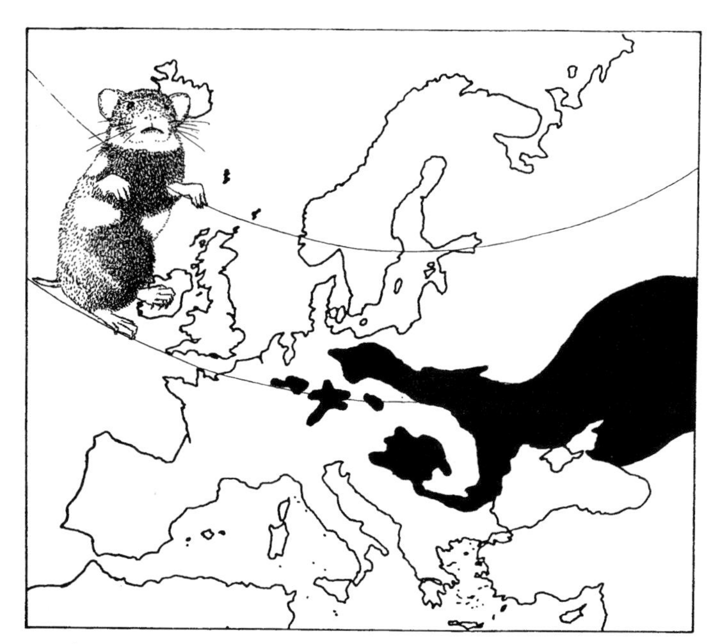

Figure 2.11 The modern distribution of the Common Hamster *Cricetus cricetus*.

As a consequence, it has been persecuted and is now rare in Western Europe, so is the subject of an active conservation campaign in the fragmented western end of its range, in Belgium, the Netherlands and western Germany (Smit and Wijngaarden 1981). In the Late Glacial, it extended its range westwards into Belgium, where Cordy (1991) lists it from six faunas of Bølling Interstadial and Younger Dryas age, and into eastern France, where it is listed from four caves by Chaline (1972). It has been recorded as a fossil in Britain, but only from much earlier times, including the Cromerian at West Runton, Norfolk, and the Glutton Stratum of Tornewton Cave, Devon, probably of Wolstonian (equivalent to Riss) age (Sutcliffe and Kowalski 1976). Again, this is a fairly rare species, and might yet be found in Late Glacial deposits in Britain.

The larger mammals have been more thoroughly studied, as they constitute the food resource for humans at archaeological cave sites in France and Spain. Gordon (1988) summarizes the ungulate/prey faunas at 50 sites belonging to the Late Glacial period in France; Reindeer were the main prey at 20 sites throughout the country, and second prey at a further seven. In some places Tarpan (at six), Saiga (one), Ibex (one) and Bison (two) were the main prey, with Aurochs, Chamois and Red Deer also important at some sites. During the cultural period termed by the archaeologists the Terminal Magdalenian, about contemporary with the Windermere Interstadial in Britain, hunters north of the Pyrenees were hunting primarily Reindeer (up to 72% of the fauna), with the Horse (Tarpan) also important; Roe and Wild Boar were almost absent, while Red Deer was scarce (up to 16% at one site). South of the Pyrenees, Red Deer was the predominant prey (up to 91% of the fauna), and Roe contributed up to 11%; Wild Boar were present at most sites, though scarce, as were Tarpan. Reindeer was only present south of the Pyrenees for very limited periods, though it did form up to 12.5% of the fauna at one site. Some sites in rocky areas did not conform to this pattern; at these the hunters specialized in killing Ibex *Capra pyrenaica* and Chamois *Rupicapra rupicapra,* and Red Deer were important secondary prey at such sites (Straus 1983). In the Azilian, roughly equivalent to the Younger Dryas in time, Reindeer had practically disappeared so far south, and Red Deer were the primary prey both in northern Spain and in southern France, with Roe, Horse, Wild Pig and Aurochs as secondary prey, and with specialized Ibex hunting continuing in appropriate sites (Straus 1986).

The end of the Würm/Devensian Glaciation is also the period when the famous cave paintings, at sites such as Lascaux in France and Altamira in Spain, were being executed. Dating of cave art is difficult, but similarities of style with similar engravings on pieces of bone or ivory preserved within datable archaeological deposits have allowed some to be given dates, and the archaeological deposits in the same caves may provide indications of dating. Altamira is believed to date from about 15 900–15 500 b.p., in the Late Glacial period (Bernaldo de Quiros 1991), and all the cave art seems to belong in the later part of the Devensian/Würm Glacial, from about 27 000 b.p. onwards. Fortunately Powers and Stringer (1975) provided a useful review of cave art from a zoological perspective, and have drawings of many examples to indicate the validity of the identifications. The most abundantly illustrated, and readily recognizable, species include Reindeer and Red Deer, Horse, Bison, Aurochs and Woolly Mammoth (Fig. 2.12). Less numerous, but equally recognizable, species include Saiga, Ibex, Wild Boar, Lion and Brown Bear. Rarer species include Musk Ox, Chamois and Hare, while probable identifications include Wolf, Cave Bear and Glutton (Wolverine). There are several very good drawings of seals, surprising in view of the inland nature of most cave sites, but Altuna (1983) has pointed out that cave art seems not simply to depict the hunters' main prey. At the cave of Ekain, Cantabria, there are 59 paintings, predominantly Horse (34) and Bison (11); other species depicted are Ibex, Red Deer, Brown Bear, Rhinoceros and fish. The excavated fauna contains over 1000 identified bones, of which 69% are Red Deer and 24% Ibex; none of the other species (Chamois, Bison/Aurochs, Reindeer, Roe Deer, Horse, Cave Bear, Wolf and Salmon) contributed more than 17 bones. The pattern at the nearby site of Tito Bustillo is similar; 83% of over 4000 bones are of Red Deer, while the two most common among 72 paintings are Horse (27) and Red Deer (23). Thus the cave paintings certainly do not simply depict the most common targets for the hunting communities who drew them, though they do generally agree well in showing the local faunas as represented in the fossil communities. The depiction of various carnivores surely reflects the dangers of the hunters' environment, and the frequent illustration of such formidable potential prey as Bison, Aurochs and Woolly Mammoth may indicate what they hoped to kill, rather than what they managed to kill.

Late Pleistocene Extinction

One of the most perplexing, and currently contentious, issues surrounding the Late Glacial faunas is the extinction of such distinctive species as Woolly Mammoth, Woolly Rhinoceros and Irish Elk. These total extinctions are obviously the most intriguing, but they are clearly associated with the loss from Western Europe of other

Figure 2.12 Cave paintings and engravings from southern Europe (after Powers and Stringer 1975). Some of the cave paintings were crude, reflecting the difficulty of the medium, but the engravings on ivory, antler and bone (such as the heads here of Musk Ox, Lynx, Saiga and Chamois) were often exquisite, showing the prowess and acute observations of the artists.

species that survive elsewhere – Musk Ox in North America, Lion, Hippopotamus and Spotted Hyaena in Africa, Saiga, Steppe Pika and Sousliks in Eastern Europe or Asia, Reindeer and lemmings in the Arctic. The conventional explanation from biologists has generally invoked the changing climate and vegetation that characterized the Pleistocene as a whole, and the end of the Last Glaciation in particular. However, there has been over the last 30 years or so a strong argument from archaeologists in particular (and with some biological support) that human hunting played a critical part. In particular, the immigration of humans into North America at the end of the Pleistocene has been argued as having a sudden impact on the relatively naive large mammal fauna of that continent, resulting in the rather sudden extinction of some 42 large (over 1 kg) mammals. This has been termed the Pleistocene Overkill Hypothesis, and owes most to the arguments of the American Paul Martin (especially Martin 1967, 1984). There is no doubt that in some parts of the world, the arrival of humans was followed very rapidly by the extinction of most of their larger prey. The two clearest examples are the extinctions of the Moas in New Zealand (13+ species of large flightless birds that became extinct in the 700 years following the arrival of the Maoris, who certainly hunted them; Cassels 1984) and the losses of the large diurnal lemurs, dwarf Hippopotamus and the Elephant Bird *Aepyornis* from Madagascar following the settlement of that island about A.D. 900 (Dewar 1984). The evidence from the main continents is less convincing. In Africa, where the fauna perhaps evolved along with Man, and was therefore less naive, only two out of 44 genera of large (over 44 kg) species became extinct. Similarly in South-east Asia, where there have been human hunters for at least half a million years, most of the large Pleistocene mammals still exist. In North America, Martin (1984) argues that 33 (73%) out of 45 genera were lost, while South America lost 46 (80%) of 58 genera and Australia 15 (94%) of 16 genera. In Europe, losses were less marked; Stuart (1991) suggests that only seven (29%) out of 24 were lost. Actually, of course, it is species, not genera, that become extinct (genera being somewhat abstract groupings made by zoologists, whereas species are groups of animals that recognize each other as potential mates and rivals). The reason for using genera in these calculations is that the fragmentary remains sometimes available to palaeontologists cannot always be readily assigned to species. As examples, both in North America and in Eurasia there is considerable doubt how many species of Horse *Equus* were present in the Late Glacial; up to four have been named in both areas, including an Ass *E. hydruntinus* and genuine horses, *E. ferus/przewalskii,* in Europe. Clearly, including species of doubtful validity increases the apparent severity of extinction. Several features of these extinctions attract attention. Larger mammals were much more severely affected than small ones; in North America, all the elephants (three species of mammoth and a mastodon), horses, camels and ground sloths disappeared, but it is hard to find a good example of a mouse, vole or shrew that is extinct. Possible small mammals extinct in North America include a vole *Synaptomys australis,* a deermouse *Peromyscus imperfectus*, a cotton-tail *Sylvilagus leonensis* and a vampire bat *Desmodus scotti* (Stuart 1991), but their taxonomic or stratigraphic position is uncertain. In Europe, one vole, *Pliomys lenki,* survived into the Upper Pleistocene in Spain, having been widespread in the Middle Pleistocene across south and central Europe, and died out at about 29 000 b.p.; other possible small mammal extinctions from Europe (a mole *Talpa magna,* hamsters *Allocricetus bursae* and *Cricetus major,* and water vole *Arvicola antiquus*) are doubtfully distinct from surviving relatives.

The best arguments for the Pleistocene Overkill Hypothesis in a continental context undoubtedly apply to North America. The current belief is that humans entered the continent from Siberia across the Bering land-bridge into Alaska about 13 000 b.p., and penetrated into the central plains about 12 000 b.p. when a corridor opened between the Cordilleran (Rocky Mountain) and Arctic ice-sheets. There are good sequences of ^{14}C dates for many of the extinct species which seem to end rather uniformly at around 10 500 b.p.; there are very few later dates for extinct species, and most of them are technically suspect for one reason or another. There is also a characteristic fluted stone point, the Clovis point, which was used by the early American hunters, and is specifically found with mammoth and mastodon specimens at several archaeological sites (Stuart 1991). Martin (1984) has argued that the geographical coincidence of these points with the distribution of the extinct elephants is itself good evidence for the Overkill Hypothesis, but one would expect hunters, their prey and their implements to coincide whether or not the prey were being overexploited. However, it must be admitted that at present the coincidence of the latest dates for extinct large mammals in North America with the arrival of humans does look convincing. What about the evidence from Europe and Asia?

Here too there are good sequences of radio-carbon dates, and even more archaeological evidence. Humans have been present for at least 500 000 years in the southern part of these continents, and the evidence of archaeological sites discussed above in Spain and France shows that humans were hunting these extinct species, as well as drawing them on cave walls, throughout the Last Glacial. Mainly, however, their prey were the smaller deer (Red Deer and Reindeer, particularly); the large, now extinct, species rarely figure. Even the famous camps made of Woolly Mammoth bones in Poland and southern Russia may have been made from bones of already dead mammoths, picked up rather than hunted deliberately; they show variable amounts of weathering, indicating that some of them had been lying around for some time (Stuart 1991). The latest dates for Woolly Mammoth are around 12 000 b.p. in Britain and France, about 12 700 b.p. in the Ukraine, and 11 000 b.p. from the Russian Plain. In Siberia, however, dates as late as 9600 b.p. have been obtained on soft tissue of specimens frozen in the tundra. Most remarkably, evidence has recently been reported of a dwarf race of Woolly Mammoth which survived on Wrangel Island in the Siberian segment of the Arctic Ocean. There were large specimens, comparable in size with typical Siberian mammoths, dated to 20 000 and 12 980 b.p., as well as a sequence of dates of dwarf mammoths between 7250 and 4010 b.p. (Vartanyan *et al.* 1993, Long *et al.* 1994). The dwarf forms were about 20% smaller – at least, their cheek teeth were, for there are no skeletons yet – and must have evolved their small size as an isolated population on the island during the 6000 years that they lingered on as the last remnants of their species. The history of the Woolly Rhinoceros is less clear, for it was always much rarer than the mammoth. The latest ^{14}C dates include several Devensian dates from Britain between 35 000 and 22 350 b.p., and a few dates from Siberia down to 27 300 b.p. Indirect dating, from archaeological sites, suggests that it may have survived to about 13 000 b.p. in southern France, and to 12 500 b.p. in Germany. There appear to be no later dates there or anywhere else, and it seems that neither it nor the mammoth survived into the Younger Dryas in Western Europe. The evidence for the Irish Elk *Megaloceros giganteus* is very similar. The Irish sites all seem to belong to the Woodgrange (equivalent to Allerød or Windermere) Interstadial, and the latest ^{14}C

date anywhere is an inprecise date from Battybetagh Bog of 10 600 b.p. There are also dates from southern Sweden at 11 490 and 11 340 b.p., from Denmark at 11 630 b.p., and from England at 12 850–10 700 b.p. These seem later than any evidence from elsewhere in Europe or Siberia, and it seems that the species had retreated to western Europe, where it was moderately abundant for a short period during the Allerød Interstadial, but was then extinguished by the cold snap of the Younger Dryas. In Ireland, there were, so far as we know, no humans at this period, and in particular there is no evidence that any of the numerous skeletons at Ballybetagh had been attacked (Barnosky 1986). On the other hand, there is evidence that the stags that died there were smaller and perhaps weaker than stags from other sites (and times?) in Ireland. The Cave Bear *Ursus spelaeus* is the one carnivore that is entirely lost, but its ecology and distribution in time are less certain than for the ungulates. A larger relative of the Brown Bear, it is common in certain Alpine cave sites of Late Pleistocene date. It appears to have been confined to Europe throughout its history, extending south as far as northern Spain and north as far as central England in the west, and across to the Caucasus in the east. The latest date from the Würm Glacial is 18 720 b.p. from Saint Michel in the French Pyrenees, and similar dates are suggested from Spain and Italy (Stuart 1991). It may have been more herbivorous than the Brown Bear, to judge from its flattened molars, and might have been confined to upland habitats; there is some evidence that populations in different areas were somewhat distinct, implying a degree of isolation of different subpopulations. It seems that the species was unable to cope with climatic changes at the end of the Pleistocene, but without better information on its ecology it is hard to explain its demise properly.

The species that died out in Europe but survive elsewhere show patterns that suggest a retreat connected with climatic and habitat change. The Reindeer, to take the most obvious example, disappeared from southern France during the Allerød Interstadial, and seems not to have returned there in the Younger Dryas. In Britain, it certainly was present in the Younger Dryas, but died out early in the Mesolithic (see Chapter 3). Both Norway Lemming and Arctic Lemming show a retreat northwards from western Europe which matches the pattern of climatic and vegetational change. Both were present in France during the Würm, but failed to return in the Younger Dryas (Chaline 1972, Marquet 1993). They did, however, get back into Belgium during the Younger Dryas, as they did into southern England. Arctic and Norway Lemming survived in the uplands of southern Poland into the Holocene (Nadachowski *et al.* 1989). The Musk Ox was always rare in western Europe, and the available dates all belong in the cluster between 18 000 and 13 000 b.p.; however, there are dates from Siberia that range from 32 000 down to 10 750 b.p., and then four dates between 3800 and 2900 b.p. from the Taimyr Peninsula. This is a classic case of a northward retreat, and moreover shows that this species survived long into the Post Glacial period (Stuart 1991).The steppe species (Pika, Saiga, sousliks) seem never to have been abundant in western Europe, and only occurred during a few short periods when, presumably, conditions were appropriately cold but dry. The species now with an African distribution disappeared much earlier. The Hippopotamus, characteristic of the Ipswichian Interglacial, only occurred in the western fringe of Europe even then, mostly in France and England, with one record from the Rhine. It seems to have lingered on in Italy into the early part of the Würm, though there are no ^{14}C dates. The Spotted Hyaena and Lion, however, survived through the Last Glaciation in southern France and northern Spain. The latest dates for the hyaena seem to be around 14 000–11 000 b.p., and there are some possible cave paintings of the species. The latest record for the Lion is a remarkable ^{14}C date of 10 670 b.p. from Lathum, Netherlands (Stuart 1991), though this fits with classical evidence that the species survived into historical times in Greece.

How does one interpret this record? It seems clear that Late Glacial habitats were not quite like either tundra or steppe environments, but some mixture of the two. Presumably the winters were too cold, and possibly the summers were too dry as well, for tree growth. However, growing seasons would have been much longer than they are now in the Arctic, and the productivity of the herbage must have been much greater. In evaluating the herbivore communities of the Bering land-bridge, Bliss and Richards (1982) argue that the habitat included more of a mosaic than the present tundra, and moreover that grasses were much more abundant (as shown by pollen analysis). They also point to the importance of willow or other shrub vegetation in the community as browse for ungulates, and macroscopic remains of various arctic willows (*Salix phylicifolia, S. reticulata, S. repens, S. polaris*) are common at British Late Devensian sites (Godwin 1975). Pollen of *Salix* is also abundant in the Late Glacial pollen record across much of north-western Europe, including Britain (Huntley and Birks 1983). Such arguments would apply even more strongly to locations at around 50°N. Such a plant community would support the large herbivores that characterize the Late Glacial, and, indeed, larger herbivores would cope better with the winter cold. These in turn would support the large predators which roamed Europe at the time, and the human hunters. The retreat of this community as the climate improved, and its split between the dry cold steppes and the wet cold Arctic, meant that some mammals also retreated to one or other habitat and others died out as their habitat disappeared.

There is no indication in Europe that human overhunting played any role in their extinction. In particular, the arrival of new populations with new tools does not seem to fit the pattern of European extinctions, in contrast to the story in North America. However, this 'Climatic Hypothesis' conclusion leaves unanswered two related questions: why did the similar changes at the ends of previous glaciations not also cause major extinctions? why did the extinct species not retreat to refuges as they must have done previously? The evidence of the Musk Ox in the Taimyr and the dwarf Mammoths on Wrangel Island suggests that they did try to retreat, but that the refuges were not big enough, or sufficiently long lasting. However, there is no satisfactory answer for the other question. Perhaps one might be forthcoming if the state of the glacial and interglacial floras at high latitudes could be compared for periods earlier in the Pleistocene. It seems that interglacials have been much shorter than glacials, in other words that cold conditions have persisted longer and therefore required of the fauna that they become increasingly adapted to cold conditions as the Pleistocene has progressed. It is possible that changes at earlier transitions were not quite like the most recent one, but it will need much more research to investigate this.

Conclusions

The Late Glacial in Britain was characterized by a mammal fauna that looks, on present-day equivalents, to be a mixture of northern Scandinavian mammals and mammals from the central steppes of Russia together with some totally extinct species. This matches a flora which was either treeless, or a mixture of open ground with birch woodland. The beetle faunas as well as direct measures of temperature confirm that it was actually warm enough for tree growth, and this must have been a productive environment, well able to support the herds of large mammals that were present. The appearance of Red Deer and Aurochs, two essentially southern species, during the warmer part of the Windermere Interstadial is not so surprising.

The archaeological specimens and the cave art both confirm that essentially northern species, notably Reindeer and Horse, were abundant as far south as southern France during the Late Glacial, and steppe species such as the Saiga appeared there in some abundance for short periods. South of the Pyrenees in northern Spain, conversely, Reindeer were rather uncommon, indeed absent most of the time, and Red Deer were the common prey of human hunters. It is surprising that Roe Deer and Wild Boar were not more common, and this prompts the double question, what was the habitat like, and did it provide the expected refuge for our present, temperate woodland, fauna?

It has been realized from pollen analysis that the deciduous woodland trees, like Hazel, Oak, Alder, Elm, Beech and Ash, species that would be the typical cover of Britain had they not been so severely cleared by Man, were largely confined to the

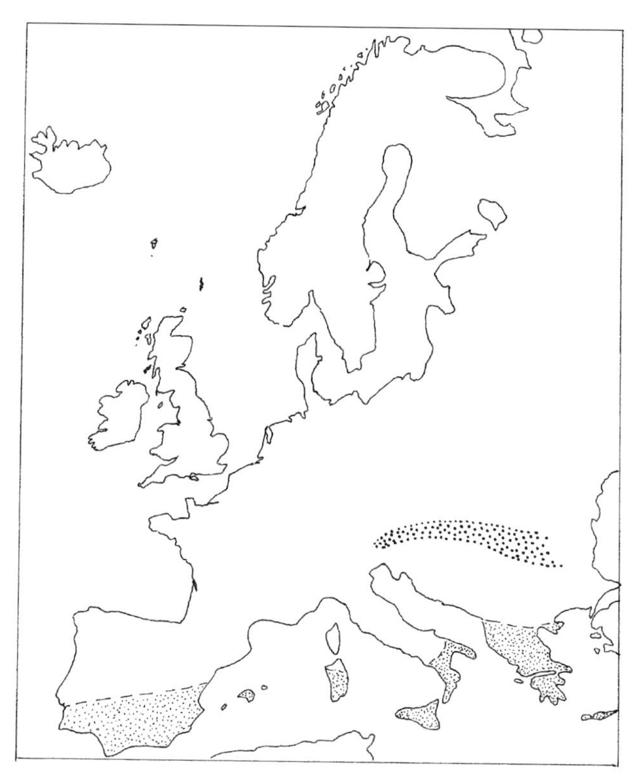

Figure 2.13 Glacial refuges in southern Europe, as demarcated by areas with more than 5% oak (fine stipple) or alder (coarse stipple) pollen at 13 000 b.p. (after Huntley and Birks 1983).

southern extremities of Europe during the Last Glaciation. The pollen atlas produced by Huntley and Birks (1983) shows Oak species confined to Spain, Italy and Greece at the glacial maximum (Fig. 2.13), and spreading northwards at varying speeds during the Post Glacial. Alder, Elm and Beech seem to have been confined to the Balkans, while Ash seems to have survived in a small refuge in northern Italy, just south of the Alps. The status of Hazel is uncertain; it was nowhere abundant during the Late Glacial, not even in the warmer interludes, yet it suddenly flourished in the Post Glacial; possibly it had a refuge in the Bay of Biscay lowlands, now flooded, or perhaps its pollen production was suppressed by all the ungulates feeding on it.

Detailed pollen curves now available for Mediterranean sites confirm that Oak, for example, was present at Ioannina in Greece throughout the last 423 000 years, that is through at least five glacial periods (Tzedakis 1993). However, the record also shows that during the glacial periods, the pollen record was dominated by pollen of herbaceous species, especially grasses and *Artemisia*, even in southern Greece; trees were present, but not forest. Hazel was only sparsely and sporadically present during this time. It seems as though the landscape was more of a parkland, with groves of trees in wetter places (Willis 1996), particularly where mountains intercepted rain-bearing winds coming in from the oceans. In these circumstances, it seems likely that many of the forest mammals, such as Wild Boar and Roe Deer, were themselves rather uncommon, and woodland specialists like the dormice may have been confined to small geographical pockets. Unfortunately at present the fossil record for such species is too poor to confirm these speculations.

CHAPTER 3

WARMING UP NICELY

Brown Bear eating bilberies

THE Pleistocene ended very abruptly. Within 50 years, the climate changed from an Arctic coolness, with mean July temperatures in central England of around 8°C, to a warm temperate climate with mean July temperatures of about 17°C. On the present map of Eurasia, that is equivalent to the difference between the Taimyr Peninsula and London. It is still a matter of some discussion how and why the climate changed so quickly, and the answer has considerable relevance to present-day concerns about global warming. Current thinking invokes major changes in the way deep currents circulate in the oceans. A sudden input of cold fresh water as the ice-cap over northern Europe started to melt could have interfered with the northward current of warm surface water (the Gulf Stream Drift) and triggered a period of instability in these currents (Rahmstorf 1995).

Whatever the basic cause, the effect is quite certain. It is confirmed by direct physical measures of the temperature using the $^{18}O/^{16}O$ ratio in shells of foraminiferans and molluscs, in ice cores and in lake carbonates (^{18}O is a rare isotope of oxygen, constituting only 0.2% of atmospheric oxygen, and is more soluble than the common ^{16}O at higher temperatures, so the ratio of the two gives a direct temperature record). Beetle faunas on land (Atkinson *et al.* 1987) and foraminiferan faunas in deep ocean sediments tell the same story; species characteristic of cold climates were rapidly replaced by species of warmer climates (Fig. 3.1). Peter Osborne's beetle faunas from West Bromwich in the Midlands are the classic example in England (Osborne 1980). His layer E, with a ^{14}C date of 10 025 b.p., has an arctic/alpine fauna including *Olophrum boreale*, *Acidota quadrata*, *Helophorus sibiricus* and *Boreaphilus henningianus*. The overlying layer F, dated to 9970 b.p., has none of these, but has instead species which occur widely across Europe, including several waterside specialists such as the weevil *Limnobaris pilistriata*.

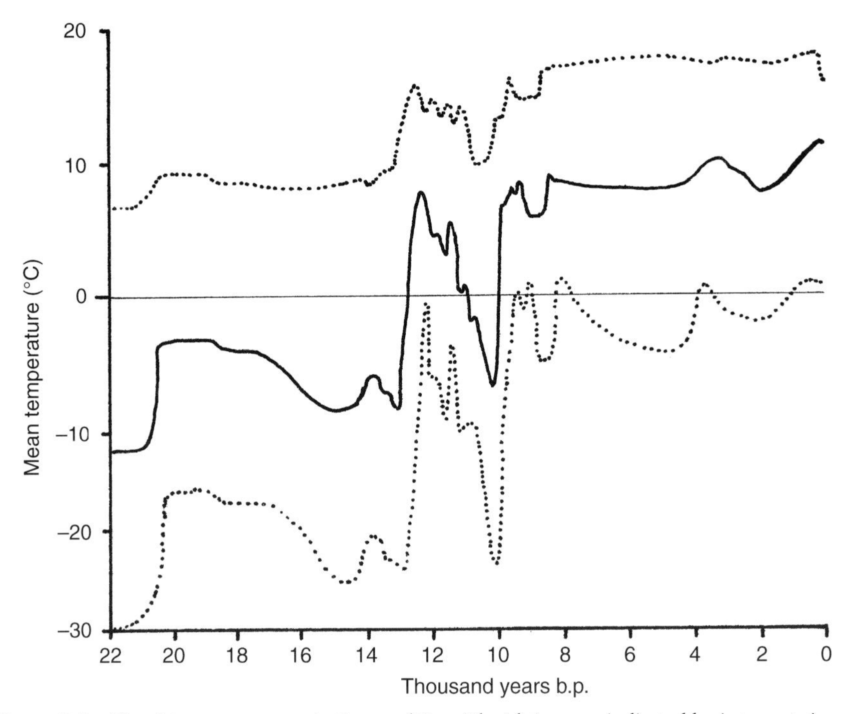

Figure 3.1 The rising temperature in Late and Post Glacial times, as indicated by interpretation of fossil beetle faunas (after Atkinson *et al.* 1987). There are only three beetle faunas prior to 15 000 b.p. and five since 8000 b.p., but 49 faunas cover the critical period in between, giving a good record of the rapidly changing conditions. Each point is the average climatic requirements of the dozen or more beetles found in each fauna, the solid line gives the average annual temperature, the dotted lines show the mean temperatures of the coldest and warmest months.

Post Glacial Mammals

It would be nice to chart the same change in the mammal fauna. We cannot. The Late Glacial (Younger Dryas) tundra fauna with Arctic Lemmings recorded, for example, at Nazeing or at Robin Hood's Cave, seems to have been very quickly replaced by a more temperate fauna. The difficulty is partly that so many good mammal faunas were excavated long ago, before the complex climatic changes, and equally complex stratigraphic record in many caves, were properly appreciated, but there is also an element of circularity in the records. If a site has lemmings or Reindeer, then it must be a Late Glacial site, and if it has instead Wood Mice or Red Deer, then it must be Post Glacial. There has to be a way out of this circle, and it is provided by carbon dating, which can now be done on small quantities of bone, without totally destroying precious specimens. What we need to know is whether in fact some of the typical mammals of Late Glacial times lingered on into the warmer climate. Some certainly did. There were Reindeer at Dead Man's Cave, Yorkshire, which date to 9940, 9850 and 9750 b.p., appreciably later than the conventional end of the Late Glacial at 10 200 b.p. Probably the Reindeer survived later in Scotland, and it is perhaps no accident that the latest date for it is 8300 b.p. from Inchnadamph in Sutherland (Clutton-Brock 1991). The Wild Horse (Tarpan) may also have lingered on into the Post Glacial on the higher, less wooded, ground in for example the Peak District (Bramwell 1977a). Though the latest date available to Clutton-Brock (1991) was from the Darent valley in Kent, at 9770 b.p., one of 9330 b.p. from Seamer Carr postdates it (see below).

However, the classic Mesolithic sites of Star Carr in Yorkshire and Thatcham in Berkshire have radio-carbon dates of 9488 b.p. and 10 050–9600 b.p., respectively, putting them within 700 (radio-carbon) years of the climatic transition (Fig. 3.2). There is also direct evidence of the vegetation around both sites, from pollen-analysis. Star Carr was predominantly birch woodland with a little pine and just the beginning of alder immigration, to judge from the pollen record (Walker and Godwin 1954, Cloutman and Smith 1988). Seeds of Hairy Birch *Betula pubescens* confirm that evidence; seeds of reed *Phragmites australis* and rhizomes of the sedge *Cladium mariscus* were also recovered, along with both White and Yellow Water-lilies and Bog-bean. The camp was evidently sited on a gravel peninsula in the reed-bed at the side of a lake, with birch scrub on the drier ground nearby.

The Star Carr fauna was predominantly large ungulates, clearly the prey of the Mesolithic hunters, but these were Red and Roe Deer, Elk (i.e. Moose), Aurochs and Wild Boar. The Reindeer and Tarpan characteristic of the Late Glacial had gone. The rest of the mammal fauna included Hedgehog, Beaver, Hare, Pine Marten, Badger, Wolf, Red Fox and Brown Bear (Fraser and King 1954, Legge and Rowley-Conwy 1988). This is the fauna of a temperate forest, and would be found in Poland or southern Sweden at the present day, if the Aurochs were not extinct. The original account of the fauna suggested that Red Deer, with 80 specimens, were much the most common target of the hunters, with Roe Deer (33) also numerous; Elk, Aurochs and Wild Boar were much less common. Exactly how many animals were represented, though, is subject to some discussion. The earlier counts were based on the abundant antlers found at the site, and indeed led to the conclusion that stags and bucks were hunted preferentially. Legge and Rowley-Conwy (1988) argue convincingly that most of the antlers were imported to the site as tools, or the material for

Figure 3.2 Mesolithic sites in Britain with mammal faunas of interest.

making them, and it is perfectly clear that, indeed, most of the antler material, of Elk as well as the smaller deer, was used as a raw material. Some of the Red Deer frontlets had holes in them for thongs, and were clearly used as head-dresses of some sort. Using counts based only on the bones (ignoring the antlers, that is), there were 26 Red Deer, 17 Roe, 16 Aurochs, 12 Elk and four Wild Boar. Thus the deer were still the most numerous prey, but not so overwhelmingly as the earlier figures suggested. The rarer species were mostly represented by one or two specimens only, though there were at least eight Beavers, as befits a lakeside site. The identity of the hare is intriguing; it is represented only by the shaft of a tibia, but Fraser and King (1954) thought it too robust to be Mountain Hare, so very tentatively suggested that it should be a Brown Hare. All the records of fossil hares and Rabbits from Britain were re-examined by Mayhew (1975), including this one. He pointed out that Mountain Hares get larger as one goes north in Europe, and at such an early post-glacial date one might expect Mountain Hares to be larger than they now are in Britain. He suggested, therefore, that the Star Carr specimen should be regarded as more probably from Mountain Hare. Needless-to-say, a tibia is insufficiently distinct to be assigned confidently to either species. Initially, the presence of Brown Bear was overlooked, but Noe-Nygaard (1983) drew attention to the presence of a characteristic axis vertebra, and Legge and Rowley-Conwy (1988) mention a jaw mixed in with those of Wild Boar. Another site nearby in the Vale of Pickering has Horse as well as Wild Boar, Aurochs, Red and Roe Deer, with a slightly earlier date of 9330 b.p. (Schadla-Hall 1988, Simmons 1995).

As the radio-carbon dates indicate, Thatcham covered a longer time span, during which the vegetation evolved from birch (Pollen Zone IV) through birch–pine (Pollen Zone V) to pine–hazel (Pollen Zone VI) woodland. The archaeological site occupied a gravel terrace alongside a lake in the valley of the River Kennet, in the Thames basin about 3 km east of Newbury (Wymer 1962). Macrofossil evidence of Hazel (nuts), birch, pine and sedges confirms in part the pollen analysis. The mammals present include most of those present at Star Carr – Red and Roe Deer, Elk, Aurochs, Wild Boar, Beaver, Fox, Wolf, Badger, Pine Marten and Hedgehog, though not Brown Bear or Hare. In addition, Common Shrew, Water Vole and Wild Cat are present, and there are a few teeth of Tarpan. More controversially, Rabbit was claimed (King 1962). This is, again, essentially the fauna of the temperate zone of western Europe, though the Tarpan might be a sign of the recent ending of colder, more open, conditions. The Rabbit is out of place on two counts. First, it is a native of the Mediterranean zone, and second is thought to have been introduced by the Normans less than a thousand years ago (see Chapter 5). Interestingly, there is plenty of evidence that Rabbits were regularly eaten by Mesolithic hunters in southern France, as well as in Spain and Italy (Jarman 1972). This led to the suggestion that perhaps the Rabbit was native to Britain, and did reach here early in the Mesolithic, as the Thatcham bones suggest, only to become extinct later as woodland became more extensive; Rabbits are not forest animals (Yalden 1982). My speculation was wasted; the Thatcham Rabbit bones have been radio-carbon dated directly and are only 270 b.p. (i.e. A.D. 1710) (Yalden 1991). Rabbits are, of course, notorious burrowers, and show no respect for archaeological sites or layers. At Thatcham, there is much less evidence of the use of antlers as material for tools than at Star Carr (there are similar quantities of flint tools and flint waste at the two sites), and the counts of bones give direct evidence of the numbers of animals killed. At least eight Red Deer and six Roe Deer were represented, along with seven Wild Boar, two Aurochsen and one Elk (Fig 3.3).

Another securely dated early Mesolithic site is Dog Hole Fissure in the famous cave area of Creswell Crags, astride the Derbyshire–Nottinghamshire border (Jenkinson 1984). A small fissure seems to have filled very quickly, for the cave deposits contain the bones of a single Wolf, Red Deer and Wild Boar spread through most of the 1.54 m depth of deposits. The additional mammals include Red Fox, Wild Cat,

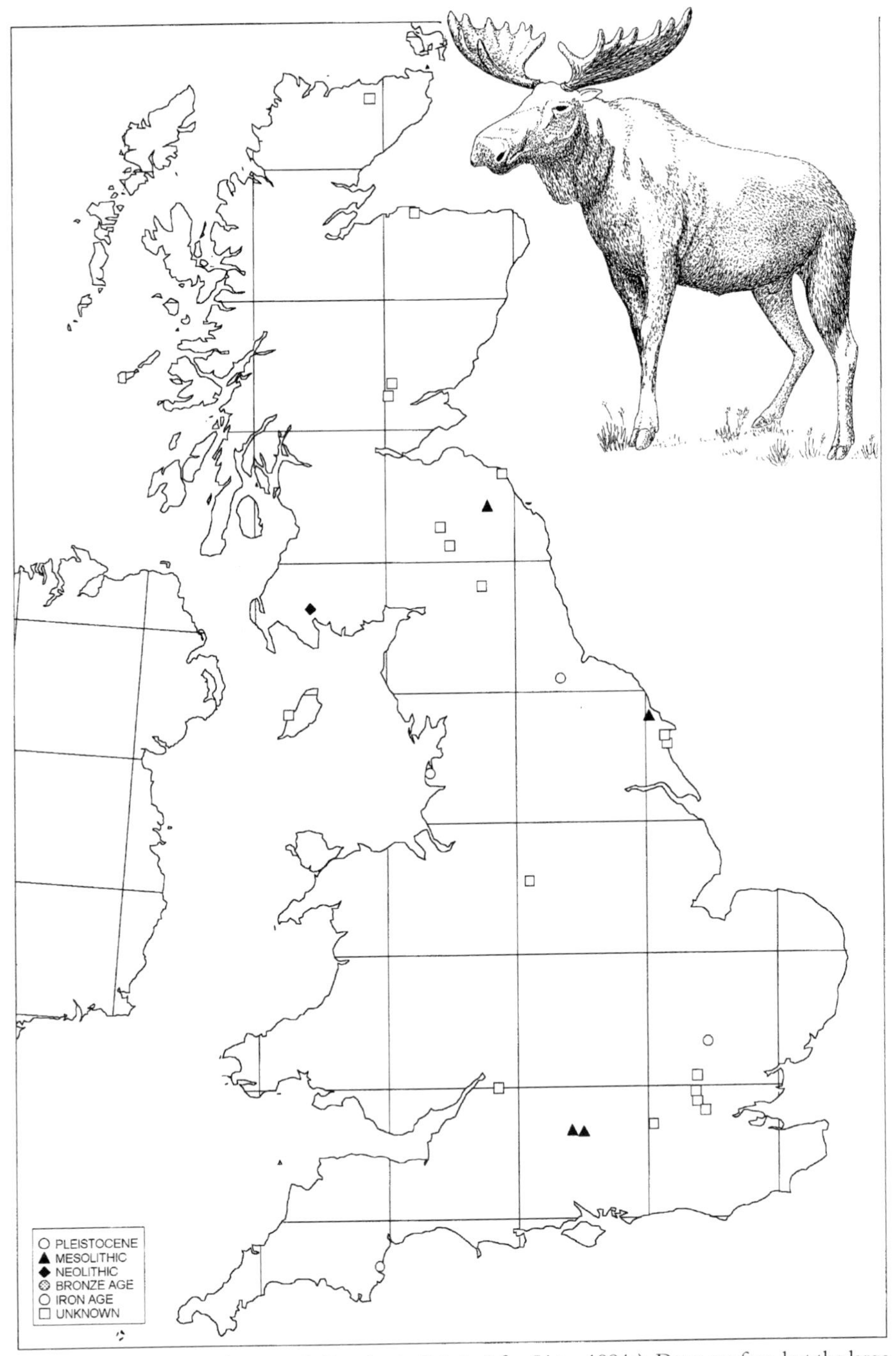

Figure 3.3 The distribution of *Alces alces* in Britain (after Lister 1984a). Dates are few, but the large proportion of northern sites (compared, for instance, with Aurochs or Brown Bear, Figs 4.7 and 4.8) might hint at a longer survival in the north of Britain.

(Mountain?) Hare, Beaver, Wood Mouse, Bank Vole, Tarpan, Aurochs, Common and Pigmy Shrew, and four bats (Barbastelle, Whiskered, Natterer's and Brown Long-eared). A radio-carbon date on some of the larger bone fragments was 9960 b.p., early within the Post Glacial period, which perhaps accords with the persistence of Tarpan and Hare. Like the other two sites, though, this is essentially a temperate woodland fauna. At Cherhill in Wiltshire, another Mesolithic fauna lacks Elk (and Horse), but has Aurochs, Wild Boar, Red and Roe Deer, as well as Beaver, probable Fox and Hare (Grigson 1983). Wawcott, west of Newbury in the Kennet valley, but not that far from Thatcham, has Elk, Aurochs, Red and Roe Deer and Wild Boar (Carter 1975), while Blashenwell in Dorset has the last four, and a date of 6450 b.p. (Rankine 1961). At Westward Ho!, apparently dated before 6585 b.p. (the date of the overlying peat), the fauna was reported to include not only Hedgehog, Red Deer and Wild Boar but also Fallow Deer; perhaps the latter was a misidentified Roe Deer (Churchill 1965, Locker 1987). Morton, in Fife, also has a fauna with Wild Boar, Aurochs, Red and Roe Deer, as well as Hedgehog and Bank Vole, and dates on charcoal of 7330–6147 b.p. (Coles, 1971).

The island of Oronsay in the Inner Hebrides provides an example of a rather different Mesolithic site, where the people lived extensively on marine resources. Shellfish remains, mostly limpets *Patella,* mussels *Mytilis* and oysters *Ostraea*, form great mounds up to 7 m high. Other marine animals are represented in these deposits including fish (especially Saithe *Pollachius virens*) and the extinct Great Auk *Pinguinus impennis*. Radio-carbon dating on charcoal and on shells and bones themselves show that these sites were in use from 6200 to 5100 b.p. (7100 to 6000 B.P.) (Mellars 1987). Mammalian prey included principally Grey Seals *Halichoerus grypus* (at least nine individuals represented in the main site of Cnoc Craig) and a somewhat dwarfed form of Red Deer (at least four animals), but Common Seal, Wild Boar, Otter and cetaceans (though only fragmentary remains, both a small dolphin or porpoise and a large baleen whale) were represented (Grigson and Mellars 1987).

What other species should have been present in Mesolithic Britain? There are few specimens of Lynx from anywhere in Britain (Jenkinson 1983), few of them very well dated, but they co-existed with Roe Deer, one of their principal prey, at several cave sites, and since the Roe Deer appears not to have entered Britain during the Late Glacial but only in the Mesolithic (Lister 1984b), the Lynx too are presumably of that date (Table 3.1, Fig. 3.4). The best dated at the time that Jenkinson was writing was that from Steetley Cave, Yorkshire, which has a Mesolithic flint implement associated with it and an appropriate pollen date. Two specimens now have ^{14}C dates, and that from Kilgreany Cave is certainly Mesolithic (see Chapter 8).

Are there any sites with a transitional fauna, indicating the change from tundra to temperate climate? Most of the large mammals are so tolerant of climate that their ranges extend across most of Europe (though the late records of Tarpan, mentioned above, may qualify). If the early Post Glacial was warm but largely tree-less, they would have colonized very quickly and readily. A better hope would be to examine the small mammals. The lemmings can be taken as good indicators of tundra conditions, while Wood Mice and Bank Voles surely indicate milder wooded conditions. There are some species that might indicate the transition. Field Voles and Water Voles occur at present well north into the tundra, and overlap geographically with both lemmings and, in most of western Europe, with Wood Mice and Bank Voles. In the

Table 3.1 The sites at which *Lynx lynx* has been reported in Britain, and the associated mammals (after Jenkinson 1983).

	Late Glacial			*Post Glacial*															
	LANGWITH CAVE	AVELINE'S HOLE	VICTORIA CAVE	CALES DALE CAVE	LYNX CAVE CLWYD E	LYNX CAVE CLWYD A	MOUGHTON FELL CAVE	KINSEY CAVE	TESDALE FISSURE	LYNX CAVE STAFFS	YEW TREE CAVE	ROBIN HOOD'S CAVE	SEWELL'S CAVE	GTR KELCO CAVE	STEETLEY CAVE	NEALE'S CAVE	GOP CAVE	INCHNADAMPH	KILGREANY CAVE
Erinaceus europaeus	X								X						X				
Sorex araneus	X														X				
Talpa europaea	X								X						X				
Crocuta crocuta																			
Felis silvestris	X			X								X			X				X
Lynx lynx	X	X	X	X	X	X	X	X	X	X	X	X	X	X	X	X	X	X	X
Mustela putorius	X	X								X									
Meles meles	X		X	X		X			X			X			X				
Canis lupus	X			X				X	X	X	X	X							X
Vulpes vulpes	X	X	X	X	X	X	X		X		X	X			X				
Ursus arctos	X	X	X				X	X	X			X		X					X
Lutra lutra	X								X									X	
Equus ferus	X	X					X					X							X
Sus scrofa	X					X			X		X				X			X	X
Cervus elaphus	X	X	X			X			X	X		X			X				X
Alces alces												X							
Capreolus capreolus	X				X	X			X		X	X			X		X		
Rangifer tarandus	X					?		X		X								X	X
Bos primigenius	X			X	X	?		X	X			X			X				
Sciurus vulgaris	X														X				
Clethrionomys glareolus	X			X					X						X				
Arvicola terrestris	X			X	X				X		X								
Lemmus lemmus	X	X																	
Dicrostonyx torquatus	X	X				?													X
Lepus timidus	X	X		X		X	X		?			X							
Ochotona pusilla		X																	

Figure 3.4 The distribution of *Lynx lynx* in Britain (after Jenkinson 1983).

sub-fossil faunas, however, they often occur with other voles, particularly the Root Vole *Microtus oeconomus* and the Narrow-skulled Vole *M. gregalis*. At present the latter occurs well to the east and north, on the wooded steppes of central Asia and on the tundra of Siberia (see Fig. 2.5). It occurred in Late Glacial Britain, along with the lemmings, but probably disappeared very soon afterwards, with them, as the climate warmed. Possibly it lingered, along with the lemmings, in the Scottish mountains, as it did in the Polish uplands (Nadachowski 1989), but we have no evidence of that.

The Root Vole is more interesting for the present discussion (see Fig. 2.6). It occurs in an isolated relict colony in northern Holland, and more extensively from northern Germany and Poland through Russia. It also was a member of the Late Glacial rodent fauna, but it seems to have persisted some time afterwards, into Pollen

Zone V at Nazeing (Hinton 1952). There are a number of small mammal faunas known from cave sites which lack lemmings, so are presumably Post Glacial, but have Root Voles, along with the other, present-day, British rodents. These seem to be the transitional faunas that one would hope to find. Unfortunately, they tend not to be the major archaeological sites that attract radio-carbon dating, so clinching evidence is so far unavailable. Among sites that belong in this possible category are the Mesolithic layers of Dowel Cave and Demen's Dale in Derbyshire (Yalden 1992). The latest date for *M. oeconomus* is the possible Bronze Age record from Nornour in the Scilly Isles (Pernetta and Handford 1970).

Composition of the Mesolithic Fauna

The Mesolithic lasted for some 5000 years in Britain, a time when the land was essentially tree-covered (Fig. 3.5) and when the sparse human population survived largely by hunting and fishing. Their impact on the vegetation is generally supposed to have been slight, though there is increasing evidence that they burned the uplands, at least, perhaps to improve the grazing and concentrate the game that they were hunting (e.g. for the Peak District, Jacobi *et al.* 1976; for the North Yorks Moors, Simmons 1995). It is tempting to suppose that most of lowland Britain looked like the Białowieża National Park in eastern Poland, the nearest surviving fragment, perhaps, of the woodland that once covered the whole of western Europe. There, a population of some 3700 Red Deer and 2700 Roe Deer live in the 580 km^2 of the Białowieża Primeval Forest along with 3400 Wild Boar, 170 Elk and the herd of reintroduced European Bison, whose numbers are restricted to 250 (Jędrzejewska *et al.* 1994, Jędrzejewski *et al.* 1993). Their predators include about 32 Wolves and 15 Lynx; formerly there were Brown Bears as well, but they have been extinct since before 1900 (Buchalczyk 1980). Among the smaller mammals, there are a few Mountain Hares, Bank Voles and Yellow-necked Mice are common, as are Pine Marten, Red Fox, Otter, American Mink (unfortunately), Polecat, Stoat and Weasel (Sidorovich *et al.* 1996). The insectivores include the Mole, Common, Pigmy and Water Shrews, along with *Sorex caecutiens* and *Neomys anomalus* which do not occur in western Europe. Similarly the rodents include *Apodemus agrarius*, *Pitymys subterraneus* and *Microtus arvalis*, *Sicista betulina* and two dormice, *Dryomys nitedula* and *Glis glis*, which have never occurred in Britain naturally, along with Harvest Mouse, Field Vole, Root Vole, Water Vole and Beaver (Pucek *et al.* 1993).

We can combine the studies of sub-fossil mammals and pollen analysis in Britain with the recent work on the mammals of Białowieża to guess at the likely nature of the Mesolithic mammal fauna of Britain. The Bison did not return to Britain in Post Glacial times, but the now totally extinct Aurochs lived a similar life. It was distributed widely throughout Europe, North Africa, including the Nile Valley, and southern Asia in the deciduous forest zone and seems to have been exterminated by forest clearance and hunting, with the last individual reputedly killed in Poland in 1627. It would have been large enough to maintain, if not create, clearings, and the additional grazing pressure from deer would also have contributed. Multiplying up from the small area of Białowieża (580 km^2) to the size of Britain (230 367 km^2) is an audacious step, but it implies that there might have been over 1 million each of Red Deer, Roe Deer

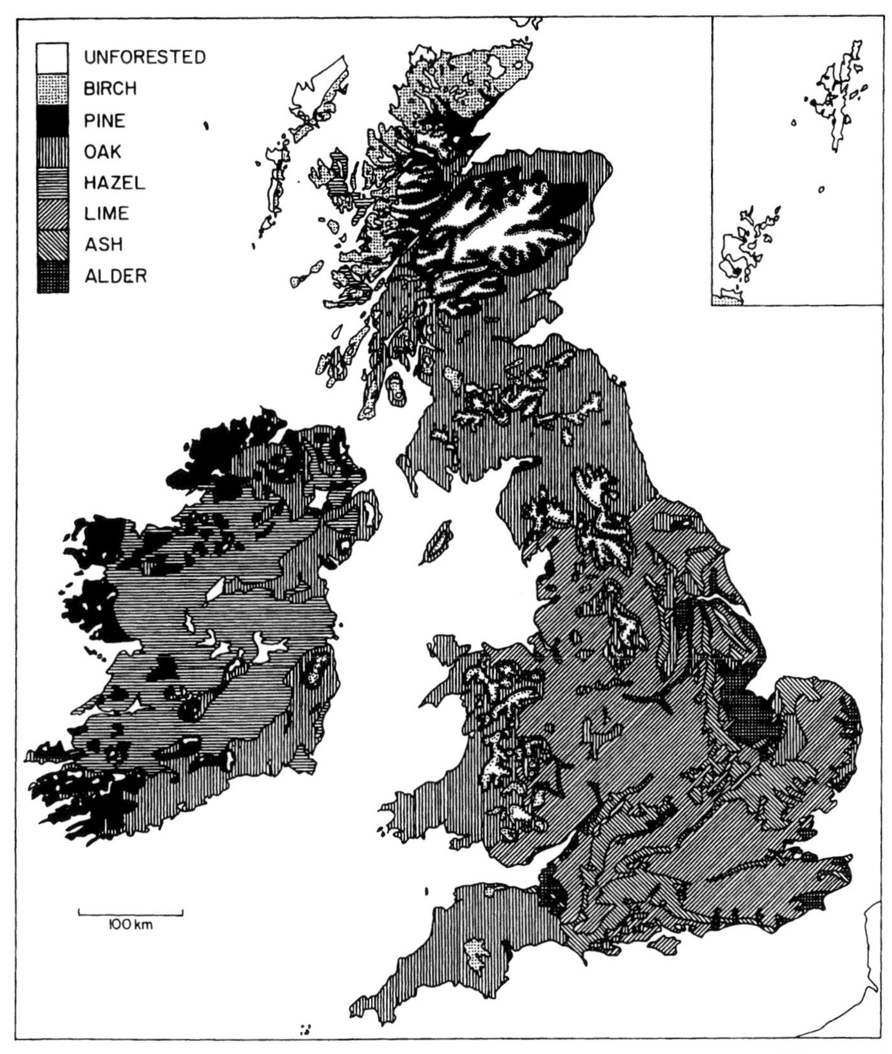

Figure 3.5 A suggested map of woodland trees predominating over Britain in the Mesolithic period, about 6000 years ago. Only the highest ground in the north of Britain would not have been covered by trees, so animals of open country (like hares) would have found little suitable habitat. (Reproduced from Bennett 1988 in *Journal of Quaternary Science* **14**, by permission of Dr K. D. Bennett and J. Wiley & Sons.)

and Wild Boar, along with 99 000 Aurochs and 67 000 Elk (Table 3.2). A similar guess suggests that there could have been about 10 000 Lynx and 20 000 Wolves preying on the ungulates, as well as perhaps 2000–3000 human hunters (Yalden 1996a). There would have been wide, meandering river valleys with lagoons and reed beds, as there are in Białowieża. Extrapolating similarly for riparian mammals from the figures given by Sidorovich *et al.* (1996), there might have been about 35 000 Beavers, 7000 Otters, 423 000 Weasels, 104 000 Polecats and perhaps half that number of Stoats. Otters living in coastal waters would have been additional to that estimate, but these numbers

Table 3.2 The populations and biomass of large mammals in Białowieża Primeval Forest (580 km^2) (data from Jędrzejewski *et al.* 1993) extrapolated to give an estimate of the Mesolithic fauna of Great Britain (230 367 km^2). (Individual masses from Jędrzejewska *et al.* 1994). The Bison did not return to Britain in the Post Glacial, but the Aurochs did, and is assumed to have contributed similar numbers and biomass (after Yalden 1996a).

Species	*Individual Mass (kg)*	*Białowieża n*	*Białowieża biomass (kg)*	*Great Britain n*	*Great Britain biomass (kg)*
Bison *Bison bonasus* (or Aurochs *Bos primigenius*)	400	250	100 000	99 250	39 700 000
Red Deer *Cervus elaphus*	100	3710	371 000	1 472 870	147 287 000
Roe Deer *Capreolus capreolus*	20	2730	54 600	1 083 000	21 676 200
Moose *Alces alces*	200	170	34 000	67 490	13 498 000
Wild Boar *Sus scrofa*	200	7375	1 475 000	1 357 000	108 619 200
Herbivore Total		10 280	833 200	4 081 160	330 780 400
Lynx *Lynx lynx*	15	19	285	7543	113 145
Wolf *Canis lupus*	32	20	640	7940	254 080

make interesting comparisons with those of the current fauna (see Chapter 9). Mountain Hares would have been limited to the uplands of Scotland, and to Ireland, by this time, and while Wood Mice and Bank Voles were more common, Field Voles must have been much more restricted in range and numbers than now. So far as we know, the Water Vole was much more abundant then; it frequently contributes 40–50% of the mammals in Mesolithic and later owl pellet faunas (Yalden 1992).

Mesolithic Extinctions

One difficulty with assessing our Mesolithic mammal fauna is that we have such a long period of time with very few faunas to represent it, and even fewer with good dating. During this time, the Lynx and Elk were thought to have become extinct (Tables 3.1, 3.3). We do not know whether hunting played a large part in this, or whether they were vulnerable to climatic and perhaps habitat change, and there are few specimens of either (Figs 3.3, 3.4) to document the arguments. It is hard to credit that habitat changes were to blame, for they both range widely across Europe, and neither deforestation (which might have affected the Elk's food supply) nor the loss of Roe Deer (the main prey of Lynx) were serious factors at this time. Perhaps the restriction of relatively large mammals to relatively small islands meant that competition was too severe, and these two were the first to disappear; and perhaps it was human hunting that tipped the balance. However, their story has been revolutionized by the recent ^{14}C dates obtained on Scottish material by Andrew Kitchener (Kitchener and Bonsall 1997). An Elk from Wigtownshire provided a date of 3925 b.p., much later than anywhere else (a supposed Roman specimen from the fort of Newstead reported by

Ewart (1911) is probably Aurochs; Andrew Kitchener, pers. comm.), but the Lynx specimen from Inchnadamph provided the greatest surprise. This was described by its discoverers as looking remarkably recent, sprawled across the top of the cave earth (Lawson 1981). Their judgement is vindicated by a date of 1770 b.p., i.e. in Roman times about 180 A.D. This is very much later than expected for Lynx to have survived in Britain; Corbet (1974) did not even discuss it as a former member of the fauna. If so conspicuous, though rare, a member of our fauna can be overlooked for so long, it raises questions about how reliably we can date the extinctions of any other species we have lost. It also provides ammunition for those who have suggested that the Lynx should be considered for reintroduction, as a relatively recent loss from our fauna (see Chapter 9).

Table 3.3 Archaeological sites for Elk *Alces alces* in Britain from Pleistocene through to Bronze Age times. Mostly based on Lister (1984a). Smith (1872) includes several other possible records, but the distinction between *Alces* and *Megaloceros* was uncertain when he wrote; even so, the preponderance of northern sites is striking.

Site	*Grid ref*	*Age*	*Source*
Gray's Thurrock, Essex	TL4533	Ipswichian?	Lister 1984a
Kent's Cavern, Devon	SX9264	31 000 b.p.	Lister 1984a
High Furlong, Lancashire	SD331387	12 400 b.p.	Hallam *et al.* 1973
Neasham, Durham	NZ340100	11 561 b.p.	Lister 1984a
Star Carr, Yorkshire	TA028810	Mesolithic	Fraser and King 1954
Thatcham, Berkshire	SU502668	9490 b.p.	Wymer 1962
Wawcott, Berkshire	SU418672	Mesolithic	Carter 1975
Peel, Isle of Man	SC2484	?	Lister 1984a
Nailsworth, Gloucestershire	ST8499	?	Lister 1984a
Whitrig Bog, Berwickshire	NT7841	7790 b.p.	Kitchener and Bonsall 1997
Auchtergaven, Aileywright, Perthshire	NO0535	?	Smith 1872
Strath Halladale, Sutherland	NC8953	?	Smith 1872
Williestruther Loch, Hawick	NT4911	?	Smith 1872
Oakwood, R. Ettrick, Selkirk	NT4225	?	Smith 1872
Coldingham, Berwickshire	NT9065	?	Reynolds 1934
Spynie, Elgin	NJ2365	?	Reynolds 1934
Barmston, E. Yorkshire	TA1659	?	Reynolds 1934
Carnaby, Bridlington, E. Yorkshire	TA1465	?	Reynolds 1934
Methven, Perthshire	NO0225	?	Reynolds 1934
Wetton, Staffordshire	SK1055	?	Reynolds 1934
Chirdon Burn, Northumberland	NY7481	?	Reynolds 1934
Broxbourne, Hertfordshire	TL3707	?	Reynolds 1934
Walthamstow, Essex	TQ3788	?	Reynolds 1934
Beckton, Essex	TQ4381	?	Reynolds 1934
Staines, Middlesex	TQ0471	?	Reynolds 1934
Ponder's End, Middlesex	TQ3695	?	Reynolds 1934
River Cree, Wigtownshire	NX4165	3925 b.p.	Kitchener and Bonsall 1997

The fate of the two earlier losses, Reindeer and Tarpan (Wild Horse), is similarly enigmatic, except that climatic and habitat changes can be more realistically invoked for them. Both deserve a second glance. There is a popular notion that Reindeer survived in Scotland to early Medieval times. This stems in part from the 12th century *Orkneyinga Saga*, a line of which reads *þat var siðr jarla naer hvert sumar at fara yrir a Katanes ok þar upp a merkr at veiða raðdryri eða hreina*. This has been translated to say that the jarls of Orkney 'used to go over to Caithness to hunt Red and Reindeer'. It is at least probable that the translation should be 'Red or Reindeer', and quite possibly a piece of poetic licence that should not be taken seriously (Dent 1974, Clutton-Brock and MacGregor 1988). The claim has been bolstered by some dubious records

Starr Carr Elk persued by Mesolithic hunters

of fragments of antler from Iron Age brochs, and by the interpretation of a Pictish symbol stone, with a deer engraved on it, near Grantown-on-Spey. These various claims have been thoroughly re-evaluated by Clutton-Brock and MacGregor (1988). The antler fragments are mostly Red Deer, or are untraceable; the Pictish engraving does show an upturned tail, like a dominant Reindeer, but its antlers are surely those of a Red Deer, with branched 'tops' quite unlike a Reindeer. There **are** Reindeer remains from Scotland, as from elsewhere in Britain and western Europe, but they date from much earlier than Medieval times; the remains from Inchnadamph, mentioned earlier, are the latest reliable record, with their date of 8300 b.p. One celebrated specimen from Rousay, Orkney, is an exception. Like the Thatcham Rabbit, this turns out to be very recent, with a ^{14}C date of 265 b.p. (Kitchener and Bonsall 1997). It is probably a souvenir of some traveller's visit to Scandinavia.

The possible survival of the Tarpan through to recent times has a similar popular hold on the romantic imagination. We do, of course, have domestic horses, and it has been tempting to suppose that they might be derived, in part or entirely, from the original wild horses of Britain. Anatomically, there is no general difference between them, such that bones from archaeological sites could be easily distinguished. One of the features of wild horses is that they have pale muzzles ('mealy mouths'), a characteristic too of such breeds as Exmoor Ponies. However, it is also a feature of wild horses, seen in all the zebras, in Przewalski's Horses, and in the cave paintings of Tarpan, that they have a short upright mane. Exmoor Ponies and other supposed survivors of the wild horses share the long, hanging, mane of domestic breeds. It is very doubtful that they are anything other than feral domestic horses, and the apparently long gap in the fossil record through Mesolithic and early Neolithic times (between 9770 and 4170 b.p.) also argues for that explanation (Clutton-Brock 1986). However, Grigson (1966, 1978) notes that odd teeth and bones do occur in archaeological sites that are in this gap; the apparent Neolithic specimens from Durrington Walls (Harcourt 1978) are one example. She remarks that the dating of those she has checked is rarely secure, but that horse remains occur frequently enough to leave some scope to the romantics.

At this time, Great Britain was, of course, not an island but part of continental Europe. The archaeological record there supports the interpretation given above. Aaris-Sorensen (1992) summarizes the story from Denmark; as the ice of the last glaciation retreated, the mammals spreading back northwards included Reindeer and Wolf by 12 500 b.p., Elk, Brown Bear and Beaver by 11 500, Aurochs, Bison and Tarpan by 10 000, and Red Deer, Roe Deer and Wild Boar by 9700 b.p., at which time he surmises that much of the rest of the present fauna (Red Fox, Badger, Pine Marten, Otter, Wild Cat and Lynx) also arrived, though their first records are a little later. In Denmark, too, the Reindeer died out early, at about 9800 b.p., and the Tarpan, as well as the Bison, a little later at around 9500 b.p. At this time, there was a land connection between Denmark and Scania (southern Sweden), broken in about 7200 b.p. as the Ancylus Lake, the freshwater precursor of the Baltic Sea, changed into the brackish Littorina Sea. This curtailed further mammal immigration into Sweden, and it may be significant that southern Sweden has an identical native mammal fauna to southern Britain, so far as we can tell. This implies a similar timing for the separation of Britain from Europe. When did that occur?

Becoming an Island

It is well known that the sea-level was greatly lowered during the glacial periods, because of the quantity of water locked in the ice caps. As the climate warmed, so ice melted and sea-levels rose. It is popularly supposed that Britain became an island by about 5000 b.p., because that is when sea-level reached its present level. This is too late, and the argument is too simple. It merits further consideration.

There are now at least three long records of rising sea-levels obtained from drilling corals in tropical parts of the world, well away from any interference by the direct weight of ice on the land surface. Two of these, from Barbados and the New Guinea area, could be affected locally by tectonic (volcanic) movements, but the third, from Tahiti, is clear of such influences, and the three records all give a very similar story (Bard *et al.* 1996). About 18 500 B.P., sea-level world-wide was at –120 m. It rose steadily to about –100 m at 14 000 B.P., jumped rather quickly at the start of the Windermere Interstadial to about –75 m, and at the start of the Post Glacial was –55 m. By the end of the Mesolithic, around 5000 b.p., it was near present sea-level (Fig. 3.6).

The impact of this rising sea-level around Britain was complicated by two factors, the weight of ice pressing down on the north of the area and the depths of existing valleys. The former affected Scotland and Ireland more than England. In fact, Britain responded to the weight of ice and its subsequent removal rather like a see-saw; the north, pushed down further, has rebounded more (the western Highlands continue to rise at about 3 mm/year) while the south, perhaps uplifted somewhat by the effect of the ice further north, has sunk, and continues to do so at about 2 mm/year. This response to the removed weight of ice is termed the isostatic uplift, while the rise due to melting ice is the eustatic uplift. In south-west Scotland, the net effect is that sea-level at the beginning of the Mesolithic was about 5 m higher than now, whereas in southern England it was indeed low, at about –50 m as the world-wide levels suggest.

The floor of the North Sea is scored by deep river channels, which carved their way down to the reduced sea-levels of glacial times. Submarine contours pick them out down to –100 m or so, again matching the world-wide picture. However, the English Channel is itself about –37 m at its shallowest and narrowest, in the Strait of Dover. World sea-levels reached –37 m by about 9000 b.p. (10 000 B.P.), which implies that the Channel would have opened, and England been cut off from France, at about that

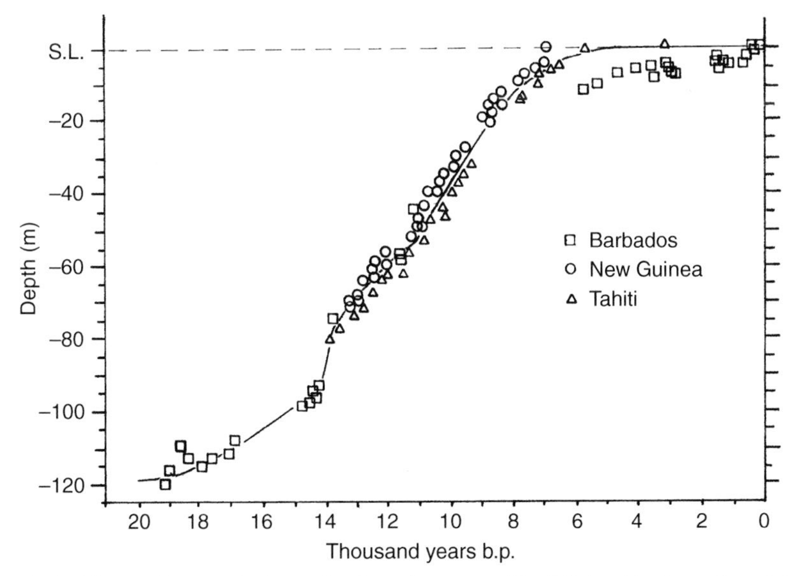

Figure 3.6 The world-wide rise in sea-levels, as documented by ^{14}C-dated borings of corals in the Bahamas, Tahiti and New Guinea (after Bard *et al.* 1996, Fairbanks 1989). Because corals grow within a few metres of the sea surface, in order to gain sufficient light for their symbiotic algae to photosynthesize, they give a good indication of sea-level.

early date (Fairbanks 1989, Bard *et al.* 1996). This would only be true, however, if the Channel had its present form and bathymetry at that time; if it has been scoured out by tides subsequently, or if the sea had to break through a chalk ridge stretching from the White Cliffs of Dover to Cap Blanc Nez, then that speculation is worthless. Fortunately, a major conference held in 1993 addressed these particular questions. It seems that the Strait of Dover was cut through the chalk ridge during an earlier glacial period than the last, as the ice of the Anglian (=Elsterian) glaciation dammed the northwards flow of the Thames/Rhine precursor rivers (Gibbard 1995). Forced to find a southward drainage, they cut a channel through the chalk ridge (Fig. 3.7). At the subsequent Hoxnian and Ipswichian Interglacials, with their higher sea levels, the northwards flow may have been resumed, but at the same time tidal action widened the Strait. In the intervening Wolstonian and Devensian Glaciations, with their lowered sea levels, the rivers may have been diverted southwards again, and certainly deepened the submarine channels. There is a buried channel, the Fosse Dangeard, in the floor of the Strait of Dover that would have been –95 m or so deep, and must represent the river valley carved out in one of those later (Wolstonian or Devensian) glaciations. Most of the erosion on the floor of the Channel has occurred during glacial times and low sea-levels, not during the high sea-levels of the interglacials (Bellamy 1995). Thus it is certain that there was already a deep valley between France and England available for flooding as the sea-level rose during the Mesolithic. It follows that when the sea-level reached –37 m, at about 9500 b.p. (Fairbanks 1989, Bard *et al.* 1996), Britain should have been cut off from Europe. This is quite early in the Post Glacial, in Pollen Zone V (Pine-Hazel), and not much later than the known faunas of Star Carr and Thatcham. It implies that there was very little time for animals of

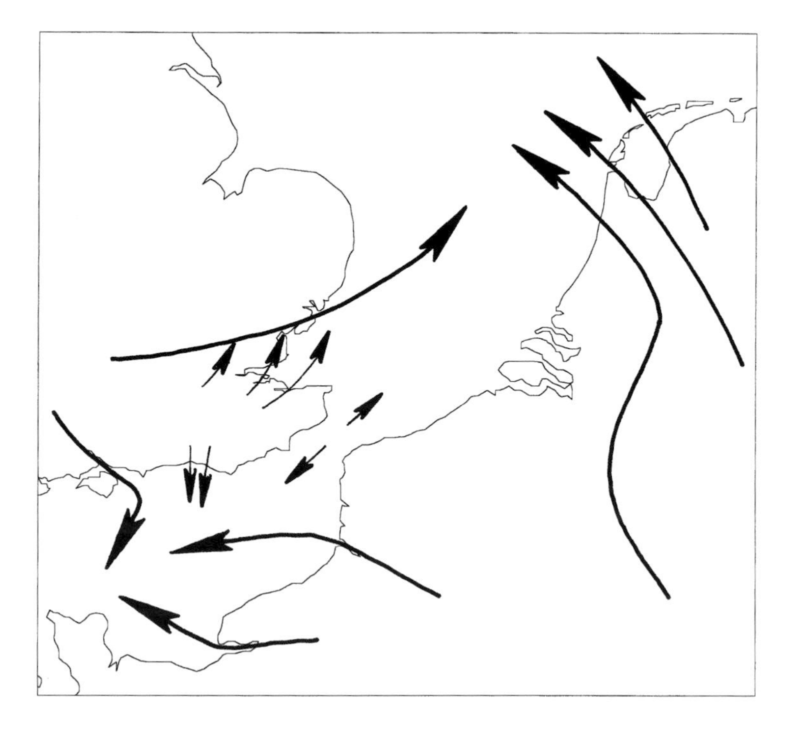

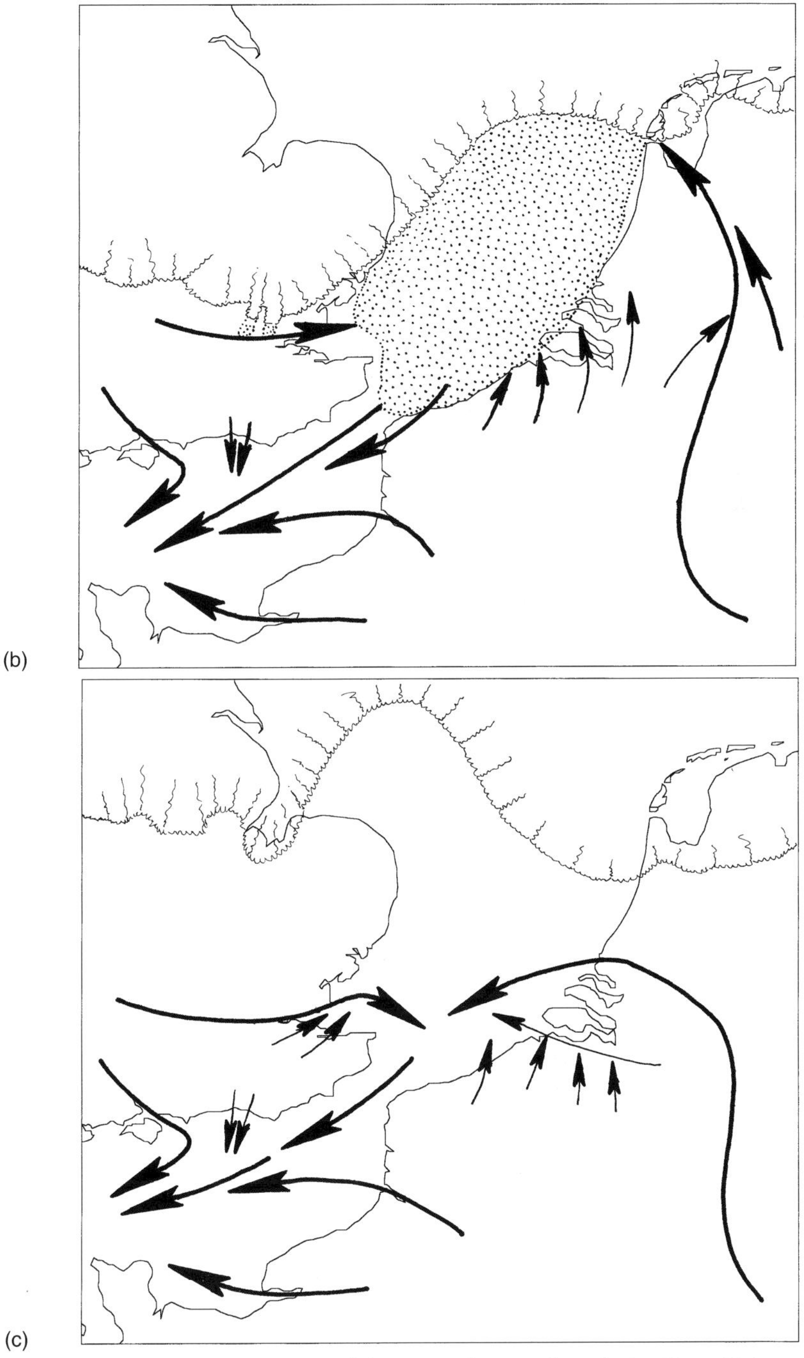

Figure 3.7 The suggested history of the Strait of Dover through the Upper Pleistocene (after Gibbard 1995). In the Cromerian Interglacial, the proto-Thames and Rhine flowed northwards (a). However that route was blocked by ice in the Anglian Glacial (b), leading to the formation of a pre-glacial lake which spilt over the chalk ridge between Dover and Wimereux, carving a channel which was further deepened in the Wolstonian Glacial (c).

Aurochs

warmer climates and conditions to get into Britain. Is there any more direct evidence that this estimate of these timings is correct?

Mollusc faunas provide one interesting test. During times of low sea-level, terrestrial and fresh-water molluscs were free to migrate into Britain from Europe. Conversely, during times of high sea-level, marine molluscs were free to migrate through the Channel into the southern North Sea. There is an early record of Cockle

Cerastoderma edule dated to 9560 b.p. from the Southern Bight, implying that brackish water, at least, had penetrated through the English Channel that far north. Full marine conditions were apparently not established until perhaps as late as 7000 b.p. (Funnell 1995, Meijer and Preece 1995). Thus there may have been a low-lying and possibly partially flooded plain persisting in the southern North Sea for some time after the Channel had opened. There are submarine peat beds below the North Sea which have been dated to Pollen Zones IV, V and VI, and with ^{14}C dates of 8500 b.p. (Godwin 1975, Mitchell 1977). These too are a little later than my estimate of 9500 b.p., but the presence of peat beneath the North Sea at later dates does not necessarily indicate that there was still a complete land-bridge so late as this. Whichever dates are preferred, it is clear that Britain became an island much earlier than the oft-quoted 5000 years ago, perhaps as early as 9500 b.p. and certainly before 7000 b.p. There is a Post Glacial mammal fauna from the floor of the North Sea, including Otter, Beaver, Wild Boar, Roe and Red Deer, Elk and Aurochs, and there are ^{14}C dates between 9300 and 8000 b.p. for this fauna (Kolfschoten and Laban 1995), but these too cannot indicate when the land-bridge was finally cut. However, it may be significant that neither peat nor fauna contribute dates as late as 7000 b.p., let alone 5500.

So what did not make it? At Cap Gris Nez in northern France, the White Cliffs of Dover are readily visible on a clear day, only 30 km away to the north. Trapping mammals there in 1961, 1963 and 1964, three British mammalogists caught 1182 small mammals, 634 specimens of which are now in the Natural History Museum, London (Yalden *et al.* 1973). Mostly these were the species which occur in southern England, too; the most numerous rodents were, in order, the Bank Vole, Wood Mouse and Field Vole, and at the time the Common Shrew was thought to be the most numerous, by far, of the insectivores. It turns out, subsequently, that there are two very similar shrews in north-western Europe, with mutually exclusive ranges, and the one in northern France (and in Jersey) is *Sorex coronatus*, the French Shrew (Meylan and Hausser 1978). Among the other British species were the Pigmy and Water Shrews, Mole, Hedgehog, Weasel and Harvest Mouse. Rather more interesting for British mammalogists, and for this chapter, were the species that are missing from England. These included two white-toothed shrews *Crocidura russula* and *C. leucodon,* the Garden Dormouse *Eliomys quercinus* and two voles, the Common Vole *Microtus arvalis*, which was actually as uncommon as it could be because only one was caught, and the Pine Vole *Pitymys subterraneus*. Why should these species occur so close to Britain, but yet be absent? Obviously the answer lies, at least in part, in the historical story just told. They are largely southern species, and the presumption is that they reached the coast after the Channel had flooded. Is there any evidence that this presumption is correct? The white-toothed shrews are as a whole a tropical group; most species are African, and some Asian, but only four species occur in Europe (Hutterer 1985). One, the tiny Etruscan Shrew *Suncus etruscus*, is confined even now to the Mediterranean region. Both *C. russula* and *C. leucodon* are close to their northern limits at Cap Gris Nez; they do just extend into Holland and northern Germany, but neither gets into Denmark (Fig. 3.8). The fourth species, *C. suaveolens*, occurs as far north as the Loire valley, but does not even reach so far as the Channel coast of France. (There is an anomaly in their distributions, that both *C. russula* and *C. suaveolens* occur in the Channel Isles, on different islands, and *C. suaveolens* also occurs on the Isles of Scilly; this problem will be discussed in Chapter 8.) Thus the evidence of

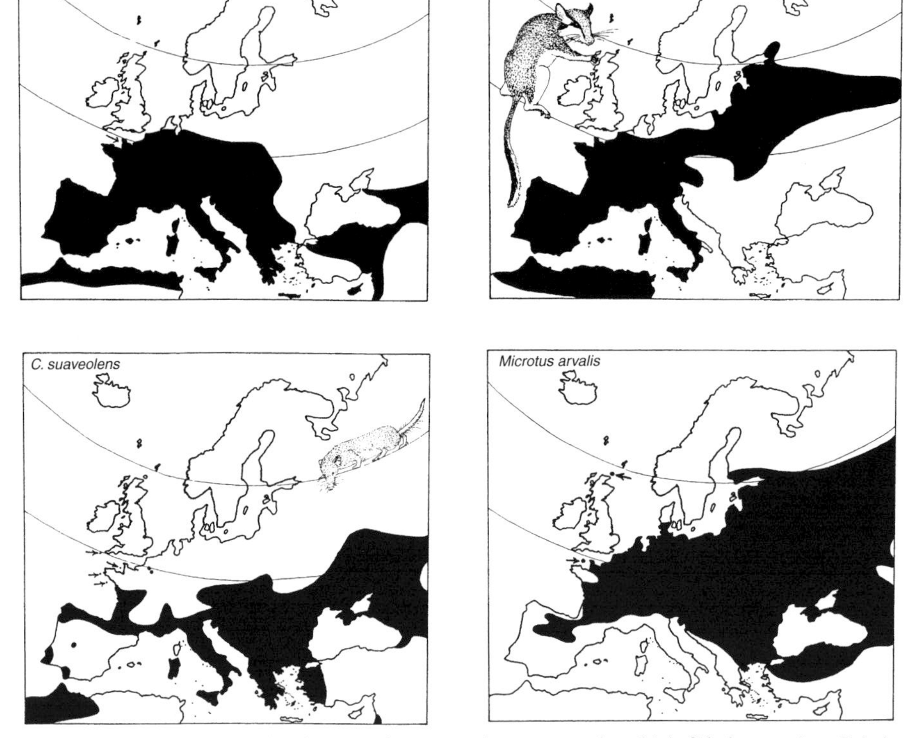

Figure 3.8 The current distribution of some southern mammals, which failed to get into Britain in the Mesolithic before the Channel was cut (after Yalden 1982).

their present distributions strongly argues for their late appearance in northern France. So too does their apparent sensitivity to cold winters (they were absent from Cap Gris Nez after the cold winter of 1962–63) and their habitat preference there, in the warmer microclimates of gardens and in the broken concrete of the wartime fortifications (Yalden *et al.* 1973). Interestingly, they benefited, and *Sorex coronatus* declined, in the hot dry summer of 1976 (Saint-Girons 1981). The subfossil record ought to provide confirmation of their history. Chaline (1972) records the presence of *Crocidura russula* and *C. leucodon* in a few sites in central and southern France that apparently date to the Late Glacial. However, the faunas look to be a mixture of temperate woodland and arctic steppe species. For example, at Flavigny-sur-Overain, Cote-d'Or, *Glis* and *Eliomys* as well as *Crocidura* co-occur with *Dicrostonyx* and *Ochotona*. It is not conceivable that these species were really contemporaneous, their habitat and climatic requirements are too different, so the faunas have been mixed up at some stage, perhaps by later, warmth-loving species burrowing down into layers that already contained the fossilized remains of arctic species. Further research on the dating of these faunas is needed, but they do suggest the presence of temperate mammals well to the south of Britain during the Late Glacial.

The Garden Dormouse is regarded by Nadachowski (1989) as a very late arrival in

Poland, perhaps at 3000 b.p. (Fig. 3.9), and this matches its rather southern pattern of distribution in Europe; like the white-toothed shrews, it is missing from Denmark and northern Germany. Similarly the Pine Vole has a rather southern distribution now, but it appeared in Poland much earlier than the Garden Dormouse, in the Late Glacial, and could have perhaps spread northwards more quickly because of its different habitat preference (dry open grassland rather than deciduous woodland); both species were present in southern France during the Late Glacial (Chaline 1972).

The situation with the Common Vole *Microtus arvalis* is rather confused because it is very difficult to differentiate reliably between the various *Microtus* species that occurred in Late Glacial or Post Glacial Europe (Hall and Yalden 1978) when these are only fragmentary specimens, perhaps isolated teeth. There have been numerous claims for the presence of *M. arvalis* in Britain at those times, for example at Ightham Fissures, Ogof-yr-Ychen on Caldey Island, Marlow, Beckford and Levaton (Sutcliffe and Kowalski 1976, Yalden 1982). The identification of these specimens as *M. arvalis* is not convincing; they can variously be explained as poorly identified or preserved specimens of *M. agrestis* or *M. oeconomus*, or as mixtures of teeth of these and *M. gregalis* (Sutcliffe and Kowalski 1976, Yalden 1982). At present, *M. arvalis* has a rather southerly distribution in Europe, like the other species in this group; it favours drier habitats and shorter grassland, especially agricultural land, in places where it occurs

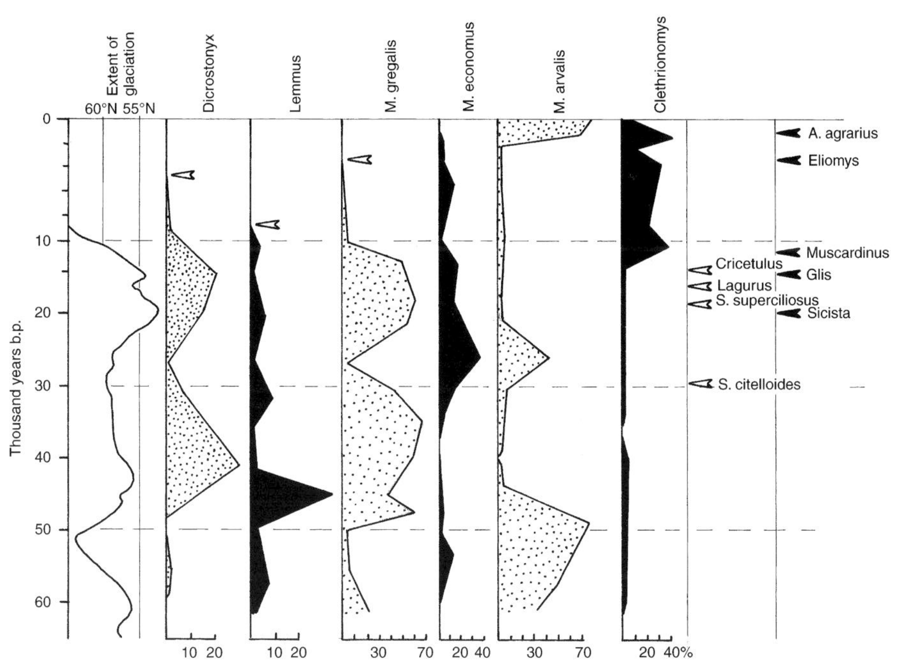

Figure 3.9 The Late and Post Glacial history of rodents in Poland, which provides a record of immigration and extinction rates at about the same latitude as Britain (after Nadachowski 1989). *Dicrostonyx* and *M. gregalis* survived well into the Post Glacial in Poland, while *M. arvalis* became common only in the agricultural period. Hollow arrows indicate some last records (the Grey Hamster *Cricetulus migratorius*, Steppe Lemming *Lagurus lagurus* and two ground squirrels *Spermophilus superciliosus* and *S. citelloides*, all steppe species) while solid arrows indicate first records.

whereas *M. agrestis* favours wetter and taller grasslands, at least in Holland where their habitat preferences have been compared (De Jonge and Dienske 1979).

There are other European species that apparently failed to reach Britain at this time. One that has caused some debate is the Stone or Beech Marten *Martes foina*. This is a common animal over much of western Europe, and frequently lives in houses, so making itself more apparent to humans than its wilder relative the Pine Marten. The two martens are very similar in size and shape, even to their skulls, but there are subtle differences in their teeth which allow good archaeological (and recent) specimens to be distinguished (Fig. 3.10); the Pine Marten usually has a yellow bib, whereas that of the Beech Marten is white, but the colour does fade on older Pine Marten specimens, both alive and dead. Such specimens have in the past triggered lively debates as to whether in fact the Beech Marten did once occur in Britain, but the evidence of all the skulls that have been examined is that it did not (Alston 1879). It has to be allowed that there are not many sub-fossil specimens of British *Martes*, and the possibility of a new specimen surprising us is always there, but the evidence from Europe is that the Beech Marten was a late immigrant to the west; Kurtén (1968) observes that it was common in the Middle East in the Late Glacial, but was probably absent from western Europe. The Edible Dormouse *Glis glis* is a characteristic member of the deciduous woodland fauna of western and southern Europe, occurring as far north as Paris and Hamburg. In Poland, it arrived early in the Late Glacial, some 2000 years before the Common Dormouse *Muscardinus avellanarius* (Nadachowski 1989). The latter did get into Britain, and into southern Sweden, while *Glis glis* did not. This discrepancy

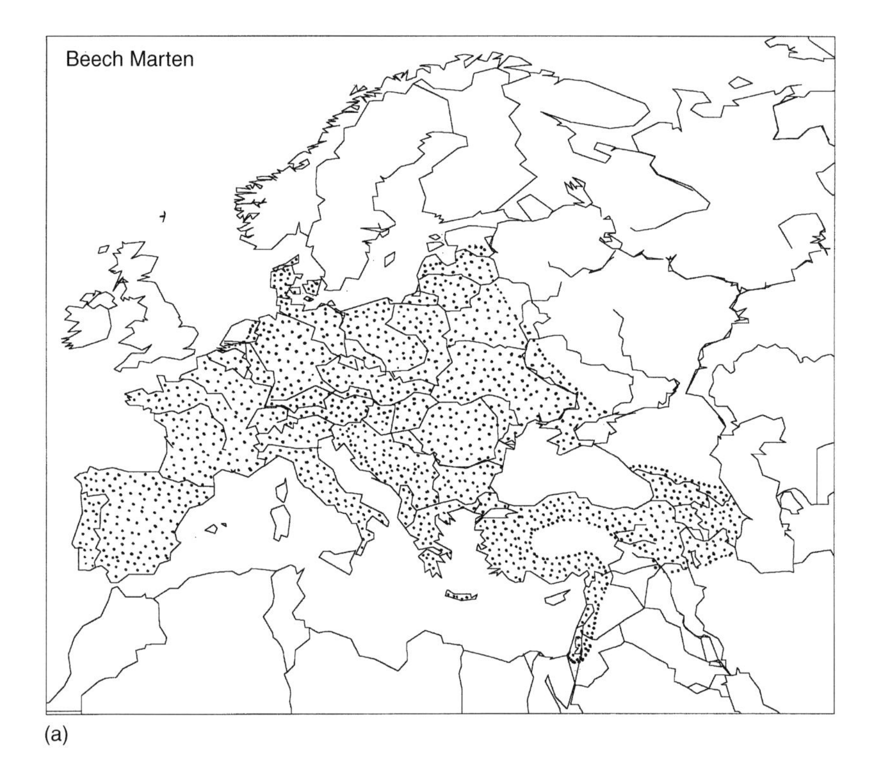

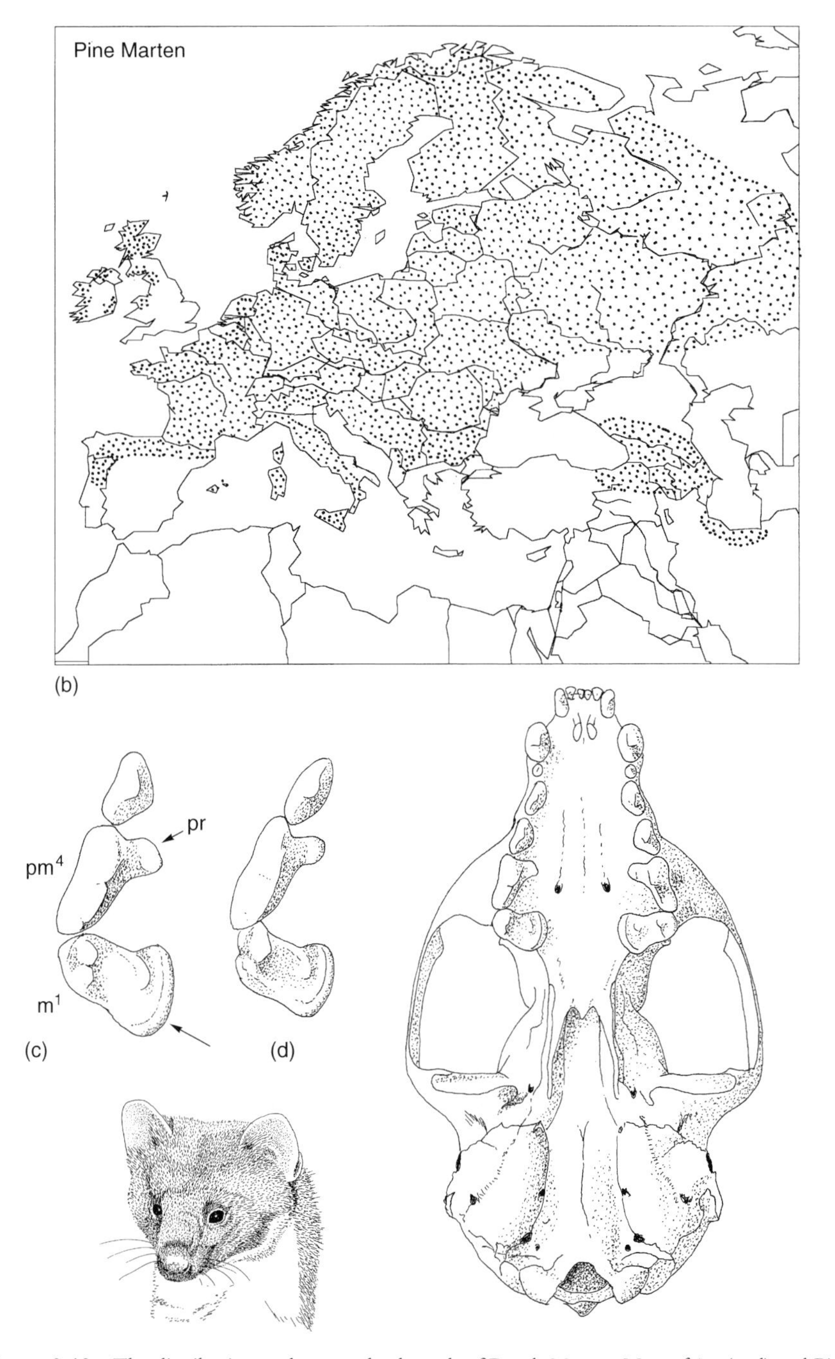

Figure 3.10 The distribution and upper cheek teeth of Beech Marten *Martes foina* (a, d) and Pine Marten *Martes martes* (b, c). The Pine Marten occurs further north, but is missing from much of the Mediterranean region and the Middle East. Skulls from cave deposits in Britain all seem to be from Pine Martens, recognizable from the different shape of the upper molar, much broader internally, and the bigger protocone (pr) on the fourth premolar.

perhaps relates to the different ecologies of the two dormice. High deciduous forest, favoured by *Glis*, did not develop fully in Britain until after 9000 b.p., and beech woodland, particularly favoured, not until 2000 b.p. (Huntley *et al.* 1989). Hazel scrub, suitable for *Muscardinus*, had appeared as early as 10 000 b.p., to judge from the pollen record. However, in southern France, *Glis*, *Eliomys* and *Muscardinus* occur together in a number of Late Glacial sites (Chaline 1972), so all three had only about 700 km to migrate back northwards from their glacial refuges to the Channel coast. The rate at which they could do so must have depended on the rate at which their woodland habitat spread.

However, a new interpretation of the recolonization of northern Europe by small mammals is suggested by the recent results of testing the mitochondrial DNA of some of the common species. The Common Shrews, Pigmy Shrews and Bank Voles of northern Europe, including Britain, seem to be genetically much closer to those of Russia and Siberia than to those of Mediterranean countries. Rather than spread northwards from what we have assumed to be their southern refuges during Glacial times, it looks as though they spread westwards from refuges in the east (Bilton *et al.*, 1998). If this result is confirmed for other species, it will indicate that we have been looking for the forerunners of our present small mammal fauna in the wrong place.

CHAPTER 4

CLEARING THE WOODLANDS

Black Rat and Roman helmet

DURING the period from about 7000 to 5000 b.p., the later part of the Mesolithic to the archaeologists, and the Atlantic Period, Pollen Zone VIIa to the palynologists, mature deciduous woodland blanketed the whole of southern Britain (Fig. 3.5). Over the next 3000 years, much of this woodland was replaced by farmland, and many of the other distinctive habitats – heather moorland, blanket bog, chalk downland – that we now regard as typical of Britain started to develop. This change began with the arrival of a farming culture that spread across Europe from the Middle East, the Neolithic (New Stone Age) culture (Fig. 4.1). It continued through the subsequent Bronze Age and Iron Age, so that when the Romans invaded, temporarily with Julius Caesar in 55 and 54 B.C., and more permanently under Emperor Claudius in A.D. 43, they invaded a farmed landscape.

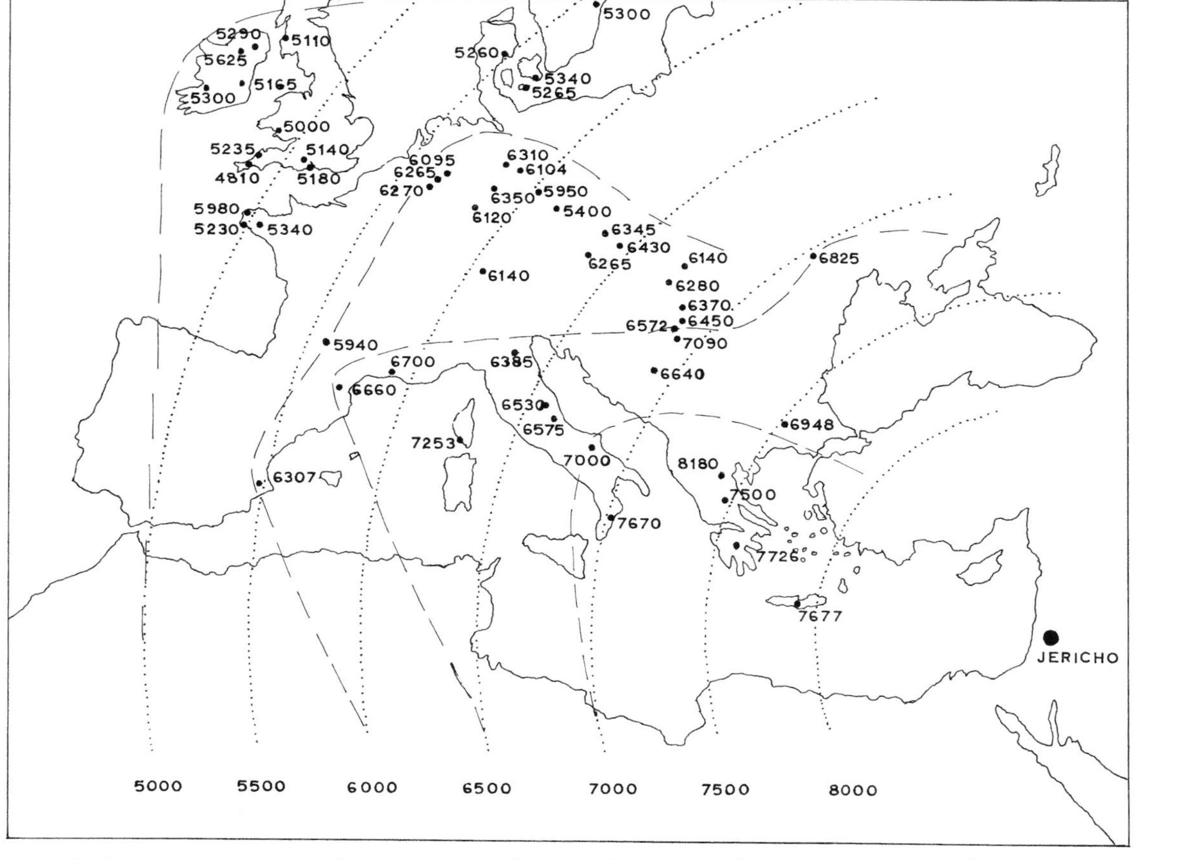

Figure 4.1 The spread of farming across Europe (after Ammerman and Cavalli-Sforza 1971). They suggest a slow rate of migration averaging about 18 km per generation (25 years). Starting from Jericho at about 9000 b.p., that would provide the arcs (dotted lines). Earliest dates (years b.p.) for cereals or associated artefacts from a number of archaeological sites across Europe show that spread through the Mediterranean was faster than this (dashed lines), but colonization northwards was slowed by the mountain ranges across central Europe.

The Appearance of Grasslands

For the palynologists, the first indication of the impending change is a drop in the proportion of Elm *Ulmus* pollen. This was noticed long before its full significance was appreciated, and used to demarcate the boundary between Pollen Zones VIIa and VIIb. As more and more pollen diagrams were published, both in Europe in general and in Britain in particular, the widespread and largely synchronous nature of this change was recognized, and a climatic interpretation was proposed: perhaps a lower winter temperature affected Elm in particular, leaving other trees unaffected. However, one tree that increased apparently in response to the decline in Elm is Ash *Fraxinus excelsior,* a much more frost-sensitive tree than Elm. Attention turned instead to human influences when it was realized that the Elm decline coincided with the first appearances of cereal pollen and also pollen of such obvious weeds of cultivation as plantain *Plantago* and Sorrel *Rumex*. The fact that Elm is among the earliest of trees to open its buds in spring and the high phosphorus and calcium content of those young leaves makes it a preferred forage for livestock, and it was suggested that Elm was selectively coppiced or lopped by Neolithic farmers in late winter. Alternatively, they may have selectively farmed those light sandy soils which also happened to be the favoured habitat of Elms, particularly the Wych Elm *Ulmus glabra*. In a detailed study of this matter, Scaife (1988) showed at Gatcombe Withy Bed on the Isle of Wight that the initial drop in Elm pollen from 15.4% to 2.3% of arboreal pollen at 4850 b.p. was followed by a longer period of 200 years or so before there was some modest recovery in the proportion of Elm pollen. Cereal pollen, as well as that of the weeds *Rumex*, *Plantago* and *Sinapis*, was continuously present from this time onwards, but the predominant pollen remained that of Oak *Quercus* throughout this period (Fig. 4.2). Thus the area remained predominantly woodland, but local interference with some of the trees, notably Elm, accompanied the creation of small fields growing crops, and supporting also the weeds of cultivation. It remains uncertain whether the interference with the Elm took the form of selective felling to create fields, selective cropping for forage, selective browsing by the accompanying livestock, or perhaps the unwitting introduction of the fungus, *Ceratocystis ulmi*, that causes Dutch Elm Disease. The beetle *Scolytus scolytus* which acts as vector of the fungus was certainly present in Britain (Girling 1988) by this time, and would have been favoured by the killing of any Elms that were felled and left lying around as logs with their bark still in place, for its larvae construct complex tunnels under the bark of dead Elms.

Initial clearances were short-lived and small; as seen in the pollen record, they appear as short-lived peaks up to 20% of Gramineae in the pollen-rain, falling back to 10% as woodland recovered. What happened next depends on the region. On the Isle of Wight, the local vegetation reverted to woodland for some 2000 years (Scaife 1982, 1988). On the other hand, on some of the downland and on Breckland, the tree cover never properly recovered, and pollen of grasses and other herbs never fell away; these areas developed early as grasslands and heathlands, and have largely remained so to the present. The downs around Winchester, for example, were dramatically cleared within a very short period around 5630 b.p., an early date, and never recovered their woodland, though further east the downs in Sussex were not cleared until the Bronze

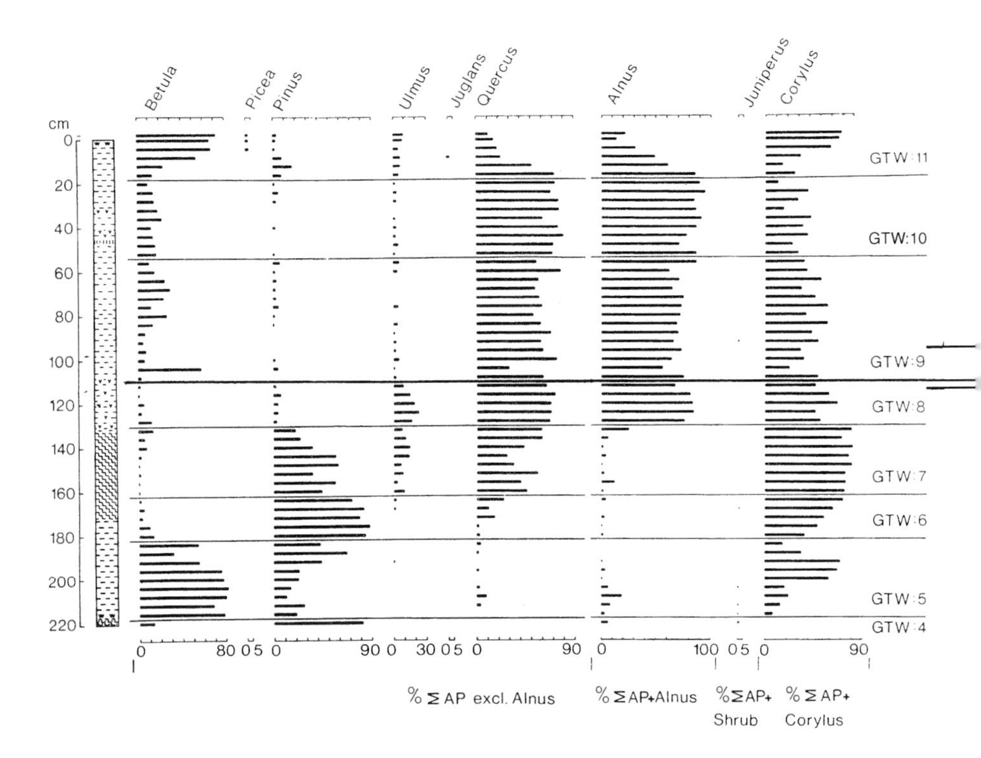

Age (Waton 1982). At Hockham Mere, on the eastern edge of Breckland, grass pollen increased to about 10% of arboreal pollen, as did heather, at about 4400 b.p., and values remained that high until Saxon times when they increased still further (Godwin 1975). Similarly, the blanket bog which covers the Peak District spread to virtually its modern extent during the period from about 5500 to 5000 b.p., though some peat started to form during the Mesolithic (see Chapter 3); charcoal, probably indicating the use of fire by hunters to improve the upland grazing, is evident for example at Robinson's Moss throughout the Mesolithic, but particularly common in an earlier phase (about 8000–7700 b.p.) and just before the time of the elm decline (5300–4700 b.p.). It is possible that fire, limiting or interfering with the upland scrub, altered the rate of evapotranspiration and initiated both the start and the spread of blanket bog (Tallis and Switsur 1990, Tallis 1991). Elsewhere, blanket bog formed at different times, reflecting the local interference of humans in the woodland cover; early in the Black Mountains of south Wales, where there is also evidence of Mesolithic activity, but much later elsewhere, for example not until Bronze Age times in the Berwyn Mountains (Tallis 1991).

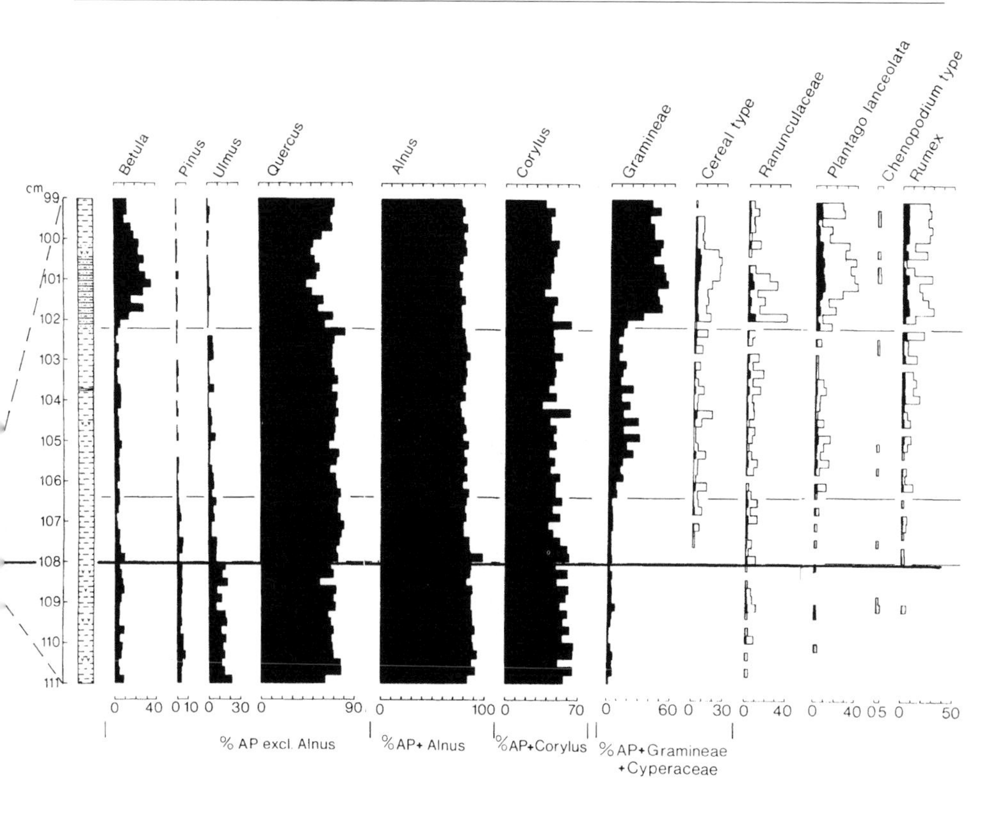

Figure 4.2 Early farming clearances in the pollen record of the Isle of Wight. The decline in Elm at 5000 b.p. (marking the boundary between pollen zones 9 and 8) seems quite sharp on the full diagram. A detailed examination in 2 cm layers shows a rather more ragged decline, matching a ragged occurrence of early cereal and weed pollens. Note that the main woodland trees, Oak and Hazel, show little evidence of change at these levels, implying that woodland remained largely intact, and that local clearance of Elm provided small patches for cultivation. (Modified from figures in Scaife 1988, by permission of Dr R. G. Scaife and the Oxford University Committee for Archaeology.)

Domestic Mammals

Whatever the immediate cause of the Elm decline, it is clear that it indicates the most significant change in the Post Glacial mammal fauna of these islands, the introduction of domestic livestock. Four species were initially involved, sheep, goats, cattle and pigs. So far as we know, horses were both domesticated and introduced rather later.

The domestic sheep is descended from the wild Urial *Ovis orientalis* found native in the mountains of western Asia, including Turkey and Iran (Schaller 1977, Uerpmann 1987). Other species of wild sheep occur further east in Asia (Argali *Ovis ammon*) and in Siberia and North America (Snow Sheep *O. nivalis*, Bighorn Sheep *O. canadensis*), but only the western populations of Urial share with the domestic sheep the same chromosome number, 2n = 54. The Urial is a grazer, adapted to the rather hot, dry conditions of the region by its tight brown coat and its limited requirement for water. Both sexes are horned, but the rams have magnificent spirals, thick at the base and curling outwards whereas the ewes have slight, rather short, conical horns (Fig. 4.3). Wild sheep have only short stumps for tails.

The earliest evidence of domestic sheep comes from various sites in the Middle East where the abundance of sheep bones increases sharply at about 9000 b.p. (Davis 1987, Clutton-Brock 1981). At Jericho, for example, the earlier layers are dominated by Mountain Gazelle and Fallow Deer bones, but the fauna changes rather sharply to one dominated by goats and cattle, with some sheep, at 9300 b.p. Perhaps the earliest domestic sheep are those from Zawi Chemi Shanidar in Iraq dated to 10 870 b.p. In Greece, there was a similar shift at 8000 b.p. from the hunting of wild ungulates, particularly Red Deer, to what must have been domestic livestock including sheep at Franchthi and at Nea Nicomedia, for sheep never occurred as wild animals in Post Glacial Europe. By 7500 b.p. sheep had spread further north-west up the Balkan peninsula. They reached the northern Alpine foothills by 6500 b.p. and Switzerland by 5500 b.p. It used to be thought that the Mouflon of Corsica (*Ovis musimon*) and

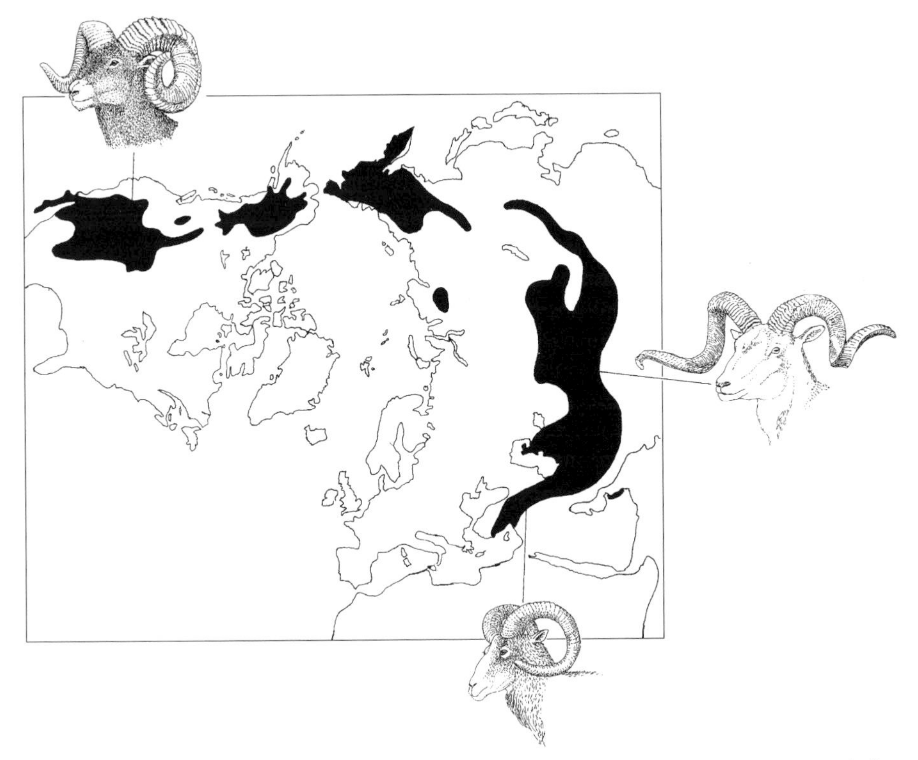

Figure 4.3 Distribution of wild sheep across the northern hemisphere (after Corbet 1978, Schaller 1977). They range from Turkey across the mountainous areas of Asia into North America, but their horn shapes vary greatly across this range.

Cyprus (*O. ophion*) were actually wild sheep; it is now realized that this is zoogeographically impossible – those islands have never been connected to their neighbouring continents at any time when sheep could have immigrated – and it is well demonstrated archaeologically that sheep were imported by early farmers, subsequently escaping to found what are therefore actually feral populations (Vigne 1992). However, they escaped so soon after their introduction, about 8000 b.p., that they were little modified by breeding from their ancestors. The earliest sheep in Britain are dated to about 5365 b.p., from a long barrow on Lambourn Downs, Berkshire (Clutton-Brock 1989). At Skara Brae in Orkney, and in other northern and western islands, sheep were the most numerous livestock at an early Neolithic date, far more abundant than in southern Britain at the same time (Noddle 1989).

The goat is similarly a native of the Middle East, descended from the Bezoar *Capra aegagrus*. This is more of a cliff-dwelling, montane species than the wild sheep, which prefer the rolling uplands, and favours browse rather than grass for its diet. Although goats and sheep have rather different ecologies, and very different fighting tactics and skull shapes, the rest of their bones are very similar. It is therefore very difficult to recognize them in archaeological sites (though Boessneck (1969) demonstrated how careful measurements could discriminate them), and often they have to be combined. This has led archaeologists to concoct such zoologically uncouth (and inaccurate) terms as 'ovicaprids' and 'ovicaprines'. (There is no animal *Ovicapra* that could serve as the base for such a name; the correct term zoologically would be caprine, derived from the tribe Caprini.) There is a further complication, that the Wild Goat is closely related to the Ibex *C. ibex*, which does occur in European mountain areas, and therefore makes it harder to recognize the arrival of domestic goats. Ibex have bluntly rounded anterior surfaces to their horns, and to the bony horn cores within them, whereas Bezoar, and their domestic descendants, have a prominent keel up the front of their horns (Fig. 4.4), so it is possible to distinguish skulls in archaeological sites. The early domestication of the goat in the Middle East is attested by the appearance of its bones in abundance at Jericho at 9300 b.p., well south of its native range. By 8000 b.p. it had appeared in south-eastern Europe, and the 'wild goats' on Crete are surely, like the sheep of Corsica and Cyprus, early feral goats. The earliest record from Britain is at Windmill Hill, Wiltshire, with a date of 4530 b.p. (Clutton-Brock 1989).

The Neolithic farmers' other two domestic ungulates, cattle and pigs, are more difficult to interpret; their history is confounded by the undoubted presence of their wild ancestors, the Aurochs *Bos primigenius* and Wild Boar *Sus scrofa*, in Britain. However, the question of the origin of domestic cattle has been quite thoroughly investigated by Grigson (1969, 1982a, 1989) and, specifically for Britain, by Noddle (1989). There is no doubt that the Aurochs is the direct ancestor of the domestic cattle of Europe. (The humped zebu cattle of India were independently domesticated, from the Asian subspecies of Aurochs *Bos primigenius namadicus*; Grigson 1969, Loftus *et al.* 1993.) The bones of Aurochs and cattle are very similar, except in size, and all fall on the same line in scatter-plots. The reduction in size is generally taken to be a sign of domestication. In the Middle East, the Aurochs bones from the early Holocene, 10 000 to 7000 b.p., at such classic archaeological sites as Jericho, Beidha and Hayonim, show no reduction in size. It has generally been accepted that the cattle from Çatal Häyük in Turkey are the earliest domestic cattle, but Grigson (1989) points out that on size they must have been wild Aurochsen. The first indications of a reduction in the size of cattle appears

Figure 4.4 Ancestral goats and ibex compared (after Schaller 1977). The Bezoar *Capra aegagrus*, the ancestor of the domestic goat, has a keel up the front of its horns, and the bone horn core is similarly keeled in cross-section. It occurs in the Middle East (black), whereas ibex, which have blunt fronts to their horns and horn cores, occur both in Europe and elsewhere in Arabia and Africa. Other species of wild goat occur further east in Asia but, unlike sheep, do not occur in North America.

during the period 6000 to 5000 b.p., in the more western of sites in the Middle East (Jericho, Ashkalon, Amuq, Fikirtepe and Qalat el Mudiq), by contrast with the more eastern contemporary sites (Matarrah, Tepe Guran, Jarmo, Hajji Firuz and Sarab). Thus cattle were domesticated at least 4000 years later than sheep and goats. By the following millennium, sites throughout the Middle East had smaller cattle, undoubtedly domestic, and the diminution in size continued into the 4th millennium (Grigson 1989). This provides the basis for an evaluation of the early cattle at Neolithic sites in Britain; they were undisputably much smaller than the wild Aurochs already present in Britain, and show no signs of any interbreeding (Noddle 1989). Admittedly sample sizes are small, but there is little overlap (Fig. 4.5), and certainly no indication of animals of intermediate size, nor of a progressive diminution in size such as occurs in the Middle East. It appears that the Neolithic farmers who settled Britain brought their cattle with them, just as they must have brought their sheep and goats. There is some suggestion, in the specimens available, that slightly different forms, what we would now call breeds, were brought into different areas of the country, perhaps in separate colonizations of southern Britain from France and northern Britain from Scandinavia, but the evidence is sparse (Noddle 1989). The earliest domestic cattle in Britain are perhaps recorded at Windmill Hill, but other Neolithic sites in Wessex such as

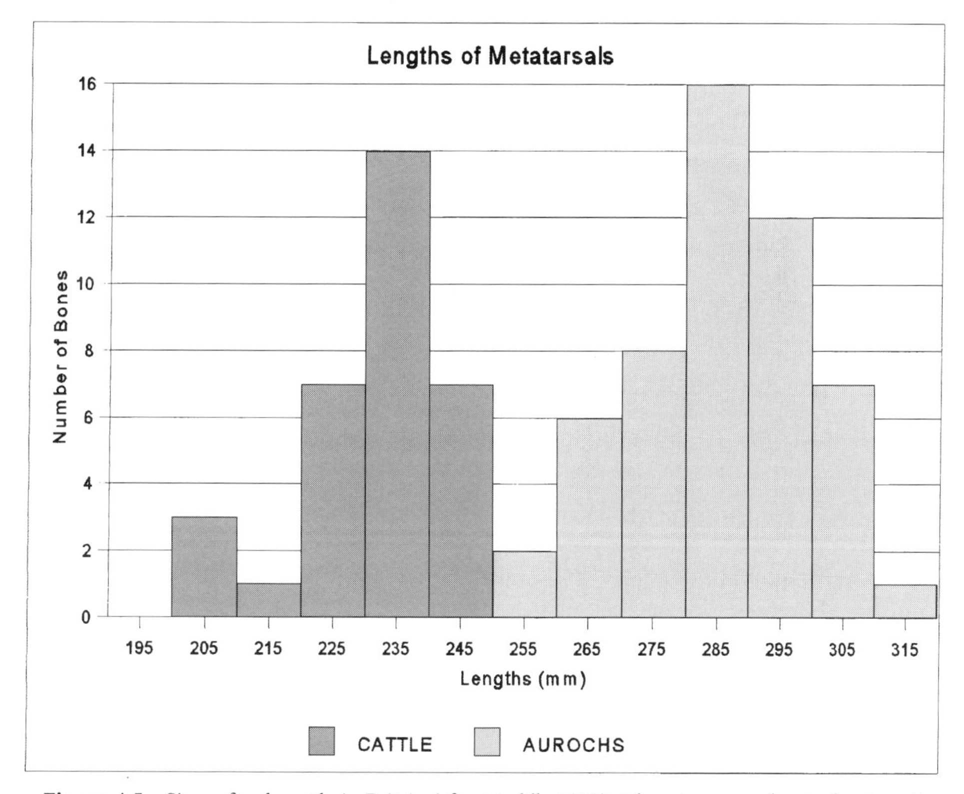

Figure 4.5 Sizes of early cattle in Britain (after Noddle 1989). There is no overlap in the size of metatarsal or metacarpal bones, suggesting that the smaller domestic cattle were introduced already domesticated, and not domesticated from or hybridized with native wild Aurochs.

Durrington Walls, Marsden, Woodhenge, Mount Pleasant and Maiden Castle also contain domestic cattle (Noddle 1989).

There appears to be little discussion, and not much evidence, of the status of domestic pigs relative to Wild Boar. Given the manner in which pigs were allowed to forage in woodland, they would almost certainly have interbred with Wild Boars (Grigson 1982b), and the distinction between the two was perhaps less profound than the difference in cattle. However, domestic pig bones and teeth are smaller than those of their wild relatives, for instance at the Neolithic sites of Windmill Hill (Grigson 1982b) and Mount Pleasant (Harcourt 1979), and they remained distinctively small through to Roman and later times (e.g. at Segontium, Noddle 1993; Caerleon, Hamilton-Dyer 1993). If the other domestic species were brought into Britain by the early farmers, so presumably were pigs. We need more evidence to confirm this. Clutton-Brock (1989) implies that the young specimen from Windmill Hill was among the earliest examples in Britain.

Horses were certainly domesticated later than the four principal ungulates, and not in the Middle East either. Wild Horses were widespread across Europe in the Late Glacial, and they seem to have lasted into the Post Glacial on the Anatolian plateau, but they appear to have been domesticated on the steppes of southern Russia, about 6000 b.p., and reappeared in archaeological sites in the Middle East at about 5000 b.p. (Uerpmann 1987). The earliest records for Britain are from Grimes Graves, Norfolk, dated to 3740 b.p., and from Newgrange in Ireland (Clutton-Brock 1989). Grigson (1966) reviews the possible Neolithic records of horses in Britain, noting 31 sites where they have been reported; she remarks that many of them are fragmentary, and that their dating and stratigraphical location is often poor. The problem of course is that horses were certainly present by Bronze Age times, and there is always the risk in archaeological sites that later bones slip into earlier deposits by various means (scavenged by Fox or Wolf, dumped into pits dug later through earlier layers) to give erroneous results. However, this large body of evidence strongly supports the idea that horses were introduced in Neolithic times, always assuming (see Chapter 3) that they had not survived through from the early Mesolithic.

One other domestic mammal was already present in Britain when the ungulates arrived, and had been present since Mesolithic times. The dog *Canis familiaris* is the domesticated Wolf *C. lupus*, and sufficiently like it to cause considerable confusion in archaeological sites. Probably the earliest convincing domestic dog is the skeleton from Ein Mallaha, Israel, dated to 11 600 b.p. (Davis and Valla 1979). This is the skeleton of a puppy, 4–5 months old, buried beside the skeleton of a human whose left hand rested on the puppy. The teeth of dogs are morphologically like those of Wolves, but slightly smaller, and the jaws of dogs tend to be shorter so that the teeth are more crowded (Clutton-Brock 1969). Dogs tend to be smaller overall, as well, but this point is obscured by the fact that Wolves are also much smaller in the Middle East than they are in Europe. However, the Mallaha puppy, and also roughly contemporaneous dog specimens from Hayonim, also in Israel, were carefully compared with a range of Wolf specimens, and are both smaller than the local Wolves, and much the same size as later dogs from Middle Eastern sites (Fig. 4.6). In the past, there has been considerable debate about the possible role of the Golden Jackal *Canis aureus* in the ancestry of the dog, or of some breeds of dog. This newer evidence rules out any

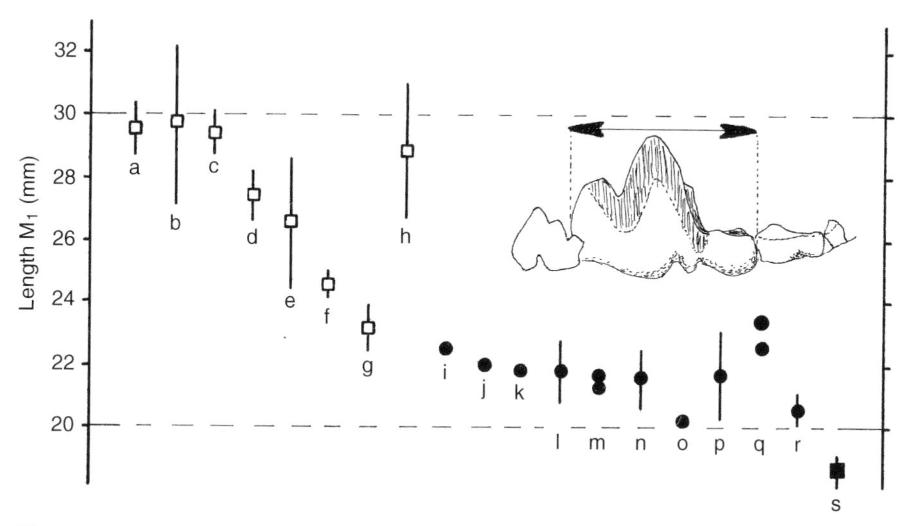

Figure 4.6 Sizes of Wolves and dogs as indicated by the length of the lower carnassial tooth (after Davis and Valla 1979). Modern Wolves (hollow squares) show a cline in size, from large in the north to smaller in the south (a, Greenland/Ellesmere Island; b, N. Europe; c, Denmark; d, Turkey; e, Zagros Mtns, Iran; f, Israel; g, Arabia), but Pleistocene Wolves from the Middle East (h) were as large as recent northern Wolves. The archaeological dogs (solid circles) from Ein Mallaha, Israel (i), Hayonim, Israel (j), Palegawra, Iraq (k), Jericho (l), Cayonu, Turkey (m), Korucutepe, Turkey (n), Arad, Israel (o), Persian-age Israel (p), modern Egypt (q) and Hungary (r) are mostly smaller than the Wolves, but all of similar sizes to each other. Golden Jackals (solid squares) are much smaller again, and below the size of these early dogs.

involvement of jackals in dog ancestry, and biochemical/genetic data confirm that only the Wolf is implicated (Davis 1987)

The earliest dogs in Britain were, surprisingly, at Star Carr, the Mesolithic encampment (see Chapter 3). This matches specimens from Germany of Late Glacial date which Musil (1984) considers to be early dogs. Thus Wolves were domesticated before the ungulates, presumably to help with hunting, but not necessarily or solely in the Middle East. Genetic analysis of dogs and Wolves from around the world confirms that they have been domesticated from, and crossed back to, Wolves in several different places; three or four different dog lineages can be recognized, but they may have been domesticated as long as 135 000 years ago (Vilà *et al.* 1997).The subsequent archaeological history of the dog in Britain has been very usefully reviewed by Harcourt (1974). The few Mesolithic specimens seem to have been moderately sized dogs, with shoulder heights of about 60 cm, the size of a Scottish Collie. In the subsequent Neolithic and Bronze Ages, there are few specimens, but they are about the same size, and not very variable. In the Iron Age, dogs become a little more common and more variable, with shoulder heights ranging from 29–58 cm, but in the Roman period dogs become much more variable, as well as more common. Shoulder heights vary from 23–72 cm, and there is clear evidence of 'toy' dogs as well as working dogs. However, dogs were not only pets and working animals; cut marks on bones from

Mesolithic Star Carr, Iron Age Highfield and Romano-British Owlesbury, as examples, suggest that they were also eaten or at least that their skins were used. In the subsequent Anglo-Saxon period, dogs became larger, on average, raising again the problem of distinguishing their bones from Wolves. Fortunately the teeth of larger dogs remain small, in relation to skull length, so skulls at least can be distinguished (Clutton-Brock 1969).

The meat economy of Britain from Neolithic through to Roman times was certainly based on domestic rather than wild ungulates. In Neolithic sites, the main stock were cattle, with sheep and pigs generally less numerous. At Windmill Hill, one of the earliest sites, 412 bones of cattle were identified, combining three different time periods, compared with 144 bones of sheep/goat and 108 bones of pig; there were

Domestic ungulates herded towards a hill-fort

only 273 bones of the other eight species present, mostly dog (Jope and Grigson 1965). Similar proportions were found at Late Neolithic Fengate; nine cattle, two sheep and ten pigs, along with one horse, one dog, and ten wild mammals (Harman 1978). At Durrington Walls, with a date of 4000–3500 b.p., pigs were the most abundant, 228 individuals being recorded along with 97 cattle, but only seven sheep and one goat. There were also four dogs and three horses, along with 24 wild mammals (Harcourt 1971). Similarly at Mount Pleasant, Dorset, there was evidence of 17 pigs, eight cattle, four sheep, one possible goat and one horse, along with only four wild mammals (Harcourt 1979). At Skara Brae in the Orkneys, sheep were perhaps more numerous; there were 17 jaws of cattle and 15 jaws of sheep in the specimens identified by Watson (1931), but only a few fragments of pigs and four shed antlers of Red

Deer; Watson remarks that the treeless Orkneys would perhaps have been better grazing lands for sheep than for cattle, and certainly poor for pigs, compared with the still relatively well-wooded Wessex countryside of early Neolithic times.

In the Bronze Age, too, cattle and sheep were the main livestock. At Grime's Graves, Legge (1992) found, using the jaws as the best index of numbers, that cattle constituted 53%, sheep/goat 39% and pigs only 5% of the 489 animals identified. Very small numbers of Red Deer, Roe Deer and Horse made up the balance. He compared these results with three other Bronze Age faunas, Runnymede and two phases at Down Farm. At Runnymede, sheep were the main stock (45%), but pigs were also important (29%) and cattle less so (only 26%). The lower-lying nature of this Thames-side site may have been more suitable for pigs, though hardly for sheep. At Down Farm, by contrast, sheep were predominant, 58% and 75% in the two faunas, and cattle important (29% and 16%), but pigs were scarce (0% and 4.5%); other species, including deer and horses, made up the balance of 12% and 4.5%.

In the Iron Age, cattle and sheep continued to be the main livestock, with smaller numbers of pig, horse and dog (Maltby 1981). Evidence for the continued presence of goats is patchy, though they were clearly present at a few sites such as Croft Ambrey and Glastonbury. The Iron Age Lake Village of Glastonbury is perhaps not the most likely place for a sheep-dominated economy, but the sample of bones identified included 3013 sheep bones, along with only 181 cattle, 58 pig, 73 horse and six goat (Dawkins and Jackson 1917). Cunliffe (1978) argues that sheep gradually became more important than cattle through the Iron Age in southern England, perhaps indicating the opening up of the downland and the need to use sheep dung to fertilize the arable land; if the sheep fed on the downs during the day and were folded at night on the fields, their dung would be left there. Presumably the Glastonbury sheep were reared nearby on the drier ground of the Mendip Hills. At Winnal Down, Hampshire, the earlier deposits (3rd century B.C.) contained about 53% cattle and 29% sheep, as well as 11% horse and 7% pig; the later Middle Iron Age fauna contained 29% cattle and 53% sheep/goat, as well as 9% horse and 10% pig (Maltby 1981).

The Roman sites have been well summarized by King (1978); he examined the bone reports from 116 sites, including a few Iron Age and Saxon sites, particularly where they were occupied during Roman times as well. A crude averaging of the figures in his Table 2 suggests, on the basis of 3815 jaws from 18 different sites, that the Roman diet included 37% cattle, 32% sheep and 31% pigs. This obscures the variation between sites and with time. Sheep tended to become less important with time; 68% of a sample of 38 earlier (1st to 2nd century) sites contained more than 30% sheep, whereas only 26% of a sample of 39 later (3rd to 4th century) sites contained so many. The sheep tended to be more numerous in southern sites, whereas cattle and/or pigs were rather more abundant at the northern and western sites, which were presumably more wooded. Cultural differences can also be detected; Roman villas produce a diet with rather more pig than contemporary native sites.

One of the largest archaeological collections of Roman bones came from Exeter (Maltby 1979). Over 18 000 bones or fragments representing at least 738 individual mammals were analysed (and a further 40 000 Medieval bones). The three domestic ungulates contributed 86% of these individuals, with cattle (245) and sheep/goat (226) equally numerous, pigs (161) slightly less so (Table 4.1). An interesting contrast is offered by the settlements of Vindolanda and Cortospitum, on the northern frontier

Table 4.1 The fauna of Roman Exeter (from Maltby 1979). Numbers are Minimum Numbers of Individuals (MNI), based on the highest count of some diagnostic element.

Date (A.D.)	*55–75*	*75–100*	*100–200*	*200–300*	*300+*	*Misc*	*Total*
Cattle	29	65	57	22	50	22	245
Sheep/Goat	30	39	72	16	47	22	226
Pig	20	29	46	19	34	13	161
Red Deer	4	4	3	2	4	–	17
Roe Deer	2	4	4	1	1	–	12
Horse	1	5	7	–	8	1	22
Hare	1	2	3	4	6	2	18
Dog	2	5	5	3	7	–	22
Cat	2	–	1	1	3	–	7
Fox	–	2	–	–	1	–	3
Otter	1	–	–	–	–	–	1
Badger	–	–	–	1	–	–	1
Hedgehog	–	–	–	–	1	–	1
Water Vole	–	1	–	–	–	–	1
Wood Mouse	–	–	–	–	–	1	1
Total	92	156	198	69	163	60	738

of Roman Britain; at both sites, domestic ungulates similarly contribute the majority of the fauna, 84% at Vindolanda and 99% at Cortospitum, but cattle are much more important than the others, 56% of the total at Vindolanda and an even larger 71% at Cortospitum (Hodgson 1968, 1977).

The Retreat of the Wild

What was the effect of this importation of domestic animals on the wild mammals of Britain? Initially, it seems likely to have been minimal. It might even have relieved the wild mammals of some of the hunting pressure they had previously sustained. Jarman (1972) has surveyed the mammal faunas of a wide range of Mesolithic and Neolithic faunas from archaeological sites across Europe. Red Deer and Wild Boar were the main prey throughout the Mesolithic and throughout Western Europe, contributing 38.5 and 17.6% of the herbivores in 52 sites with quantitative information. Aurochsen and Roe Deer also occurred frequently, but were much less important numerically, contributing 8.9 and 8.1%. In the succeeding Neolithic, wild game, particularly the Red Deer, continued to appear as an item of diet, but generally as a minor component. Jarman (1972) reviews some European sites, particularly in the Alpine area, where Red Deer continued to be important, alongside pigs and cattle; however, sheep predominated in the Mediterranean, while in north-western Europe cattle were predominant, and wild mammals were rather scarce as prey.

Among British Neolithic sites, at Mount Pleasant, Dorset, there was evidence of 31 domestic mammals, along with only two Red Deer, one Aurochs and one Wild Boar

(Harcourt 1979). The substantial fauna from Durrington Walls included 340 domestic animals, along with three Aurochsen, 14 Red Deer, two Roe Deer, one Badger, two Foxes, one Pine Marten and one Beaver (Harcourt 1971), though the Red Deer were mostly represented by picked-up antlers. At Windmill Hill, the deer were similarly represented mostly by picked-up antlers, of both Roe (11) and Red Deer (15), though another eight Red Deer had been killed. Bones of other wild species recorded there included three of Aurochs, eight Fox, 52 Wild Cat and two Badger (Jope and Grigson 1965). These sites all indicate very strongly the dependence of these early farmers on their livestock, and the unimportance to them of wild prey. The same pattern is true in the succeeding Bronze and Iron Ages; at Grimes Graves, the Bronze Age people left only four jaws of Red Deer and four of Roe Deer in a sample of 489 ungulates, mostly domestic (Legge 1992). Of 584 bone fragments at Bronze Age Fengate, only nine (four Badger, one Wolf, two Fox and two Red Deer) were of wild species (Biddick 1980). The Iron Age Lake Village of Glastonbury produced a very rich mammal fauna, including Red Deer, Roe Deer, Wild Boar, Fox, Wild Cat, Otter and Beaver, but collectively these only contributed 71 bones, against 3355 bones of domestic species (Dawkins and Jackson 1917). In Roman sites, King (1978) remarks that deer bones, mostly Red Deer, never amount to more than about 5% of the total of domestic ungulates. Deer occur in 39% of 1st century contexts, but in 53, 71 and 79% of sites in the following three centuries, suggesting that perhaps hunting was more popular in the more settled economies of later times. Roe Deer also occur widely, but even more scarcely than Red Deer. At Exeter, the Romans left the remains of only 17 Red Deer and 12 Roe Deer, a minuscule 4% of the fauna (Maltby 1979; Table 4.1), while at Vindolanda there were 11 Red Deer and nine Roe Deer, only 6% of the fauna (Hodgson 1977). Portchester Castle contained only 262 Red and 38 Roe Deer in a total of 18 923 bones (Grant 1976). A few sites claim to have Fallow Deer, including Iron Age War Ditches in Cambridgeshire and the Roman levels at Portchester Castle (Grant 1976).

If Red Deer became less important as food in the Neolithic, there is abundant evidence from both Grimes Graves, Norfolk, and Durrington Walls, Wiltshire, that they remained important as a source of tools, in the form of antler, so deer must still have been valued. It has even been argued that they were 'farmed' in some way, not hunted but husbanded to supply antlers. In particular, their antlers were used as picks. Clutton-Brock (1984) analysed samples of 283 antler picks from Grimes Graves, where they were used to mine flint from the chalk, and 332 from Durrington Walls, an encampment where they were used to dig the ditch and bank. Antler picks were used at Stonehenge, too, and have ^{14}C dates of 4410 and 4390 b.p. (Burleigh *et al.* 1982b). Clearly, Red Deer were abundant, and it has been suggested that the miners in Norfolk required up to 400 antlers each year to dig each mine shaft. However, Clutton-Brock points out that most of the picks, 82% at Grimes Graves and 87% at Durrington Walls, were cast antlers picked up after they had been shed naturally by the stags in spring; only 18% and 13% respectively were taken from stags that had been killed. On this basis, a population of some 600 deer, including 200 mature stags, from the surrounding 60 km^2 (taking modern deer densities from Poland) would have provided a self-sustaining source of picks. These were large Red Deer by modern standards, with antler diameters across the burr averaging 68 and 65 mm in the two samples; a couple of modern antlers on my wall are only 55 mm across. The full splendour of the Neolithic antlers cannot be seen, for they were broken off above the trez tine, so that the brow tine could be used as the pick; it was hammered into the chalk, indicated by battering on the burr. Some of the Grimes Graves antlers were dated directly by ^{14}C; 47 dates all cluster in a short period from 4114 to 3740 b.p. (Burleigh *et al.* 1979).

Such large Red Deer antlers are typical of Red Deer all the way through from the Late Glacial, and have in the past caused some confusion; they have been termed 'Giant Deer', and so been confused with *Megaloceros giganteus*, and they have also been considered a separate subspecies or even full species, *Cervus spelaeus*. Originally coined for specimens from Kent's Cavern, Devon, by Richard Owen in 1846, these specimens and the general problem have been usefully reviewed by Lister (1987). There is no doubt that *Cervus spelaeus* is simply a large form of the Red Deer, much larger than present-day British animals which average about 90 kg, and bigger even than modern European forest stags which attain 200 kg. However, they are within the size-range of the North American Wapiti *C. e. canadensis*, which can be 300–450 kg. The story was further confused by mixing up the bones with the genuinely larger bones of *Megaloceros*, which also occurs in Kent's Cavern, and even a few Bison bones as well. Red Deer, like many other animals, became smaller in Post-Glacial times (cf. Davis 1981 on Wild Boar and Wolf in Israel), but the present very small size of Scottish Red Deer is a reflection of their poor habitat. Taken to better conditions in New Zealand, they have become bigger again (King 1990).

The status of the wild relatives of the domestic cattle and pigs may have changed if the farmers wished to prevent their tame stock from interbreeding with their wild relatives, and even more emphatically if those wild relatives raided their crops. Aurochs declined to extinction, or nearly so, in this period. It is hard to know whether this reflects habitat change or persecution. The Aurochs was certainly a woodland species; Caesar refers to it as inhabiting the great Hercynian Forest of central Europe, which stretched from Switzerland along the Danube, and the last was killed in Poland in the Jaktorowska Forest in 1627 (Pyle 1994). Remains of Aurochs are reasonably frequent in cave sites as well as in archaeological sites of Pleistocene and earlier Holocene dates (Chapter 3). During the Neolithic, they are reported from Ireshopeburn Moor, in the northern Pennines, as well as from such Neolithic sites in southern England as Durrington Walls, Fengate, Windmill Hill and Mount Pleasant (see above). The latest dated specimen appears to be from the early Bronze Age of Charterhouse Warren Farm, Somerset, at 3245 b.p., which is the latest of six dates in the period from 4200 b.p. (Burleigh *et al.* 1982b, Clutton-Brock 1986, 1991) (Table 4.2, Fig. 4.7). Among possible later specimens are Bronze Age specimens from Maiden Castle (Armour-Chelu 1991) and one from Roman Segontium (Caernarvon), dated to the 4th century (Noddle 1993). The very large cattle bones recorded from some northern Roman sites, notably Vindolanda (Hodgson 1977), could also repay further study. There is the difficulty that domestic bulls overlap wild cows in some dimensions, and though complete specimens can usually be distinguished, the fragmentary bones in archaeological sites sometimes cannot.

The attitude to wild predators must also have changed fundamentally in the Neolithic; from being useful sources of fur and, in the case of Brown Bears, an additional source of meat, the Wolves, Lynx and Bears must have been feared as killers of livestock, particularly if severe winter weather drove them to attack coralled sheep or

Figure 4.7 Distribution of archaeological Aurochs *Bos primigenius* in Great Britain.

Table 4.2 The occurrence of Aurochs *Bos primigenius* in Britain. Dating of most records is indirect; if the accompanying fauna includes such species as Spotted Hyaena and Woolly Rhino, they are assigned to the Devensian, and if such species are absent but Horse and Reindeer are present they are assigned to the Late Glacial. Radio-carbon dates on Aurochsen are quoted directly; those prefixed c. were carried out on other species in the fauna. Middle Pleistocene sites are dated to Oxygen Isotope Stages (from marine sediments), OI 7-13.

Site	*Grid Ref.*	*Age*	*Source*
Clacton, Essex	TM1715	Hoxnian (OI 13)	Noddle 1989
Ingress Vale, Swanscombe, Kent	TQ5975	Hoxnian (OI 13)	Stuart 1982
Swanscombe, Barnfield Pit	TQ6074	Hoxnian (OI 13)	Gee 1993
Jaywick Sands, Essex	TM1513	Hoxnian?	Reynolds 1939
Grays, Essex	TQ6177	Pre-Ipswich. (OI 9)	Noddle 1989
Crayford, Kent	TQ5175	Pre-Ipswich. (OI 7)	Gee 1993
Uphall Pit, Ilford	TQ4586	Pre-Ipswich. (OI 7)	Gee 1993
Brundon, Suffolk	TL8641	Pre-Ipswich. (OI 7)	Hopwood 1939
Harkstead, Essex	TM1935	Pre-Ipswich. (OI 7)	Gee 1993
Barrington, Cambridgeshire	TL3949	Ipswichian (OI 5)	Noddle 1989
Brentford, Middlesex	TQ1778	Ipswichian (OI 5)	Gee 1993
Trafalgar Square, London	TQ2980	Ipswichian (OI 5)	Gee 1993
Swanton Morley, Norfolk	TG0117	Ipswichian (OI 5)	Stuart 1982
Hoe Grange Quarry, Derbyshire	SK2155	Ipswichian?	Gee 1993
Taplow, Berkshire	SU9182	Ipswichian?	Reynolds 1939
West Thurrock, Essex	TQ5877	Ipswichian?	Reynolds 1939
Erith, Kent	TQ5177	Ipswichian?	Reynolds 1939
Fisherton, Salisbury	SU0038	Devensian	Reynolds 1939
Pin Hole Cave, Derbyshire	SK5374	12 480 b.p.	Housley 1991
Pin Hole Cave, Derbyshire	SK5374	12 400 b.p.	Housley 1991
Aveline's Hole, Somerset	ST4758	12 380 b.p.	Housley 1991
Gough's (New) Cave, Somerset	ST467539	12 300 b.p.	Housley 1991
Gough's Cave, Somerset	ST467539	11 900 b.p.	Housley 1991
Kent's Cavern, Devon	SX9264	11 880 b.p.	Housley 1991
Pin Hole Cave, Derbyshire	SK5374	10 979 b.p.	Housley 1991
Dog Hole Fissure, Derbyshire	SK5374	c. 9960 b.p.	Jenkinson 1984
Mother Grundy's Parlour, Derbyshire	SK536743	9910 b.p.	Hedges *et al.* 1994
Star Carr, Yorkshire	TA027810	c. 9600 b.p.	Legge and Rowley-Conwy 1988
Thatcham, Berkshire	SU5167	Mesolithic	King 1962
Seamer Carr, Yorkshire	TA033819	c. 9330 b.p.	Schadla-Hall 1988
Kirkcudbright	NX6851	9074 b.p.	Burleigh *et al.* 1976
Seamer Carr, Yorkshire	TA030835	8620 b.p.	Burleigh *et al.* 1982a
Mother Grundy's Parlour, Derbyshire	SK536743	8480 b.p.	Hedges *et al.* 1994
Kildale Hall, Yorkshire	NZ609097	8270 b.p.	Burleigh *et al.* 1983
Morton, Fife	NO467257	c. 8050 b.p.	Coles 1971
Uskmouth, Monmouth	ST3283?	Mesolithic?	Noddle 1989
Wawcott, Berkshire	SU4167	Mesolithic	Carter 1975
Cherhill, Wiltshire	SU0370	Mesolithic	Grigson 1978
Wetton Mill, Staffordshire	SK0956	8847 b.p.	Grigson 1978
Netherheath Flats, Durham	NY8-3-	Mesolithic	Grigson 1978

Site	*Grid Ref.*	*Age*	*Source*
Broxbourne, Hertfordshire	TL3707	7230 b.p.	Gowlett *et al.* 1986
Mother Grundy's Parlour, Derbyshire	SK536743	6915 b.p.	Grigson 1978
Westward Ho, Devon	SS4329	c. 6585 b.p.	Grigson 1978
Blashenwell, Dorset	SY952805	6450 b.p.	Grigson 1978
Lingey Fen, Cambridgeshire	TL450585	6370 b.p.	Burleigh *et al.* 1982a
King Arthur's Cave, Herefordshire	SO546154	Mesolithic	Grigson 1978
Prestatyn, Flintshire	SJ0682	Mesolithic	Grigson 1978
East Ham, London	TQ4283	Mesolithic	Noddle 1989
Teeshead, Cumberland	NY700340	Mesolithic	Grigson 1978
Hardhill, Cumberland	NY727331	Mesolithic	Grigson 1978
Ireshopeburn Moor	NY746314	Mesolithic	Grigson 1978
Shaws, Dumfries	NX9487	Mesolithic?	Harting 1880
R. Idle, Bawtry, Yorkshire	SK6591	Mesolithic?	C. Howes, pers. comm.
Southampton Docks	SU4312	Mesolithic?	S. Hamilton-Dyer, pers. comm.
off Lowestoft	?	Mesolithic?	Reynolds 1939
off Yarmouth	?	Mesolithic?	Reynolds 1939
Dogger Bank	?	Mesolithic?	Reynolds 1939
Borth Bog, Cardiganshire	SN6290	c. 5950 b.p.	Grigson 1978
Tolpits Lane, Hertfordshire	TQ076942	5230 b.p.	Burleigh *et al.* 1982a
Pitstone, Buckinghamshire	SP933140	5520 b.p.	Burleigh *et al.* 1985
Mount Pleasant, Dorset	SY710899	Neolithic	Harcourt 1979
Durrington Walls, Wiltshire	SU152435	Neolithic	Harcourt 1971
North Marden, Sussex	SU8015	Neolithic	Browne 1986
Windmill Hill, Wiltshire	SU1083	Neolithic	Jope and Grigson 1965
Sparsholt, Oxfordshire	SU3487	Neolithic	Hamilton-Dyer 1996a
Itchen Abbas, Hampshire	SU5332	Neolithic	S. Hamilton-Dyer, pers. comm.
Littleport, Norfolk	TL5996	Neolithic	Godwin 1975
Fordyce, Banffshire	NJ5563	Neolithic	Reynolds 1939
Herne Bay, Kent	TR1768	Neolithic	Reynolds 1939
Amroth, Pembroke	SN1607	Neolithic	Reynolds 1939
Preston (Ribble), Lancashire	SD5129	Neolithic	Reynolds 1939
Avebury, Wiltshire	SU0969	Neolithic?	Reynolds 1939
Leasowe, Cheshire	SJ2791	Neolithic?	Reynolds 1939
Quy Fen, Cambridgeshire	TL5162	Neolithic?	Gee 1993
Swaffham Fen, Cambridgeshire	TL5467?	Neolithic?	Gee 1993
Newbury, Berkshire	SU4767	Neolithic?	Dawkins and Sanford 1866
West Stour, Dorset	ST790227	Neolithic?	Mansell-Pleydell 1889
Corhampton, Hampshire	SU609202	4790 b.p.	Burleigh *et al.* 1983
Lingey Fen, Cambridgeshire	TL450585	4630 b.p.	Burleigh *et al.* 1982a
Marden, Wiltshire	SU090583	Late Neolithic	Noddle 1989

Site	*Grid Ref.*	*Age*	*Source*
?Staines, Surrey	TQ0271	Late Neolithic	Robertson-Mackay 1987
Lower Mill Farm, Stanwell, Middlesex	TQ0674	Late Neolithic?	S. Hamilton-Dyer, pers. comm.
Thickthorn Down, Dorset	ST972123	Late Neolithic?	Drew and Piggott 1936
?Maiden Castle, Dorset	SY668885	4360 b.p.	Hedges *et al.* 1988
Burwell Fen, Cambridgeshire	TL565665	4200 b.p.	Clutton-Brock 1986
West Overton, Wiltshire	SU135686	4040 b.p.	Hedges *et al.* 1988
Northborough, Cambridge	TF1408	3860 b.p.	C.A.I. French, pers. comm.
Hemp Knoll, Wiltshire	SU068673	3760 b.p.	Clutton-Brock 1986
Beckford, Worcestershire	SO983363	3580 b.p.	Clutton-Brock 1986
Wilburton, Cambridgeshire	TL490750	3400 b.p.	Clutton-Brock 1986
Lowe's Farm, Cambridgeshire	TL983363	3340 b.p.	Clutton-Brock 1986
Charterhouse Warren Farm, Somerset	ST952805	3245 b.p.	Clutton-Brock 1986
Maiden Castle, Dorset	SY6688	Bronze Age	Armour-Chelu 1991
Testwood Lakes, Hampshire	SU345153	Bronze Age?	S. Hamilton-Dyer, pers. comm.
Segontium (Caernarfon)	SH4862	Roman, 4th C.	Noddle 1993
Wadworth Carr	SK5897	?	C. Howes, pers. comm.
Gainsborough	SK8189	?	C. Howes, pers. comm.
Belmore Gravel Pit, Lound	SK6986	?	C. Howes, pers. comm.
R. Idle, West Stockwith	SK6986	?	C. Howes, pers. comm.
R. Torne, near Kilham Farm	SE6602	?	C. Howes, pers. comm.
Barnwell, Cambridgeshire	TL0485	?	Reynolds 1939
Bath	ST7464	?	Reynolds 1939
Caswell Bay, Glamorgan	SS5987	?	Reynolds 1939
Caversham, Reading, Berkshire	SU7274	?	Reynolds 1939
Chatham, Kent	TQ7567	?	Reynolds 1939
Chingford, Essex	TQ3893	?	Reynolds 1939
Cropthorne, Bricklehampton, Worcester	SO9944	?	Reynolds 1939
Croyde, Devon	SS4439	?	Reynolds 1939
Derby	SK3435	?	Reynolds 1939
Didlington, Norfolk	TL7797	?	Reynolds 1939
Donnington, Newbury, Berkshire	SU4668	?	Reynolds 1939
Folkestone, Kent	TR2336	?	Reynolds 1939
Great Yeldham, Essex	TL7638	?	Reynolds 1939
Harefield, Middlesex	TQ0590	?	Reynolds 1939
Hedingham, Essex	?	?	Reynolds 1939
Hurley Bottom, Berkshire	SU8283	?	Reynolds 1939
Ipswich	TM1744	?	Reynolds 1939

Site	*Grid Ref.*	*Age*	*Source*
Isleham, 'Kent' (Cambs?)	TL6474?	?	Reynolds 1939
Jarrow Dock, Durham	c. NZ3265	?	Reynolds 1939
Kings Langley, Hertfordshire	TL0702	?	Reynolds 1939
Kirmington, Lincoln	TA1011	?	Reynolds 1939
Larkhall, Bath, Gloucestershire	ST7566	?	Reynolds 1939
Lawford, Warwickshire	TM0830	?	Reynolds 1939
Leicester	SK5904	?	Reynolds 1939
Charing Cross, London	TQ3080	?	Reynolds 1939
Chelsea, London	TQ2677	?	Reynolds 1939
Earl's Court, London	TQ2578	?	Reynolds 1939
Kensington, London	TQ2579	?	Reynolds 1939
Lambeth, London	TQ3175	?	Reynolds 1939
Westminster, London	TQ3079	?	Reynolds 1939
Locksbrook, Bath, Gloucestershire	ST7264	?	Reynolds 1939
Maidstone, Kent	TQ7656	?	Reynolds 1939
Marcham, Berkshire	SU4596	?	Reynolds 1939
Melksham, Wiltshire	ST9063	?	Reynolds 1939
Mitcham, Surrey	TQ2868	?	Reynolds 1939
Moulsham, Chelmsford, Essex	TL7105	?	Reynolds 1939
Alexandra Dock, Newport, Monmouth	c. ST3187	?	Reynolds 1939
Northampton	SP7561	?	Reynolds 1939
North Walsham, Norfolk	TG2308	?	Reynolds 1939
Northwich, Cheshire	SJ6573	?	Reynolds 1939
Nottingham	SK5741	?	Reynolds 1939
Norwich	TG2308	?	Reynolds 1939
Orton Hall, Peterborough	TL1696	?	Reynolds 1939
Oundle, Northants	TL0488	?	Reynolds 1939
Peckham, Surrey	TQ3476	?	Reynolds 1939
Romsey, Hampshire	SU3521	?	Reynolds 1939
Saffron Walden, Essex	TL5438	?	Reynolds 1939
Shustoke, Birmingham	SP2290	?	Reynolds 1939
Slade Green, Erith, Kent	TQ5276	?	Reynolds 1939
Somersham, Huntingdonshire	TL3677	?	Reynolds 1939
Stonehouse, Gloucestershire	SO8005	?	Reynolds 1939
Stroud, Gloucestershire	SO8504	?	Reynolds 1939
Sunbiggen Tarn, Westmorland	NY6707	?	Reynolds 1939
Sunbury, Middlesex	TQ1069	?	Reynolds 1939
Southampton	SU4212	?	Reynolds 1939
Tenby	SN1300	?	Reynolds 1939
Thame, Oxfordshire	SP7006	?	Reynolds 1939
Thetford, Norfolk	TL8783	?	Reynolds 1939
Tisbury, Wiltshire	ST9429	?	Reynolds 1939
Twickenham	TQ1473	?	Reynolds 1939
Twizell, Northumberland	NU1228	?	Reynolds 1939
Tyneside	?	?	Reynolds 1939
Walthamstow, Essex	TQ3788	?	Reynolds 1939
Walton-on-the-Naze, Essex	TM2521	?	Reynolds 1939

Site	*Grid Ref.*	*Age*	*Source*
Wembley Park, Middlesex	TQ1985	?	Reynolds 1939
West Wittering, Sussex	SZ7999	?	Reynolds 1939
East Wickham, Kent	TQ4576	?	Reynolds 1939
Whyteleaf, Croydon, Surrey	TQ3358	?	Reynolds 1939
Athol, Perth	NN8765	?	Reynolds 1939
Ale Water, Roxburgh	?	?	Reynolds 1939
Belhelvie Moss, Aberdeen	NJ9417	?	Reynolds 1939
Breckigo, Caithness	?	?	Reynolds 1939
Crofthead, Renfrew	?	?	Reynolds 1939
Drummond Castle, Perth	NN8418	?	Reynolds 1939
Duns, Berwick	NT7853	?	Reynolds 1939
Stockwell Street, Glasgow	NS5964	?	Reynolds 1939
Hapsburn, Roxburgh	?	?	Reynolds 1939
Jedburgh, Roxburgh	NT6520	?	Reynolds 1939
Keiss, Caithness	ND3461	?	Reynolds 1939
Kilmarnock, Ayr	NS4237	?	Reynolds 1939
Kintradwell, Sutherland	NC9107	?	Reynolds 1939
Kirkcudbright	NX6851	?	Reynolds 1939
Lanark	NS8843	?	Reynolds 1939
Lilliesleaf, Roxburgh	NT5325	?	Reynolds 1939
Linton Loch, Roxburgh	NT7726	?	Reynolds 1939
Maybole, Ayr	NS3009	?	Reynolds 1939
Mertoun, Berwick	NT6032	?	Reynolds 1939
Moulin, Perth	NN9459	?	Reynolds 1939
Muthill, Perth	NN8616	?	Reynolds 1939
Newburgh, Fife	NO2318	?	Reynolds 1939
New Galloway, Kirkcudbright	NX6377	?	Reynolds 1939
Paisley	NS4864	?	Reynolds 1939
Sandwick, Orkney	ND4389	?	Reynolds 1939
Skara Brae, Orkney	HY2318	?	Reynolds 1939
Swinton Mill, Berwick	NT8145	?	Reynolds 1939
Whitmuir Hall, Selkirk	NT4926	?	Reynolds 1939
Whitrig Bog, Berwick	NT7841	?	Reynolds 1939
Yarrow, Selkirk	NT3527	?	Reynolds 1939

cattle. The Brown Bear becomes increasingly scarce through Neolithic times (Table 4.3, Fig. 4.8), and the latest archaeological examples appear to be a jaw from Sheepen, Colchester, and bones from two other sites in the fort there, all of Roman date (Luff 1985, 1993). The latest specimens from archaeological sites with ^{14}C dates are a little earlier, from a Bronze Age barrow at Ratfyn, Wiltshire at 3500 b.p. and from northern Scotland at 2673 b.p. (Clutton-Brock 1991, Burleigh 1986). The latest specimens that happen to have been found in archaeological sites are unlikely to have actually been the last of their kind in Britain. Harting (1880) and others quote the comment of Martial that Caledonian bears were fought in the Coliseum in Rome, and Harting also quotes Welsh hunting laws that implied the bear was a beast of the chase in 8th century Wales. There are also some weak claims that place-names of Welsh, English

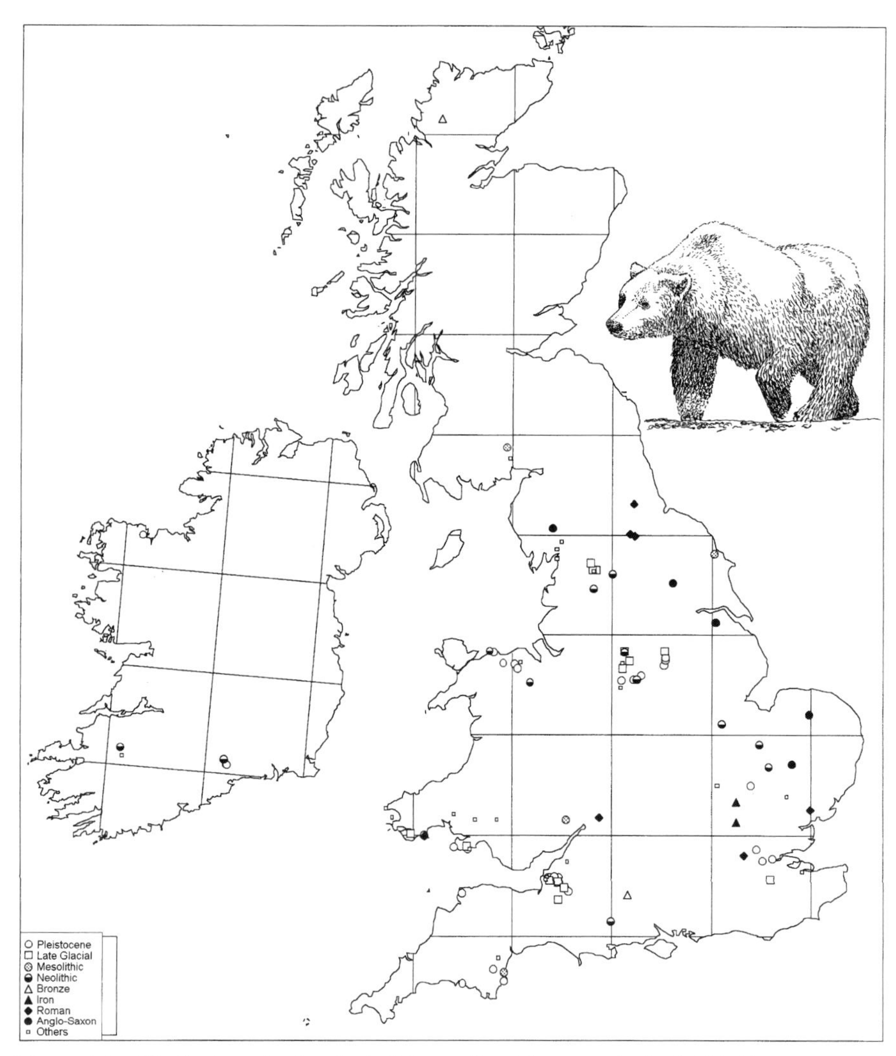

Figure 4.8 Distribution of archaeological Brown Bear *Ursus arctos* in the British Isles.

and Scottish Gaelic incorporate bears (Aybes and Yalden 1995). However, the most obvious of these is Barham in Kent, 'bear village'. It seems certain that this belonged to or was founded by someone called Bear, perhaps because he or his forefathers were huge men; it certainly cannot be taken as evidence that Brown Bears survived in Kent into Saxon times. The conventional wisdom is that Brown Bears died out in Britain in the 10th century, but evidence for their survival beyond Roman times is currently negligible; until there is evidence to the contrary, it seems safer to assume that they have been extinct for around 2000 than for 1000 years.

Actually there are a few archaeological and other indications of Brown Bears in Britain later than Roman times, but they only confuse rather than clarify the matter.

Table 4.3 The occurrence of Brown Bear *Ursus arctos* in archaeological sites in Britain. Dating of most records is indirect; if the accompanying fauna includes such species as Spotted Hyaena and Woolly Rhino, they are assigned to the Devensian, and if such species are absent but Horse and Reindeer are present they are assigned to the Late Glacial. Radio-carbon dates on Brown Bear are quoted directly; those prefixed c. were carried out on other species in the fauna. Middle Pleistocene sites are dated to Oxygen Isotope Stages (from marine sediments), OI 7-13.

Site	*Grid Ref.*	*Age*	*Source*
Heathwaite, Westmorland	SD4577	?	Jackson 1953
Whitbarrow, Westmorland	SD4486	?	Jackson 1953
Helsfell, Westmorland	SD4993	?	Jackson 1953
Thirst House, Derbyshire	SK1071	?	Jackson 1953
Llandebie Cave, Carmarthen	SN6215	?	Jackson 1953
Durdham Down, Somerset	ST5674	?	Reynolds 1906
Murston, Kent	TQ9164	?	Reynolds 1906
Bedford, Bedfordshire	TL0549	?	Reynolds 1906
Great Yealdham, Essex	TL7638	?	Reynolds 1906
Whitesand Bay, Pembroke	SM7226	?	Reynolds 1906
Grays, Essex	TQ6177	Pre-Ipswich. (OI 9)	Stuart 1982
Bleadon, Somerset	ST3456	Pre-Ipswich. (OI 7)	Reynolds 1906
Crayford, Kent	TQ5175	Pre-Ipswich. (OI 7)	Stuart 1982
Ilford, Essex	TQ4586	Pre-Ipswich. (OI 7)	Stuart 1982
Hutton Cave, Somerset	ST3658	Pre-Ipswich. (OI 7)	Jackson 1953
Cefn Caves, Denbigh	SJ0270	Ipswichian?	Neaverson 1940
Hoe Grange, Derbyshire	SK2155	Ipswichian?	Jackson 1953
Minchin Hole, Glamorgan	SS5586	Ipswichian (OI 5)	Jackson 1953
Barrington, Cambridge	TL3949	Ipswichian (OI 5)	Stuart 1982
Little Orme's Head, Caernarvon	SH8182	Devensian?	Jackson 1953
Plas Heaton Cave, Denbigh	SJ0566	Devensian	Jackson 1953
Elderbush Cave, Staffordshire	SK0954	Devensian	Bramwell 1964
Fox Hole, Derbyshire	SK1066	Devensian	Bramwell 1971
Boden's Quarry, Derbyshire	SK2959?	Devensian	Jackson 1953
Mother Grundy's Parlour, Derby.	SK5377	Devensian	Jackson 1953
Brixham, Devon	SX9255	Devensian	Reynolds 1906
Robin Hood Cave, Derbyshire	SK5374	c. 42 900 b.p.	Charles and Jacobi 1994
Castlepook Cave, Co. Cork	R603009	37 870 b.p.	Woodman *et al.* 1997
Ballynamintra Cave, Co. Waterford	X108955	35 570 b.p.	Woodman *et al.* 1997
Castlepook Cave, Co. Cork	R603009	33 310 b.p.	Woodman *et al.* 1997
Shandon Cave, Co. Waterford	X292950	32 430 b.p.	Woodman *et al.* 1997
Foley Cave, Co. Cork	R686099	26 340 b.p.	Woodman *et al.* 1997
Robin Hood Cave, Derbyshire	SK5374	28 500 b.p.	Campbell 1977
Pin Hole Cave, Cresswell	SK5377	Devensian	Jackson 1953
Langwith Cave, Derbyshire	SK5169	Devensian	Jackson 1953
Hoyle's Mouth, Pembroke	SN1100	Devensian	Jackson 1953
King Arthur's Cave, Herefordshire	SO5415	Devensian	Jackson 1953
Bowen's Parlour, Glamorgan	SS5586	Devensian?	Jackson 1953
Deborah's Den, Glamorgan	SS4188	Devensian	Jackson 1953
Worm's Head Cave, Glamorgan	SS4188	Devensian	Jackson 1953

Site	*Grid Ref.*	*Age*	*Source*
Spritstail Tor, Glamorgan	SS4943	Devensian	Jackson 1953
Nottle Tor Cave, Glamorgan	SS4943	Devensian	Jackson 1953
Picken's Hole, Somerset	ST3955	Devensian	Stuart 1982
Banwell Cave, Somerset	ST3858	Devensian	Jackson 1953
Goatchurch Cavern, Somerset	ST4758	Devensian?	Jackson 1953
Hyaena Den, Wookey Hole	ST5348	Devensian	Jackson 1953
Sanford Hill Cave, Somerset	ST4259	Devensian	Jackson 1953
Dulcote Fissure, Somerset	ST5644	Devensian?	Jackson 1953
Oreston Cave, Devon	SX5053	Devensian	Jackson 1953
Kent's Cavern, Devon	SX9264	28 720 b.p.	Campbell 1977
Pontnewydd, Denbigh	SH915710	29 000 b.p.	Hedges *et al.* 1987
Tornewton Cave, Devon	SX8167	Devensian	Grigson 1978
Victoria Cave, Yorkshire	SD8365	Late Glacial	Jackson 1953
Kinsey Cave, Yorkshire	SD8065	Late Glacial	Jackson 1953
Dead Man's Cave, Derbyshire	SK5283	Late Glacial	Jenkinson 1984
Ravenscliffe, Derbyshire	SK1773	Late Glacial	Jackson 1953
Kent's Cavern, Devon	SX9264	14 275 b.p.	Campbell 1977
Sun Hole, Somerset	ST4654	12 378 b.p.	Campbell 1977
Gough's Cave, Somerset	ST4653	c. 12 200 b.p.	Currant 1986
Fox Hole, Derbyshire	SK1066	c. 12 000 b.p.	Bramwell 1971
Elderbush Cave, Staffordshire	SK0954	Late Glacial	Bramwell 1964
Priory Farm Cave, Pembroke	SM9701	Late Glacial	Jackson 1953
Cat Hole, Glamorgan	SS5489	Late Glacial	Campbell 1977
Soldier's Hole, Somerset	ST4654	Late Glacial	Jackson 1953
Bridged Pot, Somerset	ST5248	Late Glacial	Jackson 1953
Wavering Down, Somerset	ST3755	Late Glacial	Jackson 1953
Walton, Somerset	ST4636	Late Glacial	Jackson 1953
Windy Knoll, Derbyshire	SK1283	Late Glacial?	Jackson 1953
Aveline's Hole, Somerset	ST4758	Late Glacial?	Jackson 1953
Ightham Fissures, Kent	TQ5956	Late Glacial?	Jackson 1953
Moughton Fell, Yorkshire	SD7872	Late Glacial?	Jackson 1953
Thor's Fissure, Staffordshire	SK0954	Late Glacial?	Jackson 1953
Plunkett Cave, Co. Clare	G710130	11 920 b.p.	Woodman *et al.* 1997
Red Cellar Cave, Co. Limerick	R645417	10 650 b.p.	Woodman *et al.* 1997
Star Carr, Yorkshire	TA0281	Mesolithic	Legge and Rowley-Conwy 1988
King Arthur's Cave	SO5415	Mesolithic?	Simmons and Tooley 1981
Kent's Cavern	SX9264	9100 b.p.	Campbell 1977
Donore Bog, Co. Laois	S370877	8930 b.p.	Woodman *et al.* 1997
Derrykeel Bog, Co. Offaly	N169032	8880 b.p.	Woodman *et al.* 1997
Shaws, Dumfries	NX9487	7590 b.p.	Kitchener and Bonsall 1997
Victoria Cave	SD8365	Neolithic	Dent 1974
Elbolton Hole	SE0061	Neolithic	Jackson 1953
Great Orme's Head, Caernarvon	SH7783	Neolithic	Jackson 1953
Fox Hole, Derbyshire	SK1066	Neolithic	Bramwell 1971

Site	*Grid Ref.*	*Age*	*Source*
Carrowkeel, Co. Sligo	?	Neolithic	Wijngaarden-Bakker 1986
Lough Gur, Co. Limerick	R6441	Neolithic	Wijngaarden-Bakker 1986
Greater Kelco Cave, Yorkshire	SD8146	Neolithic?	Jackson 1953
Windy Knoll, Derbyshire	SK1282	Neolithic?	Jackson 1953
Rhosddigre Cave, Denbigh	SJ1852	Neolithic?	Jackson 1953
Burwell Fen, Cambridge	TL5767	Neolithic?	Reynolds 1906
Manea Fen, Cambridge	TL4889	Neolithic?	Reynolds 1906
Barholm, Lincoln	TF1010	Late Neolithic	Harman 1993
Rain's Cave, Harborough	SK2455	Late Neolithic	Bramwell 1977
Down Farm, Cranborne Chase	ST9914	Late Neolithic	Legge 1991
Northborough, Cambridge	TF1408	3860 b.p.	C.A.I. French pers. comm.
Letchworth, Hertfordshire	TL2333	Late Neolithic	Legge et al. 1988
Ratfyn, Wiltshire	SU1541	Bronze Age	Jackson 1935
Fox Hole Cave, Derbyshire	SK1066	Bronze Age ?	Bramwell 1971
Inchnadamph	NC2717	2673 b.p.	Burleigh et al. 1976
Welwyn Garden City	TL2413	Iron Age	Schonfelder 1994
Baldock, Hertfordshire	TL2433	Iron Age	Schonfelder 1994
Little Hoyle, Pembroke	SS1199	Roman?	Jackson 1953
Sheepen, Colchester	TL9925	Roman	Luff 1985
Richmond, Yorkshire	NZ1701	Roman	Millais 1904
Catterick Bridge, Yorkshire	SE2299	Roman	Meddens 1990
London	TQ3280	Roman	Ritchie 1920
Hucclecote Villa, Gloucestershire	SO8717	Roman	Noddle 1987
Binchester, Durham	NZ2131	4th C.	L. Gidney, pers. comm.
Balkerne Lane, Colchester	TL9925	250–300 A.D.	Luff 1993
Butt Road, Colchester	TL9925	320–400 A.D.	Luff 1993
Coppergate, York	SE6052	Anglo-Scand.	O'Connor 1989
North Elmham, Norfolk	TF9820	Anglo-Saxon	Bond 1994
Elsham Wold, Lincoln	TA0312	Anglo-Saxon	Harman 1989
West Stow, Suffolk	TL8170	Anglo-Saxon	Crabtree 1989b
Eynsham Abbey, Oxford	SP4309	Anglo-Saxon	J. Mulville, pers. comm.
Colchester	TL9925	11th–14th C.	Luff 1993
Carlisle	NY4006	12th–13th C.	Stallibrass 1992

In the Anglo-Scandinavian levels of Coppergate, York, O'Connor (1989) found just the terminal phalanges – claws – of Brown Bears, and claws have also been reported from human graves elsewhere in Europe as well as in Britain. Earlier burials at Iron Age Baldock and Welwyn Garden City (Schonfelder 1994) and later, Saxon, cremations at West Stow, Elsham Wold and North Elmham (Bond 1994) all contain claws or other foot bones. The Domesday Book for Norfolk records that the City of Norwich provided annually to King Edward a bear to be baited, as well as six dogs for baiting it. There are also two much later place-names, Bearwardcote, Derbyshire and Bearwards Lane, London, where bear-keepers lived (Aybes and Yalden 1995). Furthermore, there are two medieval bone specimens, one (dated to the 11th–14th century) from Colchester, an upper jaw which had had its canine drawn (Luff 1993) and another, a lower jaw, from Carlisle (12th–13th century; Stallibrass 1992). These records all show that bears were regarded with particular esteem in later times, and traded as skins or live for baiting and perhaps as dancing bears. The use of the skins as shrouds or grave-goods implies a special reverence. Such bears give an entirely misleading impression when found in archaeological sites; they were probably brought in from remote areas, perhaps from elsewhere in Europe, and cannot be taken as evidence that Brown Bears still occurred wild in Britain at the times and places when they were buried.

The status of the Wolf from Neolithic to later times is obscured, so far as the archaeological record goes, by the fact that its domesticated descendant the dog was also present. Though skulls can be distinguished by the more crowded cheek teeth and smaller carnassial teeth of dogs, and often by their smaller size, some dogs were as large as Wolves, including those bred for hunting Wolves and some guard dogs. The temptation is therefore to identify large canid bones either tentatively as *Canis* cf. *lupus* or just as large canid. Archaeological evidence of Wolves through this period is negligible; fortunately, we know from various lines of evidence that they survived through to much later times (see Chapter 5). There has been no firm evidence whatever of Lynx at this time, until the recent dating (see Chapter 3) of the Inchnadamph specimen.

The Smaller Fauna

The retreat of the woodland and expansion of grassland during this period should have favoured the sort of mammal fauna which is now apparent – more of common grassland species, such as Field Voles, and less of woodland species such as Wood Mice. Unfortunately, archaeologists have tended to ignore small mammals, and they are greatly underrepresented unless sieving is undertaken specifically to find them. Grant (1984) checked the results of hand-collection against sieving in two season's work at Danebury; 67 out of 70 small mammal bones were missed by hand sorting and only recovered by sieving. Thus the evidence from actual archaeological assemblages is more sparse than one would like, but there are a few intriguing pointers (Table 4.4). In the Neolithic levels of Dowel Cave, Derbyshire, the small mammals include primarily Water Voles (55%) and Field Voles (34%), along with six other species, implying that grasslands already predominated (Yalden 1992). In the Bronze Age, it seems that 'sky-burials' may have attracted scavenging birds of prey to visit the sites where human bodies were exposed, and to deposit there pellets containing the remains of their previous meals. As a consequence, large accumulations of small mammals and other prey were produced. Given the frequency with which amphibian bones accompany the small mammals, perhaps the Common Buzzard *Buteo buteo* was responsible, as Tubbs (1974) remarks on the frequency with which this bird takes frogs. At a cairn on Manor Farm near Borwick, Lancashire, a small mammal fauna of 334 identified bones representing at least 36 individuals was dominated by Water Voles and Field Voles, though amphibians, mainly Common Toads *Bufo bufo*, were

Table 4.4 Small mammal faunas of archaeological sites from Neolithic to later times. (Ages abbreviated from Neolithic, Bronze Age, Iron Age, Roman, Anglo-Saxon, Medieval.) The percentage composition of each small mammal fauna is based in most cases on the Minimum Number of Individuals (MNI); ★at York, General Accident Extension site, the figures are % of 64 samples containing each species, while at Brean Down, Wigber Low, Middleton Stoney and Winnal Down they are % of identified bones. Totals include other species (*Erinaceus, Talpa, Lepus,* bats and mustelids). Sources are given in the text. S.a. *Sorex araneus*; S.m. *Sorex minutus*; N.f. *Neomys fodiens*; A.t. *Arvicola terrestris*; M.ag. *Microtus agrestis*; C.g. *Clethrionomys glareolus*; A.s. *Apodemus sylvaticus*; A.f. *Apodemus flavicollis*; Mus *Mus domesticus* ; R.r. *Rattus rattus*.

Site	*Age*	*S.a*	*S.m*	*N.f*	*A.t.*	*M.ag.*	*C.g.*	*A.s.*	*A.f*	*Mus*	*R.r.*	*Total*
Dowel Cave	Neo.	2.6	–	–	54.7	33.8	3.6	2.6	–	–	–	192
Manor Farm	B.A.	2.9	2.9	2.9	32.4	32.4	5.9	2.9	–	–	–	34
Hardendale Nab	B.A.	16	4	0.5	35	35.5	1	8	–	–	–	242
Wigber Low	B.A.	0.8	–	–	81.1	16.6	–	1.0	–	–	–	7522★
Fox Hole	B.A.	4.5	–	–	34.6	35.8	16.2	5	–	–	–	179
Brean Down	B.A.	–	0.6	–	16.7	12.7	0.6	1.3	–	58.7	0.6	150★
Snail Down, XVII	B.A.	6.0	–	–	60.6	18.2	–	3.0	–	–	–	33
Snail Down, XIX	B.A.	6.2	3.1	–	3.1	61.5	–	3.1	–	–	–	65
Dowel Cave	I.A.	7.5	–	–	40	42.5	5	5	X	–	–	40
Winnal Down	I.A.	0.8	0.3	–	67.7	28.6	–	2.4	–	0.1	–	740★
South Shields	Rom.	5.8	3.1	–	7.3	14.1	4.7	18.8	2.6	31.4	10.5	191
South Shields	Rom.	16.9	3.1	–	18.5	32.3	3.1	13.8	–	9.2	3.1	65
Wroxeter	Rom.	12.5	–	–	20.8	16.7	–	12.5	–	–	–	24
Uley	Rom.	1.2	7.3	–	8.5	58.5	4.9	2.4	–	11.0	–	82
Birdoswald	Rom.	6.4	–	3.2	32.3	29.0	12.9	3.2	–	9.7	–	31
Kingscote	Rom.	23.0	1.6	4.9	8.2	44.3	–	14.8	–	1.6	–	61
York, G.A.E. site	Rom.	1.6	–	3.1	1.6	3.1	4.7	4.7	–	60.9	–	64★
Ossom's Eyrie	Rom.	25.2	0.8	1.6	5.0	47.8	6.3	10.7	<0.1	0.8	0.3	4480
Repton	A-S	18.7	1.2	3.6	0.5	59.7	0.5	12.8	–	1.3	–	603
Middleton Stoney	Med.	18.8	2.2	2.8	3.4	20.6	0.7	36.1	–	11.5	3.8	939★
Greyfriars	Med.	10.9	7.8	10.9	1.6	4.7	3.1	20.3	1.6	26.6	9.4	64

the most abundant items (Jones *et al.* 1987). An unpublished but similar assemblage from Hardendale Nab, Cumbria, contained Water Voles and Field Voles in equal abundance, and more amphibians, especially Common Toads, than anything else; this site has a ^{14}C date of 3290 b.p. obtained on a Water Vole tibia, putting it securely in the Bronze Age (Dr S. Stallibrass, pers. comm.). Other similar sites include Wigber Low in the Peak District where Water Voles contributed 42% of the small mammal bones (Maltby 1983). The fauna from Brean Down in Somerset looks stratigraphically confused, perhaps by the Rabbit burrows that were reported, but it is nominally Late Bronze Age in date, and is also a fauna dominated by Field Voles and Water Voles (though a single House Mouse skeleton contributes most bones; Levitan 1990). Iron Age sites include Winnal Down in Hampshire, dominated by Field Vole (29%) and Water Vole (68%) bones (Maltby 1985) and the later layers of Dowel Cave, Derbyshire, similarly dominated by these two species (Yalden 1992). At Danebury, also Iron Age, Wood Mice are more numerous than Water Voles (Browne 1995). West Hill, Uley, ranging from Romano-British into Anglo-Saxon times, is dominated by Field Voles (Levitan 1993). In South Shields, the faunas from the Roman granary were interpreted by Younger (1994) as a record of the animals living in and around the site. The surprising thing is that Water Voles and Field Voles are also numerous in one of his samples, implying to me that Barn Owls had been depositing their pellets in the building. The apparent abundance of Water Voles at sites far removed from water was first remarked by Jewell (1959), examining the small mammals from Snail Down, Wiltshire. Two Bronze Age barrows there, not necessarily contemporaneous, contained rather different proportions of small mammals; Barrow XVII contained 60.6% Water Voles and 18.2% Field Voles, while Barrow XIX contained respectively 3.1% and 61.5%. Moles made up most of the rest of the samples (Table 4.4). These are clearly the faunas of open grassland, but Snail Down is in the middle of Salisbury Plain, 7 km from the River Avon. The predominance of Water Voles continues the pattern established for the Mesolithic (Chapter 3), and seems to suggest that absence of competitors provided more habitat. However, Water Voles have changed their morphology somewhat during the Post Glacial; they have become perceptably larger, and their incisors more upright (less pro-odont) since Mesolithic times (Montgomery 1975). More pro-odont molars might be associated with digging tunnels, which would perhaps have been more necessary for voles living further from the cover offered by water. The pattern suggested by all these records is that open country, dominated by the grassland voles, had supplanted the woodlands. This may be deceptive; the predators responsible for the accumulations of bones are likely to have been those that prefer to hunt over open country, even if there was little of it around, species like the Barn Owl, Buzzard and Kestrel. Thus the small mammal faunas may be telling us more about their hunting preferences than about the countryside or its small mammal communities. Woodland hunters, particularly Tawny Owls, do not produce large accumulations of pellets, and obtaining adequate samples to study their diets is hard even for modern ecologists (Southern 1954, Yalden 1985). One interesting test of the validity of such analyses is provided by Evans and Rouse (1992), who compared the results of sieving for small mammals with those for molluscs from the Neolithic deposits at Maiden Castle, Dorset. Mollusc faunas work rather like pollen analyses in chalklands (pollen is poorly preserved at such sites, but snail shells survive well, the reverse of the situation in acid peats). They found that mammal remains were rather scarce, compared with the snails, but included both grassland Field Voles and scrubland Bank Voles and Wood Mice, with the latter somewhat more abundant at times when the snail faunas also suggested a reversion to scrub as the earlier Neolithic clearances were abandoned. The fragmentary remains in this case suggested that Kestrels were the predators. Another test is provided by the Bronze Age well excavated as the Wilsford Shaft (Ashbee *et al.* 1989); again the mammal fauna was a small one, including just two Field Voles, a Bank Vole, five Wood Mice, a Pigmy Shrew and a Common Shrew. The site also yielded numerous bones of sheep and some cattle, as well as a few of pig, dog, Red Deer and Roe Deer, direct evidence of the importance of grazing mammals which was matched by the record of seeds of arable and pasture weeds, and the molluscs of open downland. The beetle fauna was dominated by dung beetles. Thus in this case the mammal fauna implies a more wooded environment than any of the other evidence; either the small mammals were wandering further or a predator dropped their remains while roosting on the well-head.

Species characteristic of woodland should have been more common in Neolithic or earlier times than they are now, and there is just a hint of this in the available evidence. The Yellow-necked Mouse *Apodemus flavicollis* is very like its more numerous and widespread relative the Wood Mouse *A. sylvaticus*. Indeed most mammalogists would have difficulty separating them, and tend to lump skulls from archaeological sites and modern owl pellets alike as *Apodemus* sp. However, the Yellow-necked Mouse has an interesting modern distribution, confined largely to south-eastern England and the counties along the Welsh border. This seems to reflect a preference for ancient deciduous woodland (Montgomery 1978, Yalden and Shore 1991), just the habitat that the Neolithic farmers started to clear. In that case, there ought to be archaeological records of *A. flavicollis* further north or in the Midlands where it is now rare or absent. There are indeed a few specimens from the Neolithic deposits in Dowel Cave, Derbyshire, from Roman age cave deposits in Ossom's Eyrie Cave, Staffordshire, from Roman Manchester (Yalden 1992) and from Roman South Shields (Younger 1994) to support that view (Fig. 4.9). The species was certainly present at Ightham Fissure, Kent, well within its modern range (named *Mus abbotti* and then renamed *Mus lewisi* by Newton 1894, 1899); its abundance there has not been fully discussed, and the faunas are of very uncertain date, though Newton referred to about 40 jaws of *sylvaticus* and only seven jaws of *abbotti (flavicollis)*. Collectively, these records do indicate a slightly wider range in former times, though not a greater abundance (absolutely, or relative to other small mammals). In well forested areas further east in Europe, notably in Białowieża, *A. flavicollis* is the common woodland mouse, to the exclusion of *A. sylvaticus*, and this might have been the situation in Britain during Mesolithic or early Neolithic times, but there is no direct evidence yet for that.

Another small mammal that shares a similar preference for deciduous woodland, and a similar rather restricted distribution, is the Dormouse *Muscardinus avellanarius*. Like the Yellow-necked Mouse, it is most common in south-eastern England and up the Severn valley, though there are small populations elsewhere. It is typically associated with Hazel, and opens the nuts in a characteristic manner, but it requires a mixture of shrubs and trees to provide its food supply through the season; it behaves ecologically more like a miniature squirrel than a mouse, though closely related to neither (Bright and Morris 1992). Unfortunately, this also means that it rarely gets caught by owls, or indeed any other predators, and is also hard to catch in conventional small mammal traps (Morris and Whitbread 1986). In 47 865 vertebrate prey from Barn Owl pellets, Glue (1974) had no Dormice (though it has been recorded occasionally since, and is taken slightly more frequently by the woodland-hunting Tawny Owl). As a consequence, it also rarely turns up in archaeological sites, either cave sites (where the small mammal faunas are often the product of Barn Owl pellets) or conventional sites. The only firmly dated record from anywhere in Britain is the specimen from The Undercliff on the Isle of Wight, with a date of 4480 b.p. (Preece 1986), and this is well within its modern range. However, there are also Neolithic specimens from Dowel Cave, Derbyshire and Romano-British specimens from Ossom's Eyrie Cave, Staffordshire, two counties where it has not been recently recorded (Bramwell *et al.* 1990, Yalden 1992). There are also undated specimens from Dog Holes, Lancashire, Pin Hole Cave, Derbyshire and Great Doward Cave, Herefordshire (Sutcliffe and Kowalski 1976). Two of these sites are also at the margin of the present range of the species, and at least hint at a broader earlier distribution (Fig. 4.10).

The species that most clearly indicate range contraction as a result of woodland clearance are two bats. It has been well demonstrated that all British bats favour woodland,

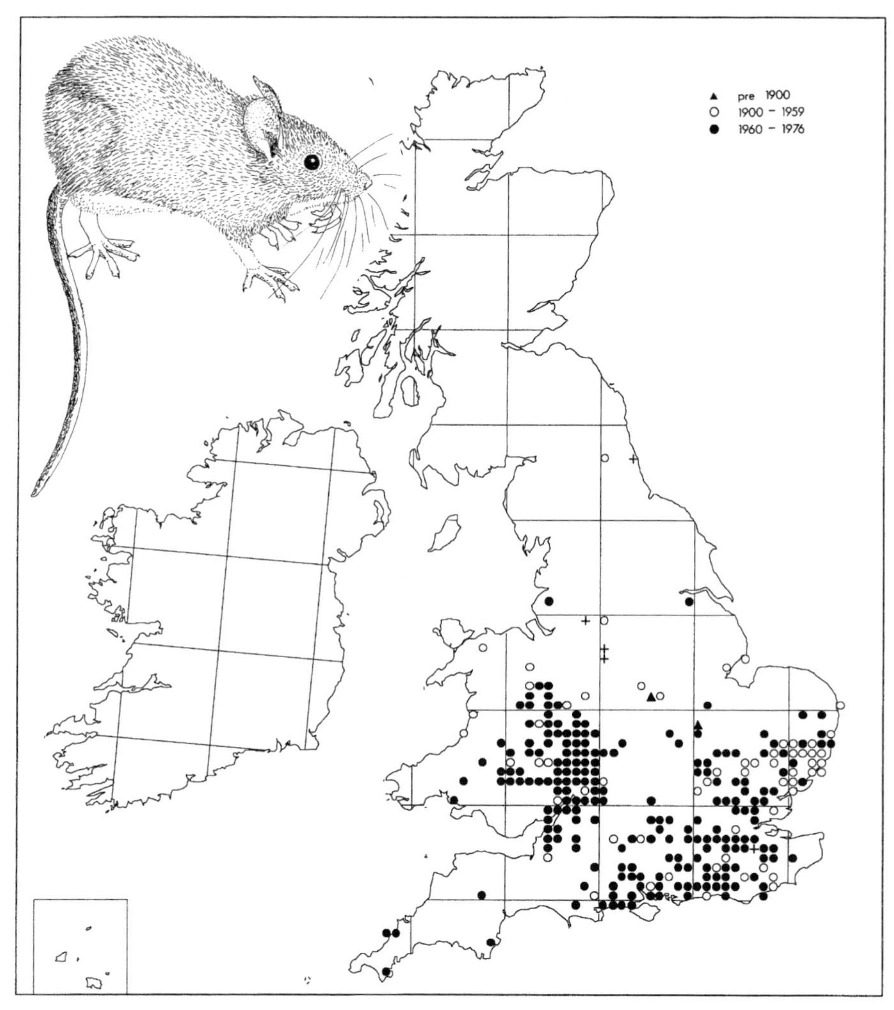

Figure 4.9 Distribution of modern and fossil (+) Yellow-necked Mice *Apodemus flavicollis* (after Arnold 1993, Yalden 1992; modern distribution updated by H.R. Arnold).

rather than grassland or moorland (Walsh and Harris 1996a, 1996b), but some species are particularly closely tied to woodland by their feeding and roosting habits. Bechstein's Bat *Myotis bechsteini* is a species that forages by flying rather slowly in and around the canopies of trees, picking off the roosting moths and flies, even earwigs, spiders and harvestman (Wolz 1993). It also roosts and breeds in holes in trees, though it usually hibernates in caves. It is probably our rarest bat, and no breeding site was known in Britain until 1996, when a site was discovered by lucky accident in the New Forest. Given its rarity, it is surprising to find that it was the commonest bat at the Neolithic site of Grimes Graves, Norfolk and at Ightham Fissures, Kent, a site of uncertain date (Yalden 1992). It is also present at Pin Hole Cave, Derbyshire. These sites are outside its present range (Fig. 4.11). The status of the Lesser Horseshoe Bat

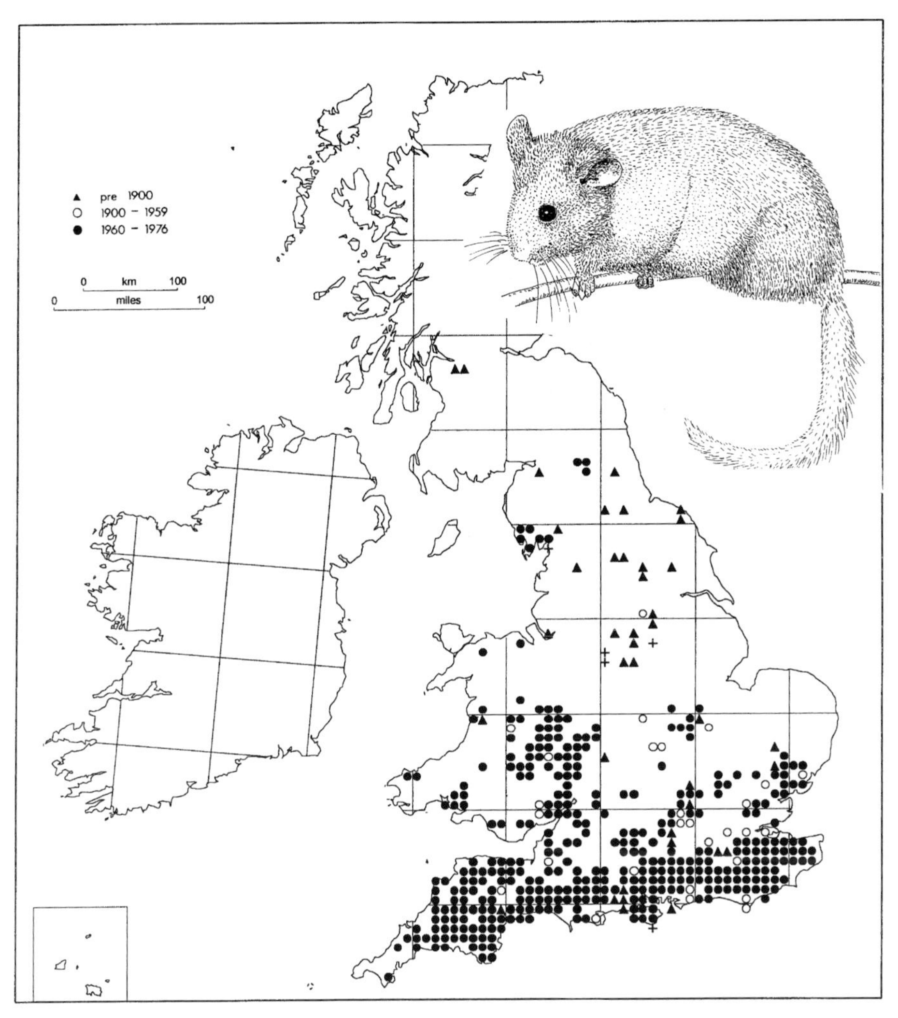

Figure 4.10 Distribution of modern and fossil (+) Dormice *Muscardinus avellanarius* (after Arnold 1993, Yalden 1992; modern distribution updated by H.R. Arnold).

Rhinolophus hipposideros is not quite so parlous now, but it has a range restricted to the south and west, probably in part by its need to hibernate hanging freely from the roofs of caves. However, it has rounded wings that facilitate hunting in woodland, and sometimes hunts from a perch, hanging from a twig and droppping down on passing insects. It has been reported from three caves (Dowel Cave, Neolithic; Ossom's Eyrie Cave, Romano-British; Wetton Mill Rock Shelter, Mesolithic?) in the Peak District and from Pin Hole Cave on the Derbyshire border with Nottinghamshire (Yalden 1992). These are sites outside its present range, though it was reputed to occur in the Peak District and further north, in Yorkshire, around the turn of this century (Fig. 4.12). Given that bats are in any case uncommon as cave fossils, these records do suggest a greater abundance and distribution for these two species in former times. The post-

Neolithic clearance of woodlands has probably affected bats in general, particularly woodland species such as these, more severely than any of the other mammals.

Introductions

As well as the domestic species, there is clear evidence during this period for the introduction of some of what we have traditionally regarded as 'wild' mammals, and suggestions that a couple of others actually belong in this category of introductions. The most clear example is the House Mouse, *Mus domesticus*, which we used to call *M. musculus*. Mice of the genus *Mus* are not native to Europe, but are an undoubted introduction from Asia. They appear to have been native to the Middle East, how-

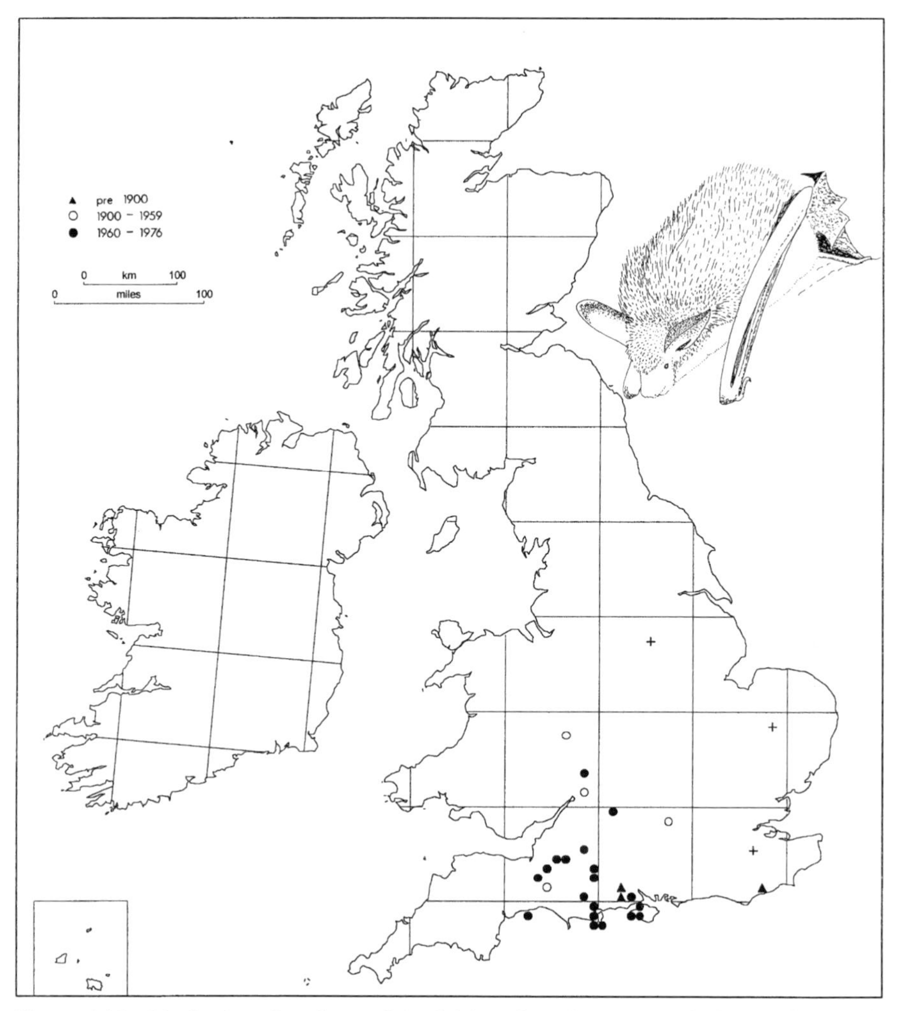

Figure 4.11 Distribution of modern and fossil (+) Bechstein's Bat *Myotis bechsteini* (after Arnold 1978, Yalden 1992; modern distribution updated by H.R. Arnold).

ever, as Tchernov (1984) quotes records from Israel throughout the Pleistocene. The earliest evidence of them as commensals may come from Early Palaeolithic (Mousterian) levels in caves such as Qafzeh, Hayonim and Tabun, Israel, where *Mus musculus* forms up to 25% of the small mammals; however, this may simply reflect the local abundance of the species at that time and place. In the Upper Palaeolithic, the species forms only 2–3% of the fauna, but then increases sharply in the succeeding Natufian (the earliest agricultural culture in the Middle East) at about 11 000 b.p. At Hayonim, it forms 48% of a large small mammal fauna, and at Mallaha 39% (Tchernov 1984). Thus the species seems to have become more numerous just at the time when cultivation of cereals and the storage of grain provided its modern habitat. The earliest records from Europe seem to be from Bronze Age sites in the Mediterranean, in Mallorca and in Spain (Brothwell 1981).

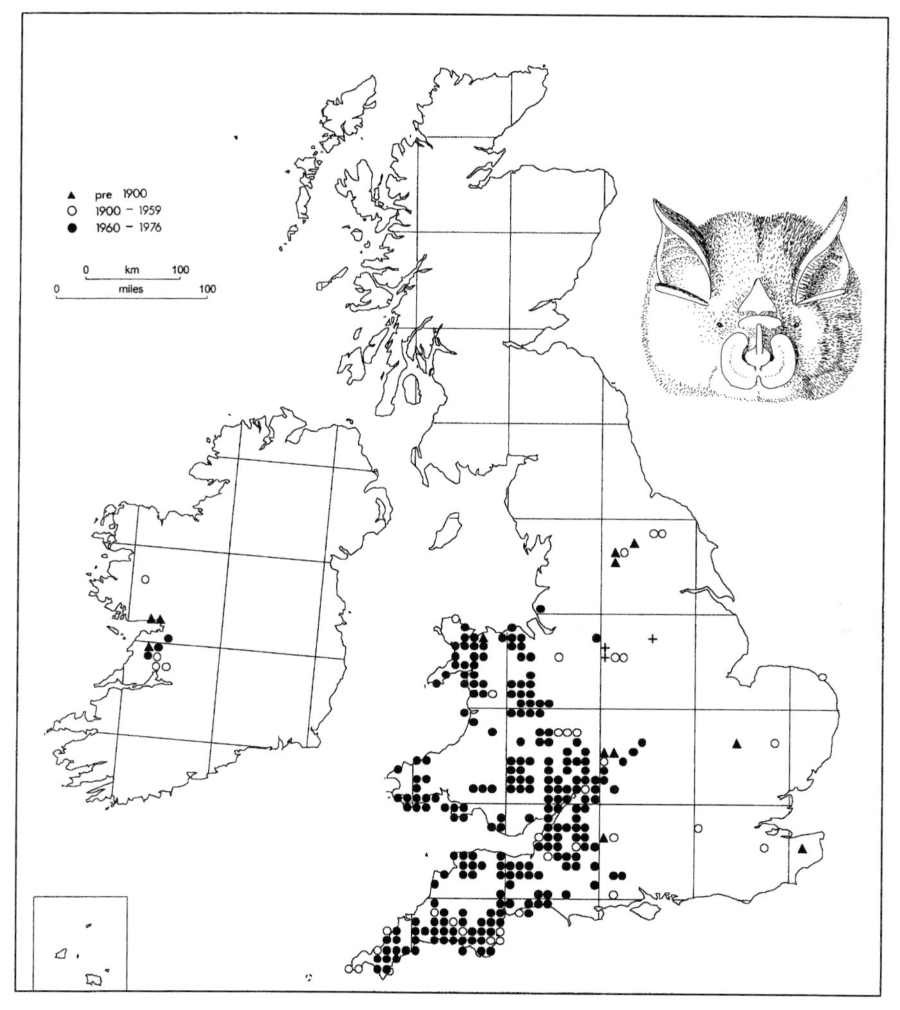

Figure 4.12 Distribution of modern and fossil (+) Lesser Horseshoe Bat *Rhinolophus hipposideros* (after Arnold 1978, Yalden 1992; modern distribution updated by H.R. Arnold).

In Britain, it was certainly present by the Iron Age, the best known record being that from Gussage All Saints, Dorset (Harcourt 1979). Other early records include Iron Age Danebury (Coy 1984, Browne 1995) and Maiden Castle (Armour-Chelu 1991), and it is possible that the species was introduced even earlier. The Late Bronze Age levels of Brean Down, Somerset have one skeleton of *Mus* (Levitan 1990), though the site is somewhat confused stratigraphically (there are Rabbit burrows, implying that faunas could have been redistributed). In Roman sites, the House Mouse is regularly present, e.g. at Winnal Down (Maltby 1985), Manchester (Yalden 1984a), York (O'Connor 1992), Uley (Levitan 1993), Birdoswald (Izard 1997) and South Shields (Younger 1994), and the species has maintained a continuous presence ever since, as evident particularly at York (Fig. 4.13). In one of Younger's samples, from a Roman granary, the House Mouse was the dominant small mammal, 31% of a total of 191 individuals. It has changed its scientific name, however. It used to be

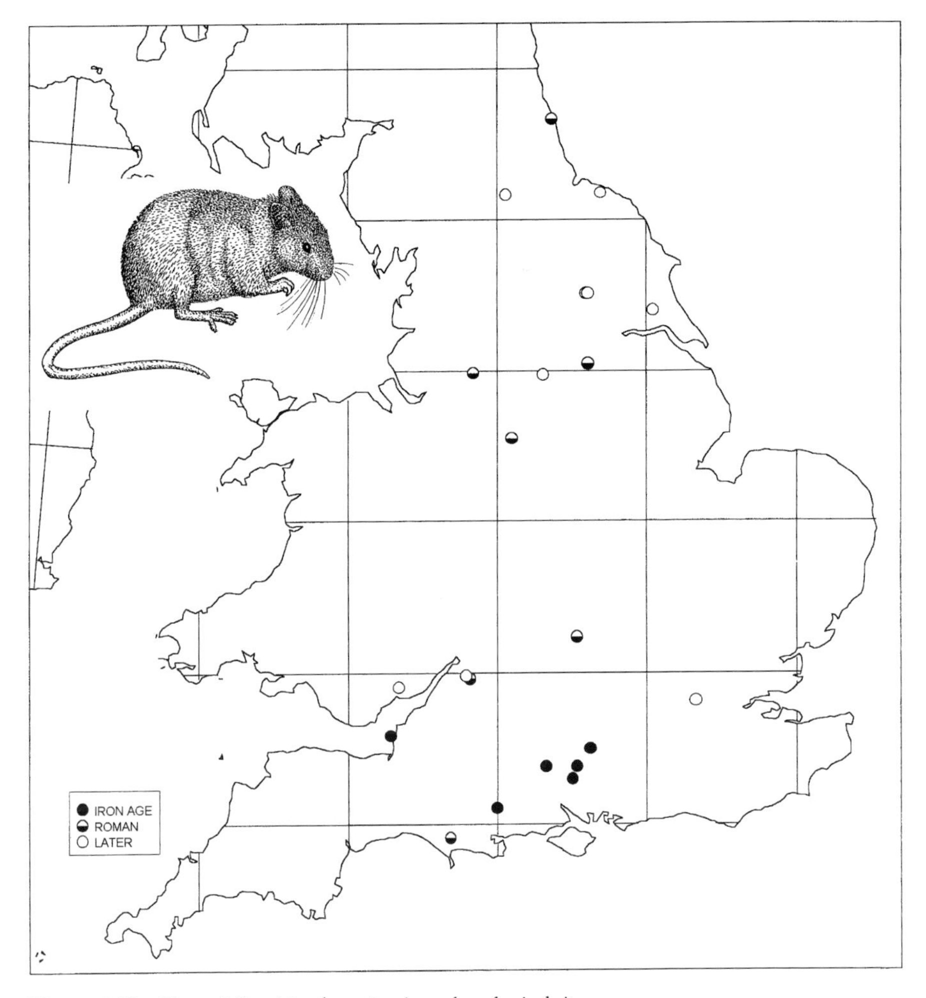

Figure 4.13 House Mice *Mus domesticus* in archaeological sites.

thought that all House Mice anywhere in Europe, and all of their descendant laboratory and fancy mice, belonged to the same species, i.e. *Mus musculus*. It became apparent in the 1970s that House Mice in Denmark divided into a north-eastern and south-western population, with a narrow hybrid zone in the middle where the two interbreed rather little, and the boundary was subsequently traced across Europe (Thaler *et al.* 1981). The north-eastern form also occurs in Sweden, and is therefore the form that Linnaeus named *M. musculus*, while the south-western form, which we have in Britain and in captivity, is strictly to be called *M. domesticus*. The two are closely related, and it is doubtful that archaeological specimens could be distinguished.

Along with the House Mouse, it seems that domestic cats arrived early in Britain. These are likely to belong to the same species as the Wild Cat *Felis silvestris*, but were domesticated in Egypt or western Asia from the African race *F. s. lybica*. The caveat is made because in the past *lybica* was regarded as a full species, but then the taxonomy of small cats was and is confused. The fact that Wild Cats in Scotland are suffering from hybridization with feral cats (see Chapter 9) emphasizes that they are closely related, probably correctly regarded as subspecies. The earliest good evidence of domestic cats is probably their arrival on Cyprus, whence there is a skull dated to about 8000 b.p. from Khirokitia (Davis 1987). Since cats are not native to Cyprus, this was presumably introduced from elsewhere in the Middle East. It is earlier than any record from Egypt, which is where the domestication of the African Wild Cat is presumed to have occurred. Again the earliest record for Britain is from Gussage All Saints, Iron Age, where the skeletons of five kittens were unearthed (Harcourt 1979). Cats are present, though scarce, at various Roman sites, for instance at Exeter (Maltby 1979), and only become more abundant in Mediaeval times (O'Connor 1992).

The Black Rat *Rattus rattus* is another Asian rodent, native to India and further south-east. In India, there are two chromosome races, one from the Deccan Peninsula with 2n = 38 and one from further north with 2n = 42 (Armitage *et al.* 1984). It is the southern form which has been introduced to Europe and subsequently around the world. There is a good sequence of archaeological records which show the species spreading to the Middle East by about 12 000 b.p. (Natufian, at Hayonim and Tabun Cave, Mt Carmel; Tchernov 1984), to Egypt by the 4th to 1st century B.C., Italy by the 2nd century B.C to 1st A.D., and to Germany by a similar time (Armitage *et al.* 1984). It used to be thought that returning Crusaders brought the Black Rat to Britain in Mediaeval times; Barrett-Hamilton and Hinton (1914, p. 582) remark 'There is no clear evidence of its presence in Europe ... prior to the Crusades (1095, 1147 and 1191) ... and there can be little doubt that it was imported from the Levant by the navies of the Crusaders'. Evidence changes. Now, archaeological specimens show clearly that it was brought here first in Roman times. The first clear evidence came from a 6 m deep well in York, securely dated as a Roman infill from a ^{14}C date on timbers at the bottom to A.D. 110 and another of A.D. 720 from charcoal at the top (Rackham 1979). A well-preserved, fairly complete skull was excavated from about 0.7 m above the dated bottom layer, in an infill containing pottery remains dated mostly to the 4th century, and it seems probable that the well was deliberately filled in; the rat skull and a few other bones found slightly lower down therefore probably also date to that century. Subsequently, the Black Rat was recorded at Roman Wroxeter, London and York (Armitage *et al.* 1984). The London specimens, from excavations in Fenchurch Street, come from the 3rd century A.D. (Fig. 4.14).

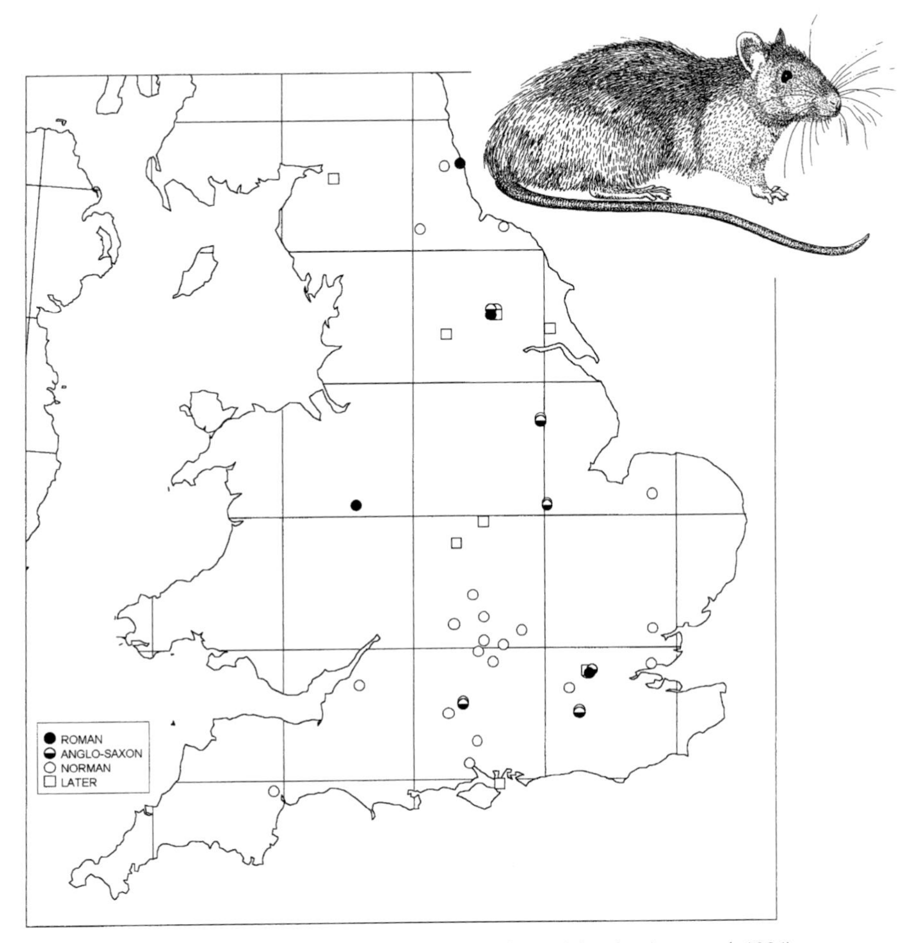

Figure 4.14 Black Rats *Rattus rattus* in archaeological sites (after Armitage *et al.* 1984).

There is evidence that the Romans imported at least one other small mammal to Britain. It is popularly known that dormice were eaten by the Romans, exploiting their habit of laying down fat to see themselves through hibernation. It is generally supposed that what we now call the Edible Dormouse *Glis glis* was the species involved, and indeed Toynbee (1973) mentions an illustration of a pair of 'squirrel-tailed dormice'. When O'Connor (1986) found parts of three skulls in Roman York, he recognized them as large dormice, and assumed they would prove to be *Glis*. It was therefore with some surprise that he later found them to be Garden Dormice *Eliomys quercinus*. Younger (1994) has subsequently found three of the same species in a granary at South Shields. One Roman description of dormice refers to them as golden, which would fit *Eliomys* better than the grey *Glis*. Presumably these were introduced as a delicacy for some feast from further south in Europe. There is nothing to suggest that such imports led to the establishment of feral populations.

Two other mammals deserve discussion as possible early introductions, though there is rather little precise information on either of them and they have traditionally been regarded as native. The Harvest Mouse *Micromys minutus* is our smallest rodent, a denizen of the stalk zone of long grasses – wheat fields, certainly, but also reed beds, canary grass, and the long grass of young forestry, hedgerows and ditches. This is a type of habitat that would perhaps have been scarce in Mesolithic times. The relevant archaeological evidence is essentially negative: the species is not recorded in Mesolithic and Neolithic small mammal faunas. The trouble is, the species is rarely recorded anywhere, partly because of its small size, which makes it easy to overlook, and partly because of its habitat. However, it does occur regularly in present-day samples of Barn Owl pellets in those parts of Britain where it is present, essentially the southern half of England. Glue (1974) recorded it in 48% of the 29 Barn Owl diets he analysed from south-east England, even though it only contributed 1% of the 7141 prey animals in those diets. It was present in similar numbers and formed a similar proportion of diets from eastern (56%, 1.1%) and south-western (46%, 0.4%) England. In the archaeological record, Harris (1979) cited only two Roman records, from Droitwich and York, while two later records, from Ossom's Eyrie Cave, Staffordshire and Viking-age Repton, Derbyshire, were given by Yalden (1992, 1996b). Another later record, from a Mediaeval site near Catterick, is given by Stallibrass (1995). If Neolithic farming created the habitats that this species required, perhaps its absence from Mesolithic to Neolithic times is genuine, not simply a reflection of its poor archaeological record. A similar problem arises with the Brown Hare *Lepus europaeus*. There is a conspicuous absence of hare bones beyond the early Mesolithic, when the Mountain Hare *L. timidus* seems to have been eliminated from England and Wales, at least, by the spread of the forests. There is, of course, an awkward problem in distinguishing the bones of these two species, but an apparent absence of hare bones of any sort from most Neolithic sites. As with the Harvest Mouse, it is at least arguable that Neolithic and later farming has created the habitat for the Brown Hare. If it was introduced, when and by whom? Caesar (*De Bello Gallico*, Book V, 12) says of the Ancient Britons 'Hares, fowl and geese they think it unlawful to eat, but rear them for pleasure and amusement', which implies that they had been introduced by Iron Age times, but perhaps not long before. There are archaeological records of hares from the Iron Age sites of Danebury Hill Fort (Coy 1984), Winnal Down (Maltby 1985), Maiden Castle (Armour-Chelu 1991) and Blunsdon St Andrews (Coy 1982), as well as from such Roman sites as Portchester Castle (Grant 1976), Exeter (Maltby 1979), Silchester (Maltby 1984), Colchester (Luff 1982), Caerleon (Hamilton-Dyer 1993), Segontium (Noddle 1993), Wroxeter (O'Connor 1987) and Hod Hill (Fraser 1968). These certainly confirm that hares, presumably Brown Hares, were present in southern England in pre-Roman and Roman times. There are some possible earlier records: from Bronze Age sites at Brean Down (Levitan 1990) and Manor Farm, Borwick (Jones *et al.* 1987), and possibly from three Neolithic sites in Wales (Gwaenysgar, Pant-y-saer, Basclodiad-y-gawres; Caseldine 1990) where hare bones occur with bones of domestic ungulates. Most intriguingly, Turk (1964) suggests that both Brown and Mountain Hare occur together in a Bronze Age site at Hartledale, Derbyshire; if these are really contemporary, they are both the latest native English Mountain Hares, and perhaps the earliest Brown Hares. There is also the puzzling record of '*Lepus timidus*' from Windmill Hill (Jope and Grigson 1965); since this was based solely on a tibia and some metapodial (foot) bones, it cannot possibly have been confidently identified as a Mountain Hare rather than a Brown Hare, but it is in any case a surprisingly early record. One difficulty with such records is that when Brown Hares were regarded as native, their presence at such sites was not remarkable; if the species was in fact introduced, the dating as well as the identification of such early specimens needs careful evaluation.

Conclusions

The 3000 years from Neolithic to Roman times saw something like the present countryside, farmland with small woodlands rather than woodland with small clearings, established over much of Britain. It saw the establishment too of something like the present-day mammal fauna, dominated by the introduced domestic stock (especially sheep and cattle) and with mammals which are typical of open grasslands such as Field Voles, Water Voles and Brown Hares as characteristic members of the fauna. However, the available archaeological mammal faunas are inadequate for determining the balance of the actual fauna of those times. Because the small mammal faunas come from pellet hoards left by predators which hunt over open country, they inevitably reflect open country conditions; they ought to contrast with faunas dominated by woodland small mammals from Mesolithic times, but do not. This is a situation which might be remedied by faunas from different types of deposit; samples trapped in natural pitfall traps, like wells and swallow holes, might give us a better view of changes over time. The status of larger wild mammals is also inadequately reflected in the available faunas. They show us that species such as Beaver, Fox, Pine Marten, Otter, Badger and Wild Cat were present, and that Red and Roe Deer were widespread. They also show that the supposed non-native species, Brown Hare and Harvest Mouse, were present by Iron Age times, but the evidence from earlier periods is not good enough to demonstrate convincingly that these really were missing and had to be introduced. At least we can be certain that the two household pests, the House Mouse and Black Rat, were introduced during this period. Some other species present in archaeological sites of this period remain enigmatic. We can be quite sure that no monkeys occurred in Britain as native mammals in the Post Glacial period; the nearest native *Macaca sylvana* are probably in the Atlas Mountains of Morocco, since those on Gibraltar seem likely recent introductions. What then should we make of the Bronze Age specimen reported by Peter Woodman from Emain Macha in Co. Armagh (Woodman 1986)? How should the fin spine of a Nile Catfish *Synodontis* at Dragonby be interpreted (May 1996)? The Romans certainly kept Fallow Deer (*damma*, as opposed to *cervus*) in parks in Italy, and depicted them in mosaics in Britain and elsewhere (e.g the mosaic from Istanbul shown as plate 71 in Toynbee 1973, see opposite). Do the occasional bones which occur in early archaeological sites of Roman, perhaps even earlier, age indicate the odd animal imported for a park or menagerie, or as venison on or off the hoof? Or are they actually intrusive bones from later deposits? What about the two records of *Eliomys*, surely imported as gastronomic treats? If they were imported, it is obvious that the Romans, and the Ancient Britons, could have imported all sorts of unexpected species. Equally uncertain is the pattern of extinction during this period; the record suggests that Elk, Aurochs and Brown

Bear were lost from our fauna during these millennia, but the possibility that one or more of them survived through to later times cannot be confidently denied. Conventional wisdom supposes that both the Brown Bear and the Aurochs should have died out in Scotland much later than the current, mostly English, archaeological record suggests (Ritchie 1920). A canine tooth from the Broch of Keiss in Caithness, of uncertain date, is suggested by Ritchie to be the latest solid evidence of Brown Bear, and there is perhaps the half-remembered Gaelic noun *magh-ghamhainn*, paw-calf. Ritchie likewise suggests that there are remains of Aurochs in some of the northern brochs. Given the surprise over Lynx, it is certainly possible that these species also lingered on, but it remains to be proved by a better archaeological record.

Fallow Deer, Roman mosaic

CHAPTER 5

THE SAXON ANGLE

Saxon ladies, ferreting

WHEN the Roman Legions were withdrawn from Britain in A.D. 410 in response to the sacking of Rome by Alaric the Goth, the Christian Romano-British culture that had been established for 200 years or so faltered and disappeared. The Dark Ages which followed certainly conceal a large cultural change. The Roman province with a Celtic native population vanished, to be replaced by a largely Germanic population and organization, at least in what became England. The people who called themselves *cymri*, who were called Britons by the Romans and *weales*, Welsh or foreigners, by the newcomers, were certainly displaced as the ruling class, and seem to have retreated to Cornwall, Wales, Cumberland or across the sea to Brittany (Morris 1973). In place of tribal areas, the counties that we still recognize came into existence, and nearly all the place-names that we now recognize for towns and villages, hills and valleys seem to have been conferred by the newcomers. Even places where some of the Britons remained were renamed Walton (*weales-ton*, foreigners' place) by the newcomers. The past was not entirely obliterated; Roman roads remained in use, and some towns retained names that were Roman. London, notably, had been established as the new Roman capital *Londinium*, in the area that is now The City, but the Saxons seem to have largely abandoned that site, establishing *Ludenwic* in the area of the Strand, but ensuring the continuity of the name. Such places as Chester and Ribchester recalled the previous existence of Roman camps (*castra)*, even if their Roman names (*Deva* and *Bremetennacum*) had been lost. On the other hand, *Eboracum,* the Roman capital of northern Britain, disappeared to be replaced by Saxon *Eoforwic,* then Norse *Jorvik,* now better known as York. Some rivers, but very few settlements, even retained Celtic names, most obviously the various Rivers Avon and Derwent.

Vince (1994) suggests that Roman life may have continued on some of the sites of former Roman towns into the 5th century, but that these were no longer functioning as towns. The Saxons seem to have settled in small rural communities, at such places as West Stow and Mucking in Suffolk, during the 5th to 6th centuries, and urban settlements seem not to have reformed until perhaps the 7th century. This accords with a gap in the record of trees felled for structural timbers (Tyers *et al.* 1994). There is some evidence of the abandonment of settled farmland, and secondary regrowth of woodland, in the post-Roman period but, as Rackham (1994) emphasizes, the countryside remained mostly a farmed, open landscape; Anglo-Saxon charters recording the boundaries of land-holdings mention 14 342 features, including 766 isolated trees, 378 hedges, 32 walls and 342 woods, as well as 629 references to furlongs, headlands and other technical terms for ploughland. The Saxons did introduce the eight-oxen plough, enabling them to plough claylands, and they did clear further woodland, for Saxon place-names referring to various sorts of clearing *(leah, hyrst, feld)* are frequent. However, they were evidently not operating in a landscape of wildwood; that had long gone.

Although the Romans wrote treatises on farming, they seem to have written little that specifically refers to Britain. The written record of animal life in Britain therefore starts with the Anglo-Saxons, in the place-names which they and their slightly later Norse rivals bestowed, and in their literature, including the Anglo-Saxon Chronicle. The latter overlaps the Norman invasion in A.D. 1066, and is therefore continuous with the written record of government and economy which comes to dominate the story of our fauna. However, that record is a hesitant one, at first, and the archaeological record continues to offer most of the evidence for several centuries more.

Place-names and Other Indications

We have a few indications from Anglo-Saxon glossaries (Wright 1884, Napier 1900) of what species were familiar to the newcomers, though they could have brought such names with them from their homelands in the North German Plain. The glossaries compare Saxon names with Latin ones, the meanings of the latter being well established by reference to classical literature. They seem to have distinguished domestic and wild pigs, sometimes translating the Latin *aper* as *wild bar* (Wild Boar) and sometimes using the term *eofor*; the Latin *verres* is translated as *tam bar* or simply *bar.* Archbishop Aelfric's Vocabulary, of the 10th century, includes *wandewurpe* (Mole), *otor, broc* (Badger), *wesle, hearma* (Ermine), *fox, wulf, bera* (Bear), *lox* (Lynx), *acwern* (Squirrel), *sisemus* (Dormouse), *mus, raet, befer, hara* (Hare), *heortbuc* (Hart, i.e. Red Stag), *hind* and *rahdeor* (Roe Deer), among others (Wright 1884). Interestingly, the Latin *urus* is translated *wesend*, implying early confusion between Aurochs and Wisent, i.e. European Bison, which was to persist into the scientific period. It seems to be a common result of the disappearance of once familiar animals that their names are transferred to a similar species. Herons became Cranes when there were no longer Cranes in the landscape, and Capercaillies were equated with Pheasants (Boisseau and

Yalden 1998, Yapp 1981). Similarly, A Pictorial Vocabulary (of the 15th century) translates *castor* as *broc*, Badger, and *fiber* as *otere*, Otter (Wright 1884), when both were correctly equated with *befer*, Beaver, in the 10th century. The confusion of the Welsh term *afanc*, which is variously translated Beaver, Otter, or some unknown monster, may have been a similar case, for it is certainly rendered Beaver in the early *Welsh–English Dictionary of 1688* (Aybes and Yalden 1995). The conclusion seems to be that Aurochs were already totally forgotten by the English of the 10th century, and that Beavers similarly had been forgotten by the 15th century.

In the Anglo-Saxon Chronicle for A.D. 937, the description of the Battle of Brunnanburgh ends by relating that the dead of the opposing Welsh, Scots and Irish army were left on the battlefield to be scavenged by the Ravens, White-tailed Eagles and Wolves. Much the same comment is made of the English dead after the Battle of Hastings in 1066 by one of the Norman commentators, and it seems likely that this was a conventional end to the saga retelling of a battle, perhaps an indication of the shame or the overwhelming nature of the defeat. However, the ending would entirely lose its point unless the listeners appreciated the significance of these animals. It seems reasonable to conclude that Wolves were familiar enough to the Saxon inhabitants of England. That conclusion is supported by the abundance of place-names which include some reference to Wolves. This evidence has to be interpreted cautiously, because it is quite clear that *Wulf*, as well as derivations from it (*Ethelwulf, Beowulf*, etc.) was a common personal name. Searle (1897) cites four known individuals called *Wulf* and 19 called *Ulf*, as well as numerous *Wulfgar, Wulfgyth, Wulfheard, Wulfhere, Wulfmar, Wulfred*, etc. Thus settlements could easily be named after people, rather than animals. The most obvious example is Wolverhampton, founded where a lady called Wulfrun, who had been given the manor in A.D. 958, donated some land for the building of a church, round which the town later developed – Wulfrun's high town. However, many of the place-names are of just the sort of wild places where the animals, but less probably specific people, might live: dales, hills and clearings in woods. In their compilation of Wolf place-names, Aybes and Yalden (1995) list 236 English places from the literature, mostly, of the English Place-name Society (EPNS). The EPNS is attempting to produce a complete record of the history and interpretation of all the place-names of England, county by county, but their listing is by no means complete as yet. Only two-thirds of the counties have so far been documented (28 are documented completely and another five have some coverage, leaving seven with no county volume), and among those which might yet yield a very interesting record for Wolves are the northern counties of Durham and Northumberland. The eastern counties are also poorly covered so far, yet should have some interesting early names, because these are counties that the Anglo-Saxons settled first (Carver 1994). The Wolf place-names tend to be associated with the higher ground of the northern counties, but are not especially associated with woodland, when analysed on a county-by-county basis. It seems that Wolves were actually exterminated fairly early from lowland Britain, where the well-wooded Weald, for instance, has provided only three names.The most numerous single category of names, 18%, refers to wolf-pits (names such as Hoopits, Woolpit), and the use of a baited pit was a long-established way to trap Wolves, whether for pest control or for their fur. Most of these are not towns but the names of fields, probably the one containing the local wolf-pit (Fig. 5.1).

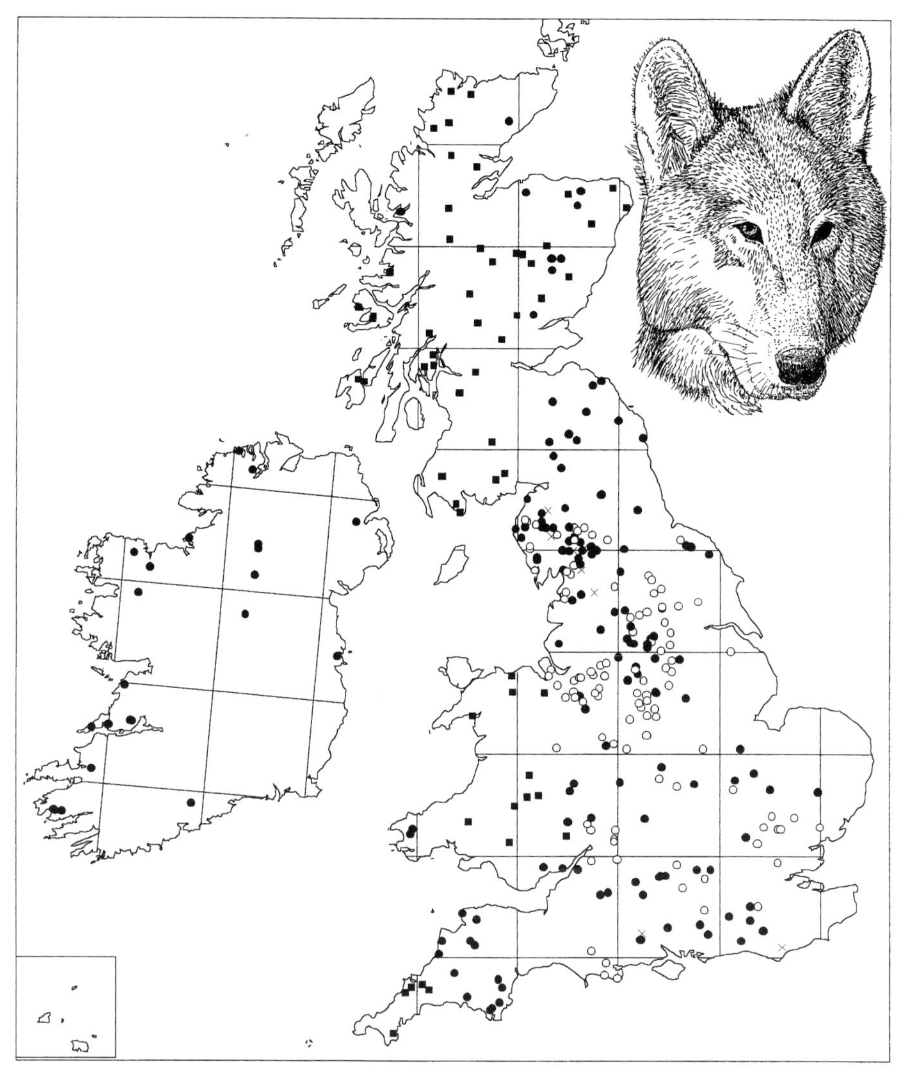

Figure 5.1 The distribution of Wolf place-names in the British Isles (after Aybes and Yalden 1995, Sara Sattar pers. comm.)

Another species that also figures in the place-name literature is the Beaver, but in contrast to the Wolf, Aybes and Yalden (1995) could find only 20 names that referred to it (Fig. 5.2); moreover, one of those (Beverkae, Fife) should probably be deleted, as a late name, conferred only in the 19th century, and not a reminder of ancient Beavers. There is the additional problem that 'to beaver' was to ferment woad, in a beaver-pit, and one or two of the relevant names could refer to this industry, rather than to the animals (Rackham 1986). However, the identification of Beverley, Yorkshire, with former rodent inhabitants of the Humber marshes and of Beverley Brook which runs into the Thames at Battersea with those of the Thames marshes

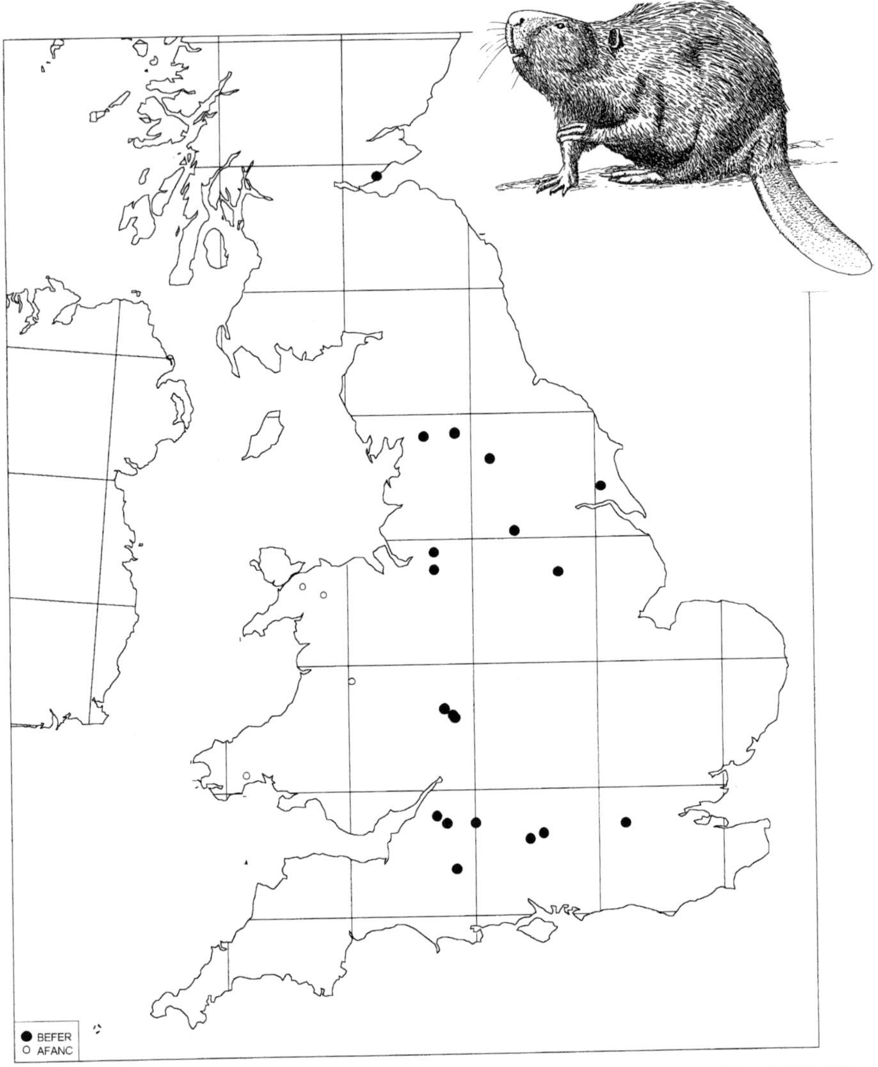

Figure 5.2 The distribution of Beaver place-names in Britain (from Aybes and Yalden 1995). There are far fewer names for this distinctive mammal than for Wolves, suggesting that it was already scarce when the Anglo-Saxon colonists were naming the countryside.

seems safe enough. Although the earlier (Neolithic–Iron Age) archaeological record has produced numerous records of Beavers in eastern England in the Fenland areas (Table 5.1), it is perhaps significant that there appear to be no relevant place-names from these areas; the plausible explanation is that they had already been cleared of their tree-cover, and were no longer suitable habitat for Beavers.

Deer also gave their names to places, and the two native species were distinguished as hart (*heort*) and hind, buck (*bucc*) and doe. However, the latter terms were also subsequently applied to the Fallow Deer when it became numerous, leading to the possibility of confusion. There are at least 185 place-names that seem to refer to the Red Deer (Harthill, Hartwell, Hindley, Hindhead, etc.: Sarah Beswick, pers. comm., Fig. 5.3), though this number assumes that *deor*, the precursor of the modern generic term deer, actually referred to Red Deer: originally it meant any wild animal (Copley 1971). Roe, mostly as names derived from *bucc*, appear in 66 place-names (Buckhill, Buckfast, but also Rogate and Reigate). There are a few names, presumably rather late ones, that may refer to Fallow Deer; Sarah Beswick identified just seven, including a couple of cases of *faelu leah*, fallow field (now Fawley) that could be fallow in the agricultural rather than cervine sense. Otters, (Wild) Cats, Foxes (often Tod-) and Badgers (Brock, but easily confused with *broc*, brook) also figure in the place-name literature; Claire Marriage found 206 Fox, 141 Badger and only 37 Otter place-names in England, as well as just 13 that refer to (Pine) Martens. There seem to be no con-

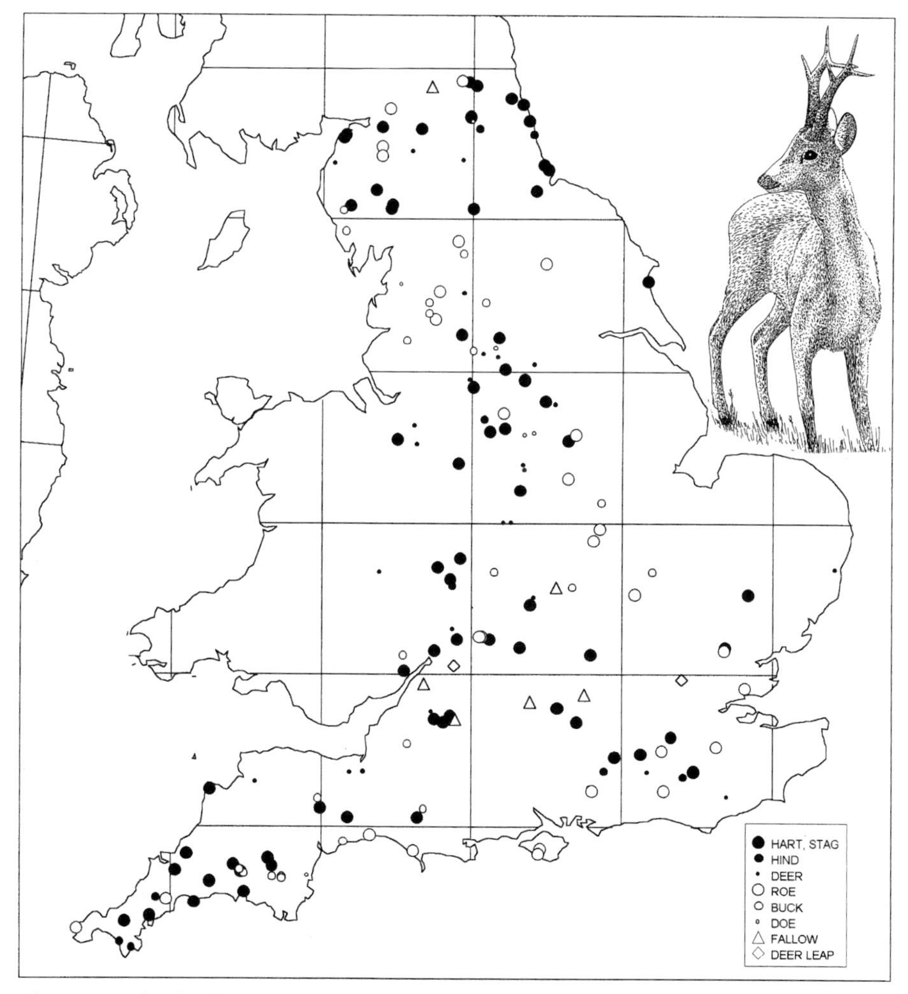

Figure 5.3 The distribution of deer place-names in England (from an unpublished project by Sara Beswick). Hart and Hind clearly refer to Red Deer, Buck and Doe are assumed to be Roe Deer.

vincing names referring to Brown Bears (Barham in Kent is presumably named after someone named Bear who founded the settlement), and possible names for Aurochs are few. Urpeth (Durham) has been suggested to mean Path of the Urus, and Strath Wradell (Sutherland) has been equated with Strath-uridale (sources in Aybes and Yalden 1995), but the scant late archaeological record reviewed in Chapter 4 makes these interpretations seem less likely.

Names from beyond the English-speaking areas are more difficult to garner, for a non-Celtic or Gaelic speaker, since there is no equivalent of the EPNS volumes for Scotland, Wales or, indeed, Ireland. There are some books that discuss place-names in these areas, but few of them are able to trace the earliest forms and dates of the names they mention in the way the EPNS volumes do. Even Cornwall, which will be covered, is so far represented only by a single EPNS volume citing some of the elements that occur in place-names there. In Wales, the Wolf is recalled in the name *bleidd*, and Aybes and Yalden (1995) suggested 15 names with this root, plus one in Hereford. The equivalent in Cornish is *bleit*, found in six place-names. In Scotland, *madaidh*, often anglicized maddie or moddie, occurs in 82 place-names that seem to imply Wolf (Sara Sattar, pers. comm.). Strictly, it translates dog, and *madaidh allaidh* is wild dog, i.e. Wolf (*madaidh ruaidh*, red dog, appears for Fox in at least six place-names). The more usual name for dog in Gaelic is *cu*, however, and *madaidh* often occurs in remote and mountainous places, suggesting that Wolf is probably the correct interpretation. Aybes and Yalden (1995) remarked on the dissimilarity between the Welsh *bleidd* and the Gaelic *madaidh*, and said that they could see no Welsh places that resembled the latter. Duncan Brown (Chairman of the Edward Lluyd Society) has since told us that an obsolete Welsh name for Fox is *madyn*, while there is an Irish word for monster or whale, *bled* (Brown 1996 and pers. comm.). The usual Gaelic for Fox is *sionnach*, which occurs in at least 17 places.

More place-names relate to the domestic species than to wild mammals, and such obvious examples as Oxford and Sheepen confirm that the Saxon settlers were farming people. A trawl of the EPNS volumes yielded at least 65 names derived from oxen and another 49 from bull, as well as 143 from other cattle names (cow, kine, calf, shipon). Sheep, including rams, tups, ewes and lambs, produced 123 names, and swine 122 names (Wayne Mason, pers. comm., Fig. 5.4). Farming seems to have sustained the population; hunting seems to have been more a sport than a means of survival. Is this a conclusion supported by the archaeological evidence?

Anglo-Saxon Archaeology

When the Dark Ages were considered to be a time of small farming communities, with no large settlements, let alone cities, it seemed understandable that there should be little archaeological evidence of their economic activities. A number of major excavations have changed both perceptions and with them our knowledge of the fauna. The Saxon site of *Hamwic* was far enough outside the centre of modern Southampton to survive unseen beneath Mediaeval farmland, but near enough to be discovered during commercial redevelopment of the modern city. In their report of the Melbourne Street site there, Bourdillon and Coy (1980) identified 48 214 bone fragments from at least 702 individual mammals and 101 birds. Sheep/goats (37.7%)

and cattle (30%) were dominant, with domestic pigs (27.4%) close behind. There was one possible Wild Boar bone, a large humerus, but no evidence of the distinctive larger teeth of Wild Boar, and only five Red Deer and two Roe Deer. The balance of the mammal fauna was made up of five horses, four dogs, 13 cats and five small mammals (Wood Mouse and Field Vole). Given the small sample of small mammals, the absence of House Mice is not significant; the cats appear to be domestic, not Wild Cats, and therefore presumably agents of pest control. Domestic ungulates therefore made up 99% of the mammal fauna, and domestic species similarly dominated the birds (mostly chicken and goose, very few wild species); farming, not hunting, was the

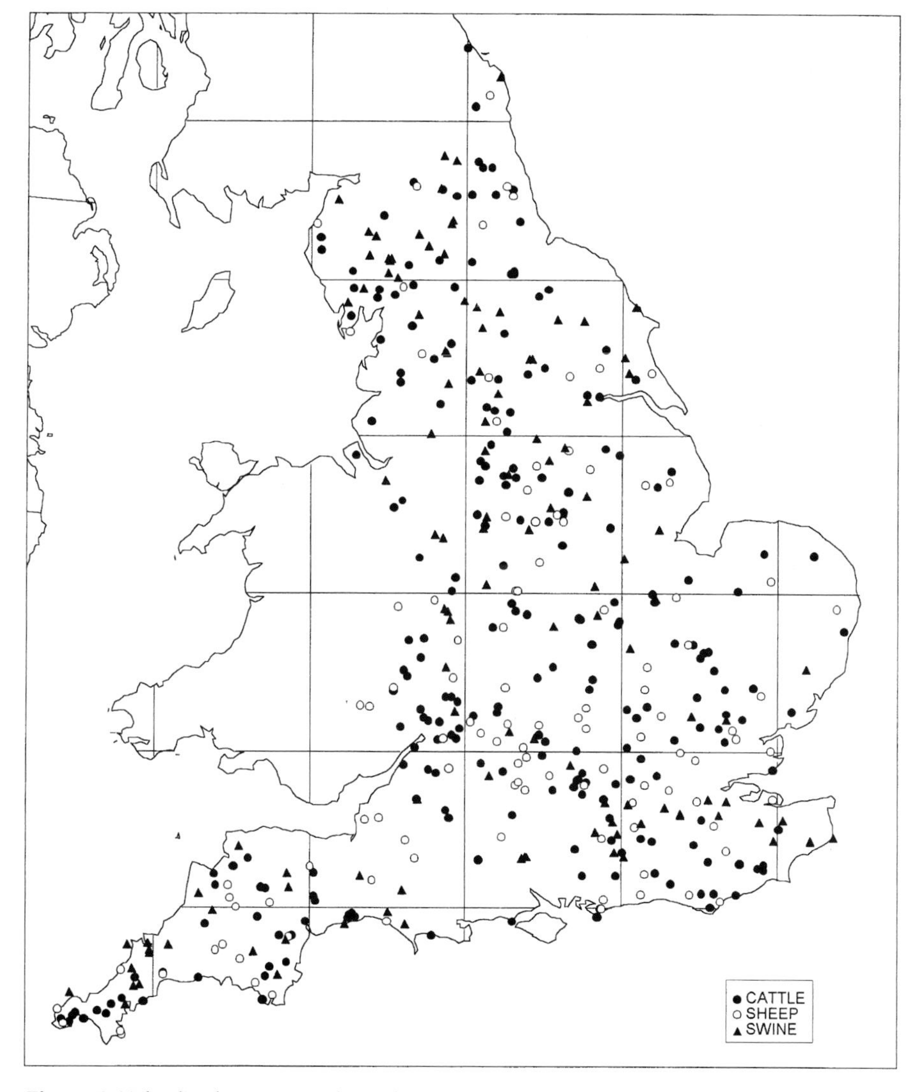

Figure 5.4 The distribution in England of place-names involving domestic animals (from an unpublished project by Wayne Mason).

main supplier of meat to this community. Two interesting additional comments are made on the fauna, one negative and one positive. First, there was no evidence for Black Rat or Rabbit, nor indeed for Fallow Deer, in this large collection; secondly, there was an indication of whale bone, just five fragments of worked bone which are certainly not identifiable to species.

The absence of Rabbit and Fallow Deer at *Hamwic* matches current thinking that these were Norman introductions, but the absence of Black Rats does not fit the revised view that they were introduced by the Romans. However, this does fit with recent results from a more northern site of similar age, the Anglian settlement of *Eorforwic*. At Fishergate, O'Connor (1991) had a fauna dated to the 8th century similarly dominated by cattle (60.9%) and sheep (25.1%) with smaller numbers of pig (9.5%). Wild mammals totalled only 2% of 15 219 identified bone fragments, and most (1404, 1.2%) were of Red Deer antlers, raw materials for tool-making rather than food. The other wild mammals offer an interesting hint of the economy and ecology of the area. There are four bones of Beaver, none with cut marks. Although there are 54 bones of House Mouse, as at *Hamwic* there are no Black Rats, despite their presence in earlier Roman York and in later Mediaeval York (O'Connor 1991, 1992; Table 5.1). A few bones of Red Fox and Pine Marten suggest hunting for fur as they do bear cut-marks, while a very few bones of Red Deer, Roe Deer and Wild Boar imply hunting for meat. The House Mouse is the commonest small mammal, but Water Shrew, Field Vole and Wood Mouse are also listed; curiously for a riverside area, no Water Voles were preserved in these levels. Another roughly contemporary site that resembles this in some respects is Repton, Derbyshire, where the Great Viking Army overwintered in A.D. 873 after sailing up the River Trent. They created a cairn there out of the ruins of a dilapidated building to cover a mass grave containing a chief and some 249 other warriors, but before that the building had been tenanted by a Barn Owl which left a useful small mammal fauna. Like the York site, there were no Black Rats. Water Voles were scarce despite the fact that the owl must have hunted over a floodplain with abundant long, wet grass; Field Voles were pre-

Table 5.1 The record of House Mice and Black Rats in archaeological excavations at York (from O'Connor 1992). The records are the presence of the species in archaeological samples from the various sites and ages. All material was sieved through a 2-mm mesh.

	Number of Samples (%)		
Site/Age	*with Mus*	*with Rattus*	*Total*
Tanner Row late 2nd–mid 3rd centuries	39 (53)	9 (12)	74
Fishergate 8th century	15 (28)	0	53
Coppergate late 9th century	17 (29)	11 (19)	59
Fishergate 13th–14th century	7 (41)	5 (29)	17
Coffee Yard 15th–18th century	9 (32)	8 (29)	28

dominant among the prey, and both Harvest Mice and Water Shrews were present (Yalden 1995). Evidently the breakdown of Roman civilization, and especially of the trade routes from the Mediterranean that must have regularly reinforced the population, meant that Anglo-Saxon Britain was too inhospitable for Black Rats to survive.

It has been suggested that the earliest Anglo-Saxon settlers must have encountered a less agricultural landscape and therefore been forced to hunt to supplement their diet. Among the earliest of sites yet investigated is West Stow, Suffolk, dating back to the 5th century and therefore close to the initial settlement. Crabtree (1989a, 1989b) has specifically considered this point. Cattle and sheep (including some goats) dominate the faunas at all three Saxon levels, as they do the preceding Roman and Iron Age faunas at this site; pigs are also important, but become less so during Saxon times. Wild mammals, including Red Deer, Roe Deer and (Brown?) Hare are present, but very rare, and there is no indication of Wild Boar (Table 5.2). The scarcity of wild mammals, no more than 0.5% of the 65 000 identified mammal bones, may indicate the much earlier clearance of what little woodland remained in this area of Breckland. However, the importance of (domestic) pigs, which one might expect to be foraging in woodland under the system of pannage that clearly was part of Anglo-Saxon farming, is very high at some of the East Anglian Saxon sites reviewed by Crabtree (1994); 29% at urban Ipswich, 30% at urban Norwich and a remarkable 68% at the rural site

Table 5.2 Archaeological records from Iron Age, Roman and Anglo-Saxon West Stow (from Crabtree 1989b). Cattle and sheep remain the most important food animals throughout these periods, deer and other game are never more than minor items of diet. The large number of unidentified bone fragments are mostly of large and small ungulates, probably also cattle and sheep, but too poorly preserved to be recognizable.

Species	*Iron Age*	*Roman*	*5th C.*	*6th C.*	*6th–7th C.*
Cattle	18.4	14.2	11.7	9.6	10.7
Sheep	1.0	1.3	3.1	2.5	2.5
Goat	<0.1	–	<0.1	<0.1	<0.1
Sheep/Goat	10.7	14.1	12.9	11.3	12.2
Pig	3.6	5.3	7.8	3.4	6.3
Horse	2.8	2.5	0.4	0.3	0.8
Dog	0.3	0.2	0.2	0.1	<0.1
Cat	<0.1	–	<0.1	0.4	–
Red Deer	0.1	0.1	<0.1	<0.1	<0.1
Roe Deer	<0.1	–	<0.1	<0.1	<0.1
Hare	<0.1	0.4	<0.1	<0.1	<0.1
Rabbit	0.3	0.2	<0.1	<0.1	<0.1
Badger	–	–	<0.1	<0.1	<0.1
Fox	<0.1	–	–	–	–
Water Vole	–	–	–	<0.1	<0.1
Mole	<0.1	–	<0.1	<0.1	–
Unidentified	62.4	61.7	63.8	71.9	67.1
Total Bones	7551	1814	21 716	50 366	4890

of Wicken Bonhunt. This last probably reflects a cultural bias, perhaps to pork being a high-status food at what may have been a royal farm.

If deer were a rare, or rarely used, resource in East Anglia, there is some evidence that they were more highly regarded elsewhere. At Porchester Castle, a site where cattle (45%), sheep (24%) and pigs (15%) still dominate the fauna, Red Deer never-the-less contributed 3% and Roe 2% of the 14 984 bones. There were also a few bones of Fallow Deer (Grant 1976); deer were more important here than they had been in preceding Roman times at the same site. This is presumed to indicate the importance of hunting at a royal hunting lodge.

Archaeological evidence for other wild mammals in Saxon England is sparse. There are a couple of Beavers, represented by six bones, from Ramsbury, Wiltshire (Coy 1980), which with the specimens from York and an (unpublished) record from Eynsham Abbey, Oxford are the only Late Saxon examples, and perhaps the last in the conventional English archaeological record (Table 5.3). There are however some additional more specialized records of Beaver specimens; a single incisor at Northampton (Harman 1979), two fragmentary bones in one of the cremations at Spong Hill (Bond

Table 5.3. Archaeological sites for Beavers in Britain from Pleistocene through to Mediaeval times. Most are records of bones and teeth, but ! indicates characteristically gnawed timber, and * archaeological finds (mounted teeth, skin bag) of artefacts made from Beavers. Middle Pleistocene sites are dated to Oxygen Isotope Stages (from marine sediments), OI 7-13. Though the Westbury-sub-Mendip fauna is mostly referrable to Oxygen Isotope Stage 13, like the Boxgrove fauna, the Beaver is a fossil derived from an earlier Pleistocene deposit (A. Currant, pers. comm.).

Site	*Grid Ref*	*Age*	*Source*
Woodbridge, Red Crag, Suffolk	TM2649	Pliocene	Newton 1891
Sutton, Red Crag, Suffolk	TM3046	Pliocene	Newton 1891
East Runton, Norfolk	TG1942	Pastonian	Sutcliffe and Kowalski 1976
Westbury-Sub-Mendip, Somerset	ST8650	Early Pleistocene	Bishop 1982
Bacton, Norfolk	TG3433	Cromerian	Sutcliffe and Kowalski 1976
Mundesley, Norfolk	TG3830	Cromerian	Sutcliffe and Kowalski 1976
West Runton, Norfolk	TG1842	Cromerian	Sutcliffe and Kowalski 1976
Kessingland, Suffolk	TM5286	Cromerian	Sutcliffe and Kowalski 1976
Happisburgh, Norfolk	TG3731	Cromerian?	Dawkins and Sanford 1866
Thorpe, Norfolk	TM4398	Cromerian?	Dawkins and Sanford 1866
Boxgrove, West Sussex	SU9007	Post-Crom. (OI 13)	Roberts 1986
Hoxne, Suffolk	TM1877	Hoxnian (OI 11)	Stuart 1982
Barnfield Pit, Swanscombe, Kent	TQ6074	Hoxnian (OI 11)	Sutcliffe and Kowalski 1976
Clacton, Essex	TM1715	Hoxnian (OI 11)	Sutcliffe and Kowalski 1976
Grays Thurrock, Essex	TQ6476	Pre-Ipswich. (OI 9)	Sutcliffe and Kowalski 1976
Ilford, Essex	TQ4486	Pre-Ipswich. (OI 7)	Sutcliffe and Kowalski 1976
Selsey, Sussex	SZ8513	Pre-Ipswich. (OI 7)	Sutcliffe and Kowalski 1976
Upnor, Kent	TQ7670	Ipswichian?	Sutcliffe and Kowalski 1976
Kents Cavern, Devonshire	SX9364	Devensian	Sutcliffe and Kowalski 1976

Site	*Grid Ref*	*Age*	*Source*
Gough's Cave, Somerset	ST467539	Late Glacial?	Currant 1986
Gough's Old Cave	ST467539	9320 b.p.	Hedges *et al.* 1987
Mother Grundy's Parlour	SK536743	Mesolithic?	Jacobi, pers. comm.
Cherhill, Wiltshire	SU0370	Mesolithic	Grigson 1983
Doghole Fissure, Cresswell, Derbyshire	SK5374	Mesolithic	Jenkinson 1984
Hartle Dale, Derbyshire	SK1680	Mesolithic	Bramwell 1977a
Thatcham, Berkshire	SU5167	Mesolithic	Wymer 1962
King Arthur's Cave, Hereford	SO546154	Mesolithic	Jackson 1953
Star Carr, Seamer, Yorkshire	TA027810	Mesolithic	Fraser and King 1954
Stanstead Abbots, Hertford	TL400106	Mesolithic?	Roberts and Roberts 1978
Middlestots Bog, Edrom, Berwickshire	NT8255	7690 b.p.	Kitchener and Conroy 1997
Linton, Borders	NT7726	6170 b.p.	Kitchener and Conroy 1997
!West Morriston Bog, Berwickshire	NT60404	6340 b.p.	Kitchener and Conroy 1997
!Runnymede	TQ018718	3770 b.p.	Ambers *et al.* 1989
Durrington Walls, Wiltshire	SU152435	c. 3900 b.p.	Harcourt 1971
Cambridge Fen	TL4760?	3079 b.p.	Montagu 1924
West Cotton, Raunds, Northants	SP976725	2900 b.p.	Hedges *et al.* 1995
Cleaves Cove, Ayrshire	NS290495	2785 b.p.	Kitchener and Conroy 1997
Burwell Fen, Cambridgeshire	TL5767	2677 b.p.	Montagu 1924
Loch of Marlee, Perthshire	NO1444	2555 b.p.	Kitchener and Conroy 1997
!Baker trackway, Westhay	ST4242	Neolithic	Coles and Coles 1986
Cresswell Resurgence, Derbyshire	SK1773	Neolithic	Bramwell 1977a
Offham, Sussex	TQ399118	Neolithic	O'Connor 1977
Chatteris, Cambridgeshire	TL3985	Neolithic?	Barrett-Hamilton and Hinton 1921
Hilgay, Norfolk	TL5795	Neolithic?	Dawkins and Sanford 1866
Hornsea, Yorkshire	TA1947	Neolithic?	Sheppard 1922
Littleport, Cambridgeshire	TL5686	Neolithic?	Montagu 1924
St. Bertram's Cave, Staffordshire	SK1054	Neolithic?	Bramwell 1977a
Reach Fen, Cambridgeshire	TL5566	Neolithic?	Montagu 1924
Newbury, Berkshire	SU4767	Neolithic?	Barrett-Hamilton and Hinton 1921
Grimes Graves, Norfolk	TL8189	Neolithic	Clarke 1922
Swaffham Fen, Cambridgeshire	TL5367	Neolithic?	Montagu 1924
Taddington Dale (Demen's Dale)	SK1671	Neolithic	Bramwell 1977b
Withernsea, Humberside	TA3427	Neolithic?	Sheppard 1903
Crossness Point, Greater London	TQ4781	Neolithic?	Dawkins and Sanford 1866
London, Royal Albert Docks	TQ4280	Neolithic?	Barrett-Hamilton and Hinton 1921
Merlins Cave, Herefordshire	SO556153	Neolithic?	Jackson 1953
Mildenhall, Suffolk	TL7074	Neolithic?	Barrett-Hamilton and Hinton 1921
Wawne, Yorkshire	TA0836	Neolithic?	Sheppard 1903
West Kennet Long Barrow, Wiltshire	SU104677	Neolithic	Piggott 1962
Whiteleaf Long Barrow, Buckinghamshire	SP8104	Neolithic	Simmons and Tooley 1981
Durrington pipeline, Wiltshire?	SU1544?	Neolithic	S. Hamilton-Dyer, pers. comm.
Tarrant Keynston, Dorset	ST914037	Neolithic?	Mansell-Pleydell 1895

Site	*Grid Ref*	*Age*	*Source*
Barholm, Lincolnshire	TF1010	Late Neolithic	Harman 1993
Staines, Surrey	TQ0271	Late Neolithic	Robertson-Mackay 1987
Lower Mill Farm, Stanwell	TQ0673	Late Neolithic?	S. Hamilton-Dyer, pers. comm.
Brean Down?	ST2858	Bronze Age	Levitan 1990
Caldicot, Gwent	ST4888	Bronze Age	Hamilton-Dyer 1997
Testwood Lakes, Hampshire	SU345153	Bronze Age?	S. Hamilton-Dyer, pers. comm.
Glastonbury, Somerset	ST4938	Iron Age	Dawkins and Jackson 1917
Coneybury Hill	SU134414	Iron Age	Coy and Maltby 1987
Micklemoor Hill, West Harling, Norfolk	TM975857	Iron Age	Clark and Fell 1953
Meare, Somerset	ST4541	Iron Age	Gray 1966
Meare, Somerset	ST4541	Iron Age	Cornwall and Coles 1987
Mingies Ditch, Oxford	SP391059	Iron Age	Wilson 1993
St Minver, Cornwall	SW9677	Iron Age	Coy 1987
Haddenham	TL4675	Iron Age	Evans and Serjeantson 1988
Ulrome, Yorkshire	TA1656	Iron Age	Sheppard 1903
Hedsor, Buckinghamshire	SU9086	Roman	Barrett-Hamilton and Hinton 1921
Shapwick, Wimbourne, Dorset	ST9301	Roman	N. Grace, pers. comm.
Edinburgh Castle	NT251736	c. 1680 b.p.	Kitchener and Conroy 1997
Ramsbury, Wiltshire	SU2771	Saxon	Coy 1980
Northampton, St. Peter's Street	SP7561	mid Saxon	Harman 1979
Spong Hill, North Elmham, Norfolk	TF9820	Saxon	Bond 1994
*Ducklington, Oxford	SP3507	Saxon	Wilson 1992
*Wigber Low, Derbyshire	SK205513	Saxon	Wilson 1992
*Lechlade	SU2199	Saxon	Wilson 1992
*Sutton Hoo, Suffolk	TM2849	Saxon	Bruce-Mitford 1975
Aylesbury, Walton	SP8213	Saxon	Noddle 1976
Theale, Reading, Berkshire	SU6371	Saxon	Hamilton-Dyer 1996b
Wirral Park Farm, Glastonbury	ST4938	900 A.D.	Darvill and Coy 1985
York, Fishergate	SE6052	Anglian	O'Connor 1991
Eynsham Abbey, Oxford	SP4309	Late Saxon	J. Mulville, pers. comm.
Winchester	SU4829	Mediaeval	O'Connor 1977
Ardrossan, Ayrshire	NS239422	?	Kitchener and Conroy 1997
Hay Wood Rock Shelter, Somerset	ST1323	?	Sutcliffe and Kowalski 1976
Lincolnshire fens, Lincolnshire	TM4559	?	Harting 1880
Pifflehead Wood, Newtondale	SE8395	?	Simms 1972
Romsey, Hampshire	SU3521	?	Harting 1880
Staple Howe, Yorkshire	SE8974	?	Sutcliffe and Kowalski 1976
Whenby Lake, Yorkshire	SE6170	?	Simms 1972
Ressendale, Sedbergh, Westmorland	SD6692	?	Macpherson 1892
Wolesey Palace, Winchester, Hampshire	SU4829		
Bed of River Avon, Chippenham, Wiltshire	ST9173		
Bed of River Avon, Melkshan, Wiltshire	ST9064		

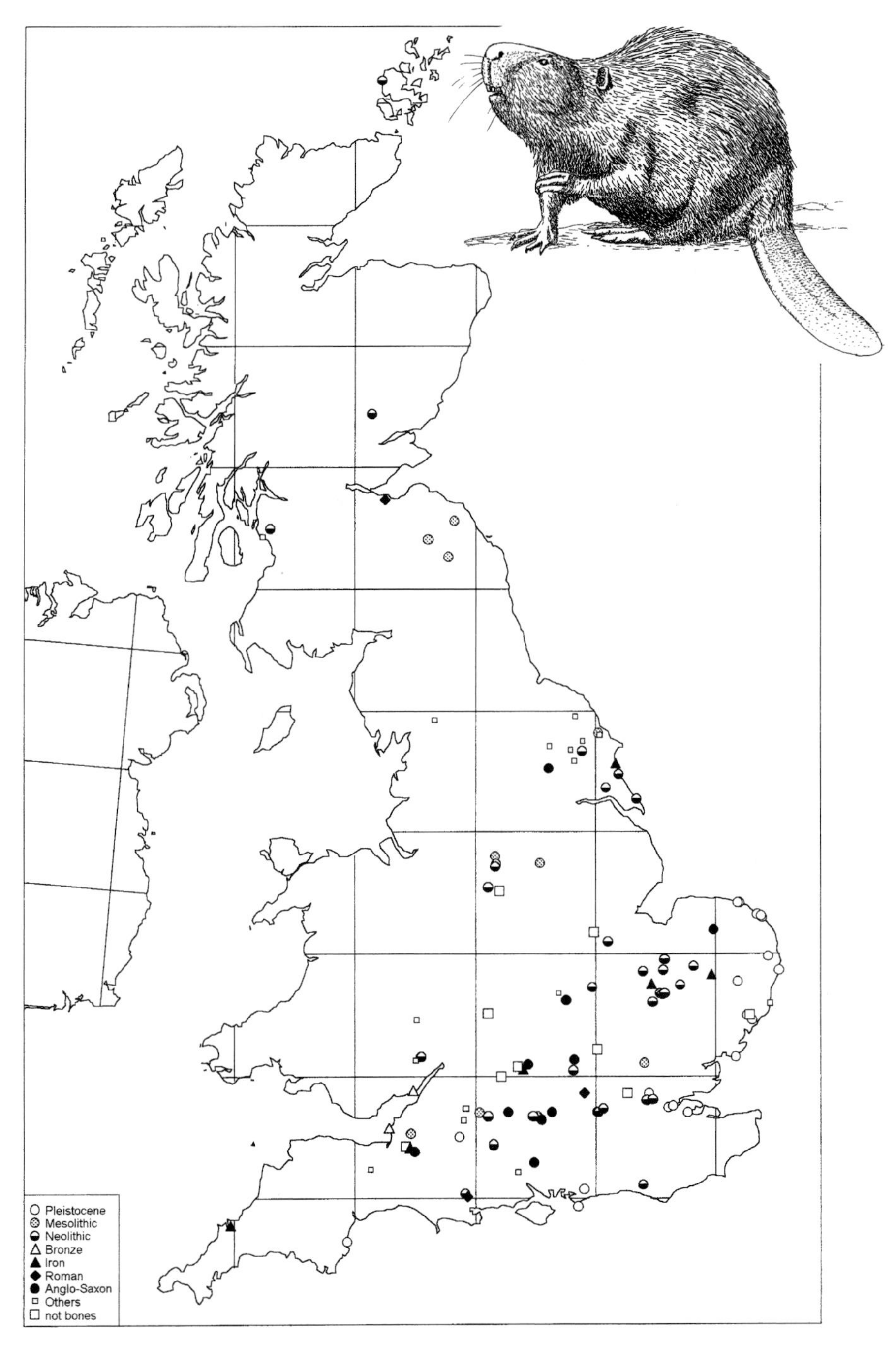

Figure 5.5 The archaeological and documentary record of the Beaver *Castor fiber* in Britain (from Aybes and Yalden 1995, Kitchener and Conroy 1997, and data in Table 5.3). Note the abundance of sites in eastern Britain, where there appear to be no place-names (Fig. 5.2).

1994), in the form of a Beaver-skin bag with the Sutton Hoo lyre (Bruce-Mitford 1975) and as teeth mounted as brooches in a number of burials (Wilson 1992) (Fig. 5.5). Hares are present at most Saxon sites, including Portchester Castle (Grant 1976), West Stow (Crabtree 1989a), Coppergate, York (O'Connor 1989), Brandon (Crabtree and Campana 1991), Flixborough (Dobney and Mills 1994), Hartlepool (Rackham 1989) and North Elmham (Noddle 1980). Fox and Badger appear at Portchester Castle, West Stow and Ramsbury, Badger and Otter at Brandon, and Fox and Otter at North Elmham. Fox is also recorded at Canterbury Cathedral (Driver 1990) and Ipswich (Jones and Serjeantson 1983), Pine Marten at Andover (S. Hamilton-Dyer, pers. comm.). None of these is surprising, and adds little to our understanding of Saxon mammal faunas. There is little evidence for any of the larger mammals that would have been hunted as game or vermin; Wild Boar and Wolf must have been present, but yield little by way of archaeological evidence (Fig. 5.6, Table 5.4) (Wild Boar is present at Anglo-Scandinavian Coppergate, York; O'Connor 1989). The Brown Bear was perhaps already extinct, but the evidence of skins from Saxon sites discussed in Chapter 4 could indicate a rare late survival, with appropriate symbolic status.

Table 5.4 Archaeological sites for the Wolf *Canis lupus* in Britain. The later record is poor because of the uncertainty whether large canid remains are Wolf or Dog. Despite this, there are enough records to indicate the wide spread in time and geographically, which matches the place-name evidence (see Chapter 5). Records from Pliocene to Cromerian belong strictly to *Canis mosbachensis*, the smaller forerunner of *C. lupus*. Dating is rarely certain for these records; sites where Wolves occur with species such as Spotted Hyaena and Woolly Rhinoceros are assigned to the Devensian, those with Reindeer and Mammoth to Late Glacial, and those with Post Glacial species such as Roe and Red Deer to Mesolithic if lacking domestic species, or to Neolithic if domestic ungulates are also present.

Site	*Grid Ref*	*Age*	*Source*
Boyton, Suffolk	TM3747	Pliocene	
Woodbridge, Suffolk	TM2649	Pliocene	
Sidestrand, Norfolk	TG2539	Cromerian	Turner 1995
West Runton, Norfolk	TG1842	Cromerian	Stuart 1982
Boxgrove, West Sussex	SU9007	Post-Cromerian (OI 13)	Roberts 1986
Westbury-sub-Mendip, Somerset	ST8650	Post-Cromerian (OI 13)	Bishop 1982
Ingress Vale, Kent	TQ5975	Hoxnian (OI 11)	Stuart 1982
Swanscombe, Kent	TQ6074	Hoxnian (OI 11)	Stuart 1982
Grays, Essex	TQ6177	Pre-Ipswichian (OI 9)	Harting 1880
Brundon, Suffolk	TL8543	Pre-Ipswichian (OI 7)	Hopwood 1939
Stoke Tunnel, Suffolk	TM1-4-	Pre-Ipswichian (OI 7)	Stuart 1982
Valley of Roding, Ilford, Essex	TQ4486	Pre-Ipswichian (OI 7)	Harting 1880
Slade Green, Erith, Kent	TQ5177	Pre-Ipswichian (OI 7)	Reynolds 1909

Site	*Grid Ref*	*Age*	*Source*
Bleadon, Avon	ST3457	Pre-Ipswichian (OI 7)	Harting 1880
Hutton Cavern, Somerset	ST3458	Pre-Ipswichian (OI 7)	Harting 1880
Crayford, Kent	TQ5275	Pre-Ipswichian (OI 6/7)	Reynolds 1909
Ipswich, Suffolk	TM1644	Ipswichian?	Reynolds 1909
Bacon Hole, Glamorgan	SS5686	Ipswichian	Reynolds 1909
Barrington, Cambridge	TL3949	Ipswichian	Stuart 1982
Beilsbeck, Yorkshire	SE7944	Ipswichian?	Harting 1880
Durdham Down, Somerset	ST5674	Ipswichian	Reynolds 1909
Kirkdale Cave, Yorkshire	SE677856	Ipswichian	Boylan 1981
Bracklesham, West Sussex	SZ8096	Ipswichian?	Harting 1880
Cefn near St Asaph, Clwyd	SJ0261	Ipswichian	Reynolds 1909
Hoe Grange, Longcliff, Derbyshire	SK2155	Ipswichian?	Reynolds 1909
Minchin Hole, Glamorgan	SS555869	Ipswichian	Harting 1880
Ravenscliff, Glamorgan	SS5586	Ipswichian	Harting 1880
Tornewton Cave, Devon	SX8167	Ipswichian	Turner 1995
Joint Mitnor, Buckfastleigh, Devon	SX743664	Ipswichian	Turner 1995
Tewkesbury, Gloucestershire	SO8932	Ipswichian?	Harting 1880
Thame, Oxfordshire	SP7005	Ipswichian?	Harting 1880
Pontnewydd, Clwyd	SJ006714	225 000 b.p.	Turner 1995
Shandon Cave, Co. Waterford	X292950	27 500 b.p.	Woodman *et al.* 1997
Castlepook Cave, Co. Cork	R603009	23 470 b.p.	Woodman *et al.* 1997
Brixham, Bench Cave, Devonshire	SX9255	Devensian	Harting 1880
Broxbourne, Hertfordshire	TL3607	Devensian	
Black Rock Quarries, Pembroke	SN1100	Devensian	Jackson 1953
Banwell, Somerset	ST3858	Devensian	Harting 1880
Sanford Cave, Avon	ST3959	Devensian	Harting 1880
Bosco's Hole, Glamorgan	SS5586	Devensian	Harting 1880
Clwyd Vale, Cae Gwyn, Clwyd	SJ088727	Devensian	Reynolds 1909
Crow Hole, Glamorgan	SS5586	Devensian	Harting 1880
Deborah's Den, Glamorgan	SS4188	Devensian	Harting 1880
Elderbush Cave, Staffordshire	SK098549	Devensian	Jackson 1953
Ffynnon Beuno, Clwyd	SJ083723	Devensian	Reynolds 1909
Fisherton, Wiltshire	SU0038	Devensian	Reynolds 1909
Hyaena Den, Wookey Hole	ST532481	Devensian	Jackson 1953
Isleworth, Middlesex	TQ1675	Devensian	Stuart 1982
Kents Cavern, Devonshire	SX934641	Devensian	Harting 1880
King Arthur's Cave, Herefordshire	SO546154	Devensian	Jackson 1953
Langwith Bassett, Derbyshire	SK517695	Devensian	Reynolds 1909
Long Hole, Glamorgan	SS4684?	Devensian	Harting 1880
Mother Grundy's Parlour, Cresswell, Derby.	SK535742	Devensian	Jenkinson 1984
Murston, Sittingbourne, Kent	TQ9163?	Devensian	Harting 1880
Oreston Cave, Devonshire	SX5053	Devensian?	Harting 1880
Paviland, Glamorgan	SS437859	Devensian	Harting 1880

Site	*Grid Ref*	*Age*	*Source*
Picken's Hole, Somerset	ST3955	Devensian	Stuart 1982
Pin Hole Cave, Cresswell	SK533742	Devensian	Jenkinson 1984
Plas Heaton, Clwyd	SJ031691	Devensian	Jackson 1953
Rhinoceros Hole, Wookey, Somerset	ST532479	Devensian	Proctor *et al.* 1996
Robin Hood's Cave, Creswell	SK535742	Devensian	Charles and Jacobi 1994
Spritsail Tor, Glamorgan	SS4493	Devensian	Harting 1880
Tattershall Castle, Lincolnshire	TF2157	Devensian	Stuart 1982
Torbryan, Devon	SX815674	Devensian	Reynolds 1909
Uphill, Somerset	ST316583	Devensian	Harting 1880
Waterhouses, Staffordshire	SK0355	Devensian	Reynolds 1909
Windsor, Berkshire	SU9676	Devensian	Harting 1880
Wookey Hole, Somerset	ST5347	Devensian	Harting 1880
Wretton, Norfolk	TF6800	Devensian	Stuart 1982
Yealmpton Cave, Devonshire	SX5751	Devensian	Reynolds 1909
Fox Hole Cave, Derbyshire	SK100663	c. 12 000 b.p.	Bramwell 1971
Plunkett Cave, Co. Sligo	G710130	11 150 b.p.	Woodman *et al.* 1997
Aveline's Hole, Somerset	ST476588	Late Glacial?	Jackson 1953
Bridged Pot, Ebbor Gorge	ST526488	Late Glacial	Jackson 1953
Gough's Cavern, Cheddar Gorge	ST468540	Late Glacial	Jackson 1953
Wetton Mill, Staffordshire	SK095561	Late Glacial	Bramwell 1976
Soldier's Hole, Cheddar Gorge	ST468540	Late Glacial	Jackson 1953
Ightham, Kent	TQ6056	Late Glacial?	Reynolds 1909
Kinsey Cave, Yorkshire	SD804657	Late Glacial?	Jackson 1953
Lynx Cave, Staffordshire	SK106540	Late Glacial?	Jackson 1953
Sun Hole, Cheddar Gorge	ST467541	Late Glacial	Collcutt *et al.* 1981
Thor's Fissure, Staffordshire	SK098549	Late Glacial	Jackson 1953
Walton, nr Clevedon, Somerset	ST4274	Late Glacial	Reynolds 1909
Wavering Down, Somerset	ST3755	Late Glacial	Jackson 1953
Windy Knoll, Derbyshire	SK1282	Late Glacial?	Harting 1880
Dog Hole Fissure, Cresswell Crags	SK5374	Mesolithic	Jenkinson 1984
Star Carr, Yorks E	TA0281	Mesolithic	Fraser and King 1954
Fox Holes, Clapdale, Yorkshire	SD756714	Mesolithic?	Jackson 1953
Yew Tree Cave, Pleasley Vale	SK516649	Mesolithic?	Jackson 1953
Fox Hole Cave, Derbyshire	SK100663	Neolithic	Bramwell 1971
Helsfell, Cumbria	SD4893	Neolithic?	Macpherson 1892
Burwell Fen, Cambridgeshire	TL5767	Neolithic	
Mount Pleasant, Dorset	SY710899	Neolithic	Harcourt 1979
Haverbrack Bank Pot	SD482802	Neolithic?	Jackson 1953
Heaning Wood Bone Cave	SD267748	Neolithic?	(Cave Guide)
Hightown Peat Beds,	SD2903	Neolithic?	Neaverson 1940–43
Kirkhead Cavern	SD391756	Neolithic?	(Cave Guide)
Merlewood Cave	SD411789	Neolithic?	Jackson 1953
Newbury, Berkshire	SU4767	Neolithic?	Reynolds 1909
Rhyl Peat Beds	SJ0181	Neolithic?	Neaverson 1940–43
Teesdale, Durham	NY9325	Neolithic?	Reynolds 1909
Ulrome, York	TA1656	Iron Age	Sheppard 1903
?Catterick, Yorkshire	SE2299	Roman	Simms 1972

Site	*Grid Ref*	*Age*	*Source*
Elderbush Cave, Co. Clare	R323739	1730 b.p.	Woodman *et al.* 1997
Ramsbury, Wiltshire	SU2771	Saxon	Coy 1980
Pevensey, East Sussex	TQ6404	Mediaeval	Harting 1880
Rattray, Aberdeen	NK0857?	Mediaeval	M. Gorman, pers. comm.
Arnside Knott, Cumbria	SD4577	?	
Bed of River Avon, Chippenham, Wiltshire	ST9173	?	
Denver Sluice, Norfolk	TF5901	?	Harting 1880
Mildenhall Fen, Suffolk	TL6678	?	
Lions Wall, Colchester, Essex	TL9925	?	
Vale of Kennet, Wiltshire	?	?	Harting 1880
Wythemail, Northamptonshire	SP8471	?	
Fundon Well, West Sussex	TQ1208	?	
Gomeldon, Wiltshire	SU1935	?	

Beaver

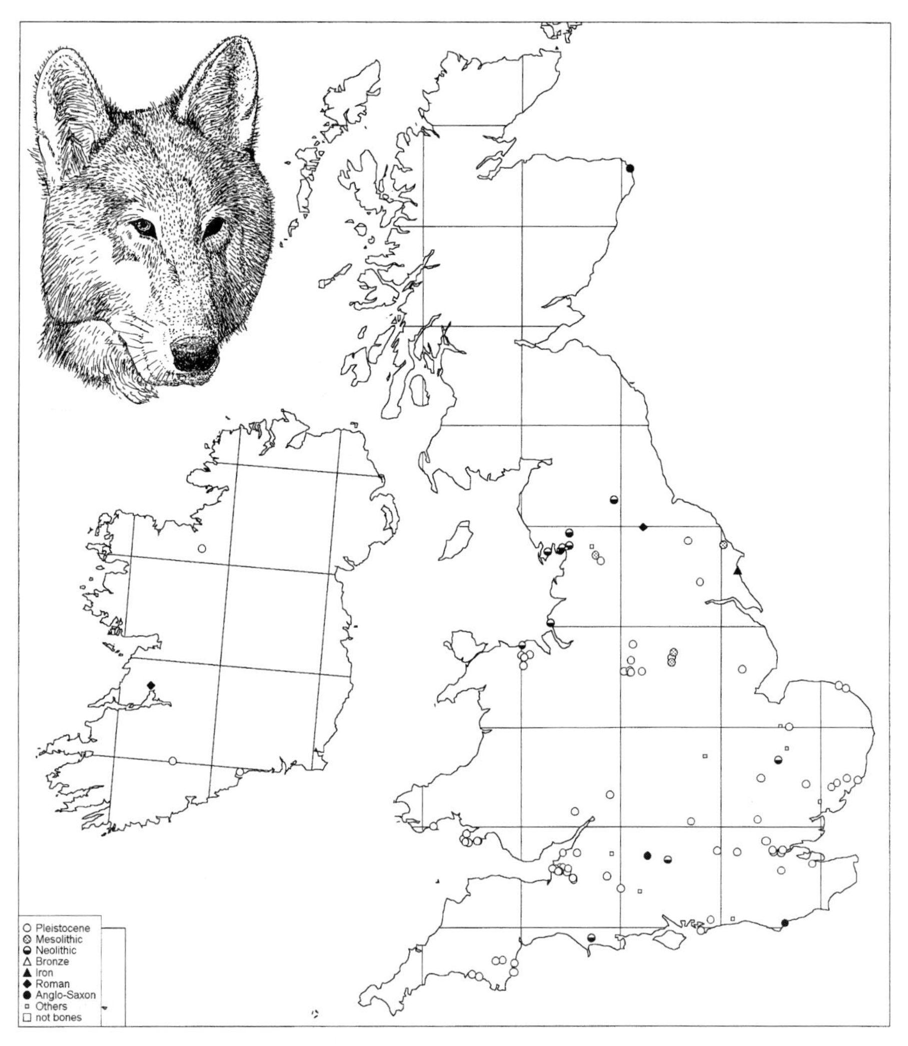

Figure 5.6 The archaeological and documentary record of the Wolf *Canis lupus* in the British Isles. Though the animal must have been common, there are few late archaeological records, due primarily to the difficulty of distinguishing its remains with certainty from those of domestic dogs.

Norse and Norman Influence

By late Saxon times, certainly out of the Dark Ages and into the historical period, we know that Norse settlers threatened the emergent Christian communities of Saxon England, establishing Danish kingdoms in Mercia and Northumbria for part of the 10th century, and bestowing Norse, rather than Saxon, names on settlements and parts of the landscape in much of northern Britain. Thus wolf-derived place-names in Cumbria are more likely to be something like Uldale (Old Norse *ulfr*, *dalr*) than

Woodale (Aybes and Yalden 1995). However, the Norsemen who settled first in Normandy, and then some 150 years later invaded England at Hastings in 1066, had a more dramatic effect on the fauna. They also established a kingdom in Sicily and southern Italy, which meant that they had contact with the hunting and park-keeping traditions of southern Europe, perhaps a carry-over from the Roman traditions of keeping parks stocked with wild animals. As Conqueror, William also had a direct interest in establishing just what his new kingdom was worth for taxation purposes, so instituted the Domesday Survey of 1086. This is a remarkably complete record, unique for its time, and gives us an invaluable view of the state of England at that time. It does not cover the four northern counties of Westmorland, Cumberland, Durham and Northumberland, not then part of England, and even Lancashire was clearly a border region, not properly surveyed, but tagged onto the Cheshire survey along with parts of Flintshire and Denbighshire. As such a complete record, it has been minutely examined by historical geographers, notably by H.C. Darby and his colleagues. Their five regional volumes are summarized in Darby (1977). Rackham (1980, 1986) has re-examined particularly the record for woodland, one of the main subjects for the inquisition.

The enquirers were to ask: the name of each place, who owned the land, 'now' (i.e. 1086) and in the time of King Edward (i.e. the Confessor, prior to the invasion)? How many hides (areas of land, nominally 120 acres) and ploughs (both those in lordship and the men's)? How many villagers, cottagers and slaves, how many free men and Freemen? How much woodland, meadow and pasture? How many mills and fish-ponds? How much has been added or removed? What was and is the total value? How much has or had each free man or Freeman: all threefold, before 1066, when given by King William, and now; and whether more could be had than at present (Morris 1975).

Clearly this is a tax return, not an ecological survey, and it tells us little directly of the fauna of the country. Despite this, it certainly does contain indirectly a great deal of value. It confirms that England was poorly wooded; Rackham (1990) calculates that perhaps there was 15% woodland cover. There were already many regions with virtually no woodland, and others that were much more wooded than any present county. This applied even within counties; western Cheshire, including the Wirral, had very little woodland, but most of the eastern fringe of the county was completely wooded (Yalden 1987). Rackham (1990) draws a distinction in modern England between the Planned Countryside and the Ancient Countryside. The former, a large swathe across the Midlands from Dorset to Norfolk, Lincolnshire and the East Riding of Yorkshire, where there was little woodland in the Domesday Survey, was laid out in straight hedges and ditches on the pre-existing open field system by the Enclosure Acts of the 18th and 19th centuries. By contrast, the Ancient Countryside was created piece-meal any time between the Neolithic and Mediaeval times; it has an irregular patchwork of fields and woods, crooked lanes and hedges. The distinction is clearly visible by the time of the Domesday Survey, and has repercussions still on the mammal fauna. It has already been remarked (Chapter 4) that Dormice and Yellow-necked Mice have a very odd distribution in this country, largely missing from the Midlands, but more evident in the south-east and up the Severn Valley. It is quite clear that this is a pattern of distribution that reflects that of the Ancient Countryside (Figs 4.7 and 4.8). It explains one detail of their distribution which used to puzzle me as a student;

why are neither to be found in that so thoroughly documented study site, Wytham Wood near Oxford? Answer, because it lies right in the middle of the Planned Countryside and, despite its appearance as ancient deciduous woodland, is in fact a piece of secondary woodland that developed in the Middle Ages on former common land (Rackham 1976).

The Domesday Survey may say little directly about wild mammals, but there are some interesting hints about some of the husbandry of feral ones. The Anglo-Saxons did not have hunting estates or forests; Gilbert (1979) suggests that the Roman law of game as *res nullius*, belonging to whoever killed it irrespective of whose land it was on, was not supplanted until the Frankish Empire in about the 7th century. Hunting law which limits the right of hunting on other's land, or which states that game, or some of it, might belong to a landowner or sovereign, clearly indicates that *res nullius* has been supplanted. So do the existence of penalties for infringing that right. If this was a notion foreign to the Saxons, it is not surprising that the Anglo-Saxon Chronicle (for 1087) observes with some concern in writing his obituary that King William loved the deer as though they were his sons, and threatened severe penalties for hunting 'his' deer!

> He made great protection for the game
> And imposed laws for the same
> That who so slew hart or hind
> Should be made blind
>
> He preserved the harts and boars
> And loved the stags as much
> As if he were their father
> Moreover, for the hares did he decree
> That they should go free
> (Whitelock 1961)

Forests were places set aside for hunting of game, and moreover for royalty, not just anyone, to hunt. Other activities, such as gathering firewood, were not necessarily prohibited, but in order to protect the deer and their grazing, there were usually restrictions on the levels or times of grazing, and on cutting of timber. Generally, Red Deer, Roe Deer and Wild Boar were the species protected, but Hares were honorary deer in the Forest of Somerton. In Domesday, there are a few indications of forest; the New Forest in Hampshire is clearly indicated, as a recent creation that did not exist in the time of King Edward. Rackham (1980, 1990) suggests that 25 are either explicitly or implicitly mentioned. These include Windsor, Savernake and Delamere Forests, the Forest of Dean and Wychwood. Most were created rather later, reaching a maximum in England of around 143 (Fig. 5.7). Of these, 90 were the king's and the rest private. Corporal or capital punishments for offences against the king's venison were apparently carried over from Carolingian French law; they were rescinded by Henry III in 1217 (Shaw 1956). In Scotland, Gilbert (1979) suggests that hunting law was not introduced until the 12th century. David, Earl of Huntingdon and brother of Alexander I, King of Scotland in 1107–24, must have seen forest law in action in Huntingdon, all of which was included in Huntingdon Forest, and which neighboured Rockingham, Whittlewood, Salcey and Rutland Forests as well. When he became King of Scotland himself in 1124, he seems to have introduced the system

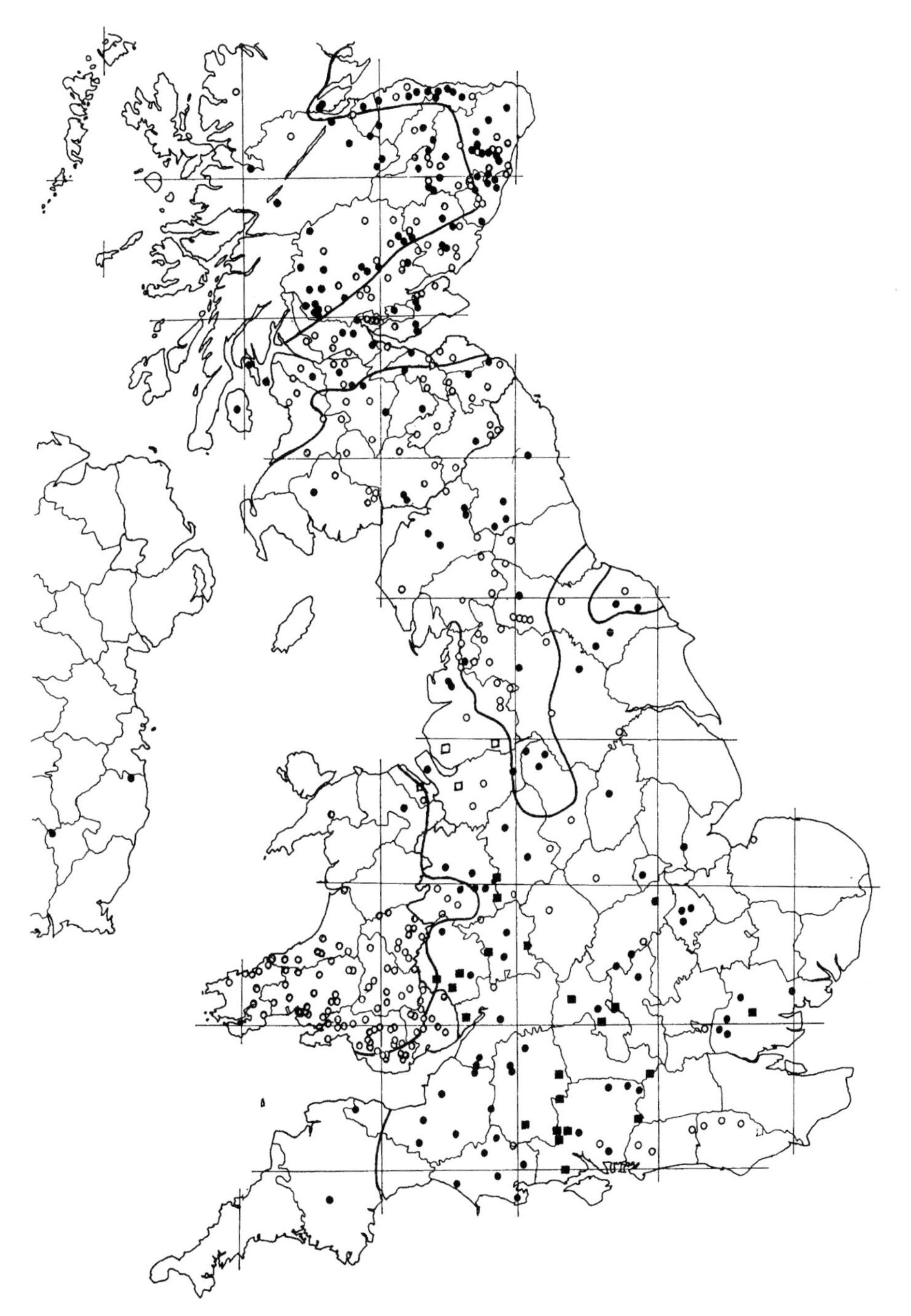

Figure 5.7 The distribution of forests in Domesday and later England (Fig. 6.5 from Rackham 1986 *History of the Countryside*, based on Anderson 1967, Gilbert 1979, Rees 1933, Linnard 1982 and Rackham 1980; by permission of Dr O. Rackham).

there. He was himself a regular huntsman throughout his lifetime, hunting especially in southern Scotland, but he also granted hunting rights to many of his nobles. Gilbert (1979) traces the extension northwards of the forest system over the next four centuries.

Hunting could either involve driving or chasing game. A drive would involve large numbers of beaters driving game towards the huntsmen, and could result in large numbers of deer being killed. Probably this practice explains a somewhat enigmatic entry in some of the Domesday Survey returns, particularly in the western counties, which mention hays. Cheshire seems to be the only county that might have a complete record of these, and they are all included in the returns for woodland, along with hawk's eyries. There are 99 hays listed for Cheshire, and in Kingsley and Weaverham they are specifically referred to as *haii capreolorum*, hays for or of Roe Deer. Kingsley's woodland is reported to have been put into the earl's forest (i.e. Delamere). It seems that these hays were hedged enclosures into which deer were driven as a means of hunting or harvesting them. Their existence implies that habitat for deer in general, Roe Deer in particular, was still abundant at least in places (Yalden 1987). Other counties where hays are mentioned include Shropshire, Hereford, Gloucester, Worcester and Warwick, but the surprising feature is their apparent absence from other counties. Place-names apparently founded on hays occur elsewhere, yet no hays are listed in their Domesday returns.

Chasing game required the use of dogs as well as horses, and implied the targeting of single animals. It was regarded as a purer form of hunting. At least three different techniques were used to pursue deer; stalking, coursing and *par force* hunting (Gilbert 1979). Coursing involved the use of greyhounds with the hunters following on horseback, stalking required instead a quiet pursuit with a scent-following hound, usually by a single huntsman. *Par force* hunting was a complex and ritualized technique, involving pairs of scent-following hounds stationed to hunt in relay along the likely path of the deer. The evidence for this form of hunting actually being practised is thin, and Gilbert (1979) concludes that driving deer was much the most usual form of hunting.

Parks

Another Norman introduction was probably the park, as a piece of private land which was surrounded by a ditch and bank with a fence (pale) in which to enclose semi-wild animals. For the Normans, it was a place to keep deer, as a source of readily available venison. It is possible that parks were already in existence in Anglo-Saxon times, for Rackham (1990) observes that the deerhay at Ongar, Essex, mentioned in a will of 1045, was later the site of the Great Park of Ongar. In Domesday, 35 parks are mentioned, including this one, but there is no mention of Fallow Deer (*contra* Lever 1977); if they refer to animals at all, it is simply as *parcus bestiarum silvaticum*, a park for wild animals. Though the notion of parks as fenced or walled areas to keep in semi-wild animals goes back at least to the Romans, it certainly did not survive in Britain through Saxon times. The animals kept initially in such parks would have been largely Red Deer, perhaps also Roe Deer, Wild Boar and 'wild' cattle, but Roe Deer are poor animals for parks, being territorial and largely confined to woodland. A deer that is more of a communal, herd dwelling, species, and one moreover that grazes as much as it browses, would obviously be much more suitable. This is, of course, exactly what the Normans introduced, in the Fallow Deer, now the most widespread and abundant deer in lowland England. This is smaller than Red Deer, so a smaller park could contain a larger herd of Fallow than Red Deer. (In the modern Tatton Park, which has both species, the Red Deer calves have a poorer survival rate than Fallow fawns, and their adult body weights have suffered more from overcrowding; Birtles 1998). The creation of parks was a popular one in Mediaeval England, and Rackham estimates that there were around 3200 in England by about A.D. 1300, covering perhaps 2% of the country (Fig. 5.8). Like forests, parks often retained their other land-uses, and while usually they contained some woodland, they also usually contained grasslands (launds or lawns).

The Fallow Deer is of somewhat uncertain origin. It occurred in Britain during the Hoxnian Interglacial, at Clacton and Swanscombe, and then again during the Ipswichian Interglacial. In the Post Glacial, there is scant evidence for it even in Mediterranean areas of Europe, and none whatever in north-western Europe (Lister 1984b). There is a record of Fallow Deer from the Mesolithic site of Westward Ho! (Churchill 1965), but no detail is given, and it is not really credible as a record; an alternative identification suggests Roe Deer have been misidentified (Levitan and Locker 1987). Fallow Deer were undoubtedly present in Turkey and Uerpmann (1987) considers *Dama dama* to have been confined to Anatolia, replaced elsewhere in the Middle East by the Persian Fallow Deer *Dama mesopotamica*. Fallow Deer appear in the Neolithic of Greece and Bulgaria. So do sheep, and the possibility that Fallow Deer were also transported by early travellers is very real. In Britain, their introduction has been variously ascribed to the Phoenicians, Gauls, Ancient Britons and, more plausibly, to the Romans. Fitter (1959) toys with the idea that they may even have been native. Lever (1977) confidently asserts that they were well established in Britain in Roman times. There are indeed a few records from Roman and Saxon sites (see above and Table 5.5), but no evidence that they were established here. The *cervi palmati* that surprised spectators to Gordian's amphitheatrical performance in A.D. 238 (Toynbee 1973) can hardly have been the very familiar Fallow Deer, which the Romans referred to as *damma*; it seems likely that they were Elk, but whether any of them really came from Britain may be doubtful. The idea that the 'bucks' mentioned by Archbishop Aelfric's huntsman were Fallow (Lever 1977) ignores the probability that they were Roe; the Saxon *rann* and *raegan*, Roe Buck and Roe Doe, are mistranslated *dammas* and *capreas* (see Wright 1884). Many sites with large samples of Roman bones have no Fallow Deer (e.g. Sheepen, Colchester, Luff 1982; Exeter, Maltby 1979; Caerleon, Hamilton-Dyer 1993), though, as noted in Chapter 4, deer generally are scarce at Roman sites, so Fallow Deer might be missed. At sites where they do occur (such as Iron Age Glastonbury, Roman Abingdon and Wroxeter, Roman and Saxon Porchester Castle, Saxon Hereford and Ipswich; Table 5.5), there are also certain Fallow Deer in the later, Mediaeval, layers, raising the possibility of contamination. It is also possible that salted venison, imported from France or Italy, produced a few genuine Fallow bones of Roman and Saxon date. What the archaeological record certainly shows is an abundance and regularity of Fallow Deer remains in Mediaeval sites. At Launceston Castle, founded in 1067 to suppress Cornish rebellion and yielding bones spanning the following centuries, Fallow Deer are only just

Table 5.5 Archaeological sites for Fallow Deer. The validity of the identifications and/or dating of these records is not endorsed by their inclusion in this table; some of them may well be misidentified bones of other artiodactyls, and others may be later specimens contaminating earlier horizons. It would require a major research programme to disentangle these. However, the record from Westward Ho! is considered by Levitan & Locker (1987) to be Roe Deer. The ^{14}C dated specimen from the Darent gravels emphasizes the difficulties of being certain of dating derived solely from stratigraphical context.

Site	*Grid Ref*	*Age*	*Source*
?Westward Ho!	SS4329	Mesolithic	Churchill 1965
Glastonbury	ST5039	Iron Age	Darvill and Coy 1985
War Ditches, Cambridgeshire	TL4857	Iron Age	
Abingdon	SU4997	Roman	Wilson 1975
Catterick Bridge	SE2299	Roman	Meddens 1990
Wroxeter	SJ5608	Roman	Meddens 1987
Barnsley Park	SP0806	Roman	Webster *et al.* 1985
Portchester Castle	SU6105	Roman	Grant 1976
Ipswich	TM1644	Anglo-Saxon	Jones and Serjeantson 1983
Hereford	SO5140	Anglo-Saxon	Noddle 1985
Ipswich	TM1644	Anglo-Saxon	Crabtree 1994
Portchester Castle	SU6105	Anglo-Saxon	Grant 1976
Cheddar Palace	ST457532	Anglo-Saxon	Higgs *et al.* 1979
Barking Abbey	TQ4384	Anglo-Saxon	Rackham 1994
North Elmham Park	TF9820	Mediaeval	Noddle 1980
Hereford	SO5140	Mediaeval	Noddle 1985
Canterbury	TR1557	Mediaeval	Driver 1990
Faccombe Netherton, Hampshire	SU3857	Norman–Mediaeval	Sadler 1990
Clarendon Palace	ST1730	Norman–Mediaeval	King 1988
Cheddar Palace	ST457532	10–13th C.	Higgs *et al.* 1979
Odiham Castle, Hampshire	SU727521	12th C.	S. Hamilton-Dyer, pers. comm.
London, Baynard's Castle	TQ3280?	14th C.	Armitage 1982
London, Baynard's Castle	TQ3280?	16th C.	Armitage 1982
Aylesbury	SP8113	Mediaeval	Allen and Dalwood 1983
Bath	ST7464	Mediaeval	Cunliffe 1979
Copt Hay, Tetsworth, Oxfordshire	SP6802	Mediaeval	Pernetta 1973
Trowbridge, Wiltshire	ST8557	Mediaeval	Bourdillon 1993
King's Lynn, Norfolk	TF6120	Mediaeval	Clarke and Carter 1977
Coventry	SP3378	Mediaeval	Noddle 1978
Jarrow	NZ3465	Mediaeval	Noddle 1987
York, Petergate	SE6052	Mediaeval	Ryder 1971
Oxford	SP5106	Mediaeval	Wilson *et al.* 1989
Launceston	SX330846	Mediaeval	Albarella and Davis 1996
Exeter	SX9292	Mediaeval	Maltby 1979
Sandal Castle	SE433418	Mediaeval	Mayes and Butler 1983
Okehampton Castle	SX5895	Mediaeval	Maltby 1982
Prudhoe Castle	NZ0962	Mediaeval	Albarella and Davis 1996
King's Langley, Hertfordshire	TL064025	Mediaeval	Locker 1977
Lodge Farm, Kingston Lacey, Dorset	ST974021	Mediaeval	Locker 1994
Bishops Palace, Sonning, Reading	SU7575	Mediaeval	S. Hamilton-Dyer, pers. comm.
Friar Street, Reading	SU7272	Mediaeval	S. Hamilton-Dyer, pers. comm.

Eton, Windsor	SU9678	Mediaeval	S. Hamilton-Dyer, pers. comm.
Sir Ch. Wren's House Hotel, Windsor	SU9676	Mediaeval	S. Hamilton-Dyer, pers. comm.
Broughton Mains, Hampshire	SU3132	Late Mediaeval	S. Hamilton-Dyer, pers. comm.
Darent Gravels, Kent	TQ525550	150 b.p.	Burleigh *et al.* 1982a
Staines, Middlesex	TQ0471	?	Reynolds 1934

Figure 5.8 The distribution of parks in Mediaeval Britain (Fig. 6.1 from Rackham 1986 *History of the Countryside*, based on Cantor 1983, Gilbert 1979, Rees 1933 and Linnard 1982, by permission of Dr. O. Rackham).

represented in the 12th century, but become much more numerous than Red and Roe Deer in the late 13th century, a relationship which persists through to modern times (Albarella and Davis 1996). The pattern at Sandal Castle, near Wakefield, is very similar (Mayes and Butler 1983). Overall at Launceston, there are 439 Fallow Deer bones, mostly 15th century, compared with 49 Red and 53 Roe Deer bones, indicating very clearly its prevalence over the two native species by Mediaeval times. This is a moderately good match to the numbers which Henry III obtained for his feasts; Rackham (1980) calculates an average annual take of 607 Fallow Deer, 159 Red Deer and 45 Roe Deer, as well as 88 Wild Swine, over 6 years between 1231 and 1272. Other sites which span Anglo-Saxon to Norman times similarly show Fallow Deer entering the fauna only in later, Norman, times, for example Southampton (Noddle 1975), London (Armitage 1982), Faccombe Nettleton (Sadler 1990) and Trowbridge (Bourdillon 1993). Later Mediaeval sites regularly contain Fallow Deer bones, for example North Elmham Park (Noddle 1980), Hereford (Noddle 1985), Canterbury (Driver 1990), Aylesbury (Allen and Dalwood 1983) and Coventry (Noddle 1978).

The notion that Fallow Deer spent some time as domesticated animals is supported by the very low level of genetic diversity seen in British and indeed European populations (Pemberton and Smith 1985, Pemberton 1993). This confirms that they have passed through a population bottle-neck. It contrasts with the obvious colour variation (menil, dark, white) that characterizes British park and wild populations, but visual polymorphism of that sort is exactly what humans tend to select, whereas the hidden genetic protein polymorphism is easily lost.

At Launceston, the hind limb bones of Fallow Deer are much more frequent than the rest of the skeleton, implying that haunches of meat, rather than entire deer, were being brought to the Castle. The bias applies also to Red Deer, but not to Roe Deer, so far as the smaller sample sizes allow evaluation. Albarella and Davis (1996) review similar biased representation of hind limbs from a number of other Mediaeval castles, including Barnard Castle (London), Sandal Castle (West Yorkshire), Okehampton (Devon) and Prudhoe Castle (Northumberland), a useful summary because most of these sites have not been fully published. This seems to contrast with the evidence presented by Rackham (1986, Ch. 6 heading) for the orders sent out by Henry III for his Christmas dinner in 1251, which specified the transport of hinds, Roe Deer and Wild Boar to York. Live animals would surely travel better than haunches of meat in Mediaeval England, leaving uncertain the reason for the bias at the castles; perhaps the deer were killed and butchered nearby, but only the best joints were consumed in the castles.

In Scotland, the earliest mention of Fallow Deer is in 1288–1290 (Gilbert 1979). In England, they seem to have been introduced slightly earlier, as might be expected; Chapman and Chapman (1975) review the early records. They misinterpret the *Colloquies of Aelfric* as mentioning the hunting of *dammas* during the 11th century in the Essex forests (see above), but remark the distinction in the New Forest between the *canes damericos currentes* with which William Briwer was allowed to chase bucks in 1223 and the *canes cervericius currentes* mentioned in 1227, which were clearly hounds for chasing Red Deer. They also mention an agreement over hunting in Charnwood Forest and Bradgate Park, Leicestershire, dated 1247, which refers to bucks and does. These could be either Fallow or Roe, but the dates for the hunting season for the bucks, 1 August to 14 September, indicates that they were Fallow Deer, because Roe

Buck, with their different breeding season, could be hunted to 29 September. Thus documentary evidence reinforces the archaeological evidence, indicating that Fallow were a Norman introduction.

The presence of other species in parks is less commonly documented. Wild Boar must have been even more difficult to fence in than Fallow Deer, and they could not have been readily persuaded to imprison themselves by jumping into a park over a deer-leap, which was one way of stocking a deer-park. Harting (1880) implies that parks at Cornbury, Oxfordshire in 1339, Savernake in 1539–1543 and Chartley in 1593 contained Wild Boars, and supposes in part on this evidence that Wild Boars did not become extinct in Britain until the end of the 16th century. Rackham (1986) argues, contrarily, that the last Wild Boar seem to have been in the Forest of Dean and in the Forest of Pickering in Yorkshire during the 13th century. He notes that Henry III had 200 from Dean and 100 from Pickering for his Christmas dinner in 1251; his Queen Eleanor ate 50 sows and 20 boars at Christmas 1253, and the King 80 in 1257. In 1260 Henry III ordered a dozen killed from the Forest of Dean, and Rackham suggests these as the last genuine Wild Boar in Britain. A slightly later record is the 80 taken from Pendle Forest to Pontefract Castle in 1295, which yielded a revenue to the Forest of £3 6s 1d (Shaw 1956). However, Wild Boars were clearly regarded as important beasts of the chase, particularly in Europe, and there must have been a great temptation to reintroduce them as a status symbol.

The white Wild Park Cattle were a similar status symbol, but their origin poses an even greater problem. In size and horn shape, they are not particularly closer to Aurochs, their wild ancestor, than any other domestic cattle, and there seems no reason to suppose that they are in any real sense 'Wild Cattle'. However, they have clearly been enclosed in various parks for many centuries, and their original purpose and status remains uncertain. One plausible, but certainly unproven, suggestion is that they represent the survivors of Roman bulls bred in relation to one of the religious cults for sacrificial purposes. It is hard to see such practices surviving through the Dark Ages, and a Norman influence, originating at about the time that parks were introduced as a concept, seems much more likely. Harting (1880) lists 19 parks which at one time or another had contained 'Wild Cattle', but only six still contained them when he wrote, and only three (Cadzow, Chillingham and Chartley) kept them into recent times (Whitehead 1953, 1963). Some were polled, others horned; some had red-brown ears, others black ears; manes were typical of bulls in some herds, but apparently not others; the park cattle seem a fairly heterogeneous collection. Chartley was described in 1593 as containing Red and Fallow Deer, Wild Swine and 'wild beasts', i.e. cattle. The Chillingham herd may date back as far as 1292 when 'a park with wild animals' is mentioned, but the cattle themselves are first mentioned in 1646 (Hall and Hall 1988).They are probably the only pure-bred herd, the others have been crossed with such breeds as Longhorn and Highland Cattle to maintain their numbers at low points. The Chillingham cattle are very homogeneous genetically, as would be expected in a herd that was reduced to 13 individuals in 1947. They retain the domestic cattle trait of having no fixed breeding season, and calves are born in every month of the year. Conversely, they suffer the natural mortality rate, of around 50% of calves dying in their first year, that wild cattle might have experienced (Hall and Hall 1988). Harting is at pains to indicate that the parks containing 'Wild Cattle' were established on the sites of ancient hunting forests that might have contained genuine wild cattle,

Medieval Boar hunt

but the evidence is not convincing. Wild Cattle, i.e. Aurochsen, were not white, they were much larger than park cattle, and they had surely become extinct in Britain by Roman times. A great deal of romantic speculation inspired much 18th and 19th century writing about the extinct large mammals of Britain, including these.

WARRENS

The Normans seem also to have introduced another important member of the present-day fauna, the Rabbit. The earliest Anglo-Saxon vocabularies do not contain the Latin *cuniculus* or a native equivalent, but by the 15th century it is translated *conynge* or *conninge* (Wright 1884), which became coney; a warren was a coneygarth. The term Rabbit, derived from mediaeval French *rabette*, was used until the 18th century only for the juveniles, which were a culinary delicacy (Tittensor and Tittensor 1985). This implies that Rabbits were not familiar to the early Anglo-Saxon settlers, and certainly matches the fact that warrens are not mentioned in the Domesday Book. The care with which early Rabbits and Rabbit warrens were treated by the Normans, in legislation and in husbandry, matches the implication that they were recently introduced.

The earliest definite mention of a Rabbit warren seems to be that on the Isles of Scilly in 1176 (Sheail 1971). Rabbits were caught by ferrets in 1272, and 2000 skins were exported from Lundy in 1274 (Thompson and Worden 1956). His Christmas dinner in 1251 saw Henry III and his court consume 450 Rabbits (Rackham 1986). In the following centuries, Rabbits appear regularly on the menus of state banquets, and were highly valued; they appeared on the menu in 1309 at Canterbury, at the coronation feast for Henry IV in 1399, and the installation feasts of the Archbishops of Canterbury in 1443 and of York in 1465 (Thompson and Worden 1956).

The early warrens were often installed on off-shore islands, where presumably the Rabbits were safer from both human and animal predators. Lundy, Skokholm, Scilly, the Farnes and the Isle of May are all mentioned as early warrens, and the warren on the Isle of Wight had a custodian in 1225. On the mainland, there was a royal coneygarth at Guildford in 1235. Others were within parks: the king stocked his park at Windsor with Rabbits in 1244 (Tittensor and Tittensor 1985), and by the early 17th century most estates had a functional Rabbit warren. Initially, the right of warren was a right to hunt, and have the exclusive right to hunt, 'lesser game', an ill-defined term which seems to have included Hares, Foxes and Martens (as opposed to 'greater game', deer and boar). In practice, it seems that Rabbit warrens were established on private land without the need for Royal permission. Rabbit warrens were highly institutionalized production units, often in sandy or free-draining land, with artificial pipes or burrows to encourage the Rabbits to breed, and fenced against predators. Often they were established in artificially constructed raised mounds, to ensure the drainage. The earliest example in Scotland seems to have been at Crail in 1264–66; further Royal warrens were reported in the late 13th century at Perth and near Crammond. Ecclesiastical and lay landholders with warrens are reported in the 14th century, and an Act of Parliament of 1503/4 encouraged the construction of warrens, along with parks, dovecots, fish ponds and orchards (Gilbert 1979).

The Rabbit appears not to have been native outside of Iberia and southern France. It was recorded at several Mesolithic sites there (Jarman 1972; see Chapter 3), but seems to have been unknown to the ancient Greeks, though they knew the (Brown) Hare, *lagos*. The Greek historian Polybius is said to make the first mention of Rabbits, in the second century B.C.; referring to Corsica, he says that there were no hares, but burrowing animals called *kunikloi* which resembled small hares. Roman writers describe the Rabbit, *cuniculus*, as common in Spain, and it seems to have been imported to Italy before A.D. 230, when it was well established. The Romans apparently domesticated Rabbits, and kept them in walled enclosures - *leporaria* (Thompson and Worden 1956). The subsequent history of Rabbits in Europe, and their precise method of introduction to Britain, is uncertain, but the Norman Kingdom in Sicily would have provided likely contact with and sources of Rabbits.

The archaeological record is of course confused, even wrecked, by the distinguishing, and distinguished, ability of Rabbits to burrow. Their unfortunate appearance in the supposed Mesolithic layers at Thatcham (Chapter 3) is only the most dramatic example. They are reported for example in 'Bronze Age' deposits at Brean Down (Levitan 1990) and at Anglo-Saxon West Stow (Crabtree 1989a), as well as in many other cave and open sites. In many cases, the excavators dismiss these as obvious contamination; at Brean Down the burrows of Rabbits penetrating the Bronze Age deposits were pointedly mentioned. Occasionally, the reverse is carefully noted; Bourdillon and Coy (1980) specifically checked that there were no Rabbit bones among the 45 704 Anglo-Saxon bones identified from Southampton (*Hamwic*). Several other carefully excavated Anglo-Saxon sites with large samples of bones are similarly devoid of Rabbit, including middle Saxon faunas from Maiden Lane and Jubilee Hall, London (West 1993) and North Elmham Park (Noddle 1980), Hereford (Noddle 1985), Ipswich (Jones and Serjeantson 1983) and late Saxon Canterbury (Driver 1990). On the other hand, Rabbit bones do appear in the Mediaeval levels of many sites, including some of those where they are absent from earlier levels (e.g. Southampton (Noddle 1975), North Elmham Park, Hereford, Canterbury). At Launceston, Rabbit bones are absent from the earliest levels, but appear in the late 13th century deposits, perhaps 150 years later than Fallow Deer there (Albarella and Davis 1996). Allowing for the undoubted difficulty of interpreting the archaeological record of a species that is well able to conduct its own archaeological excavations, the documentary and archaeological records seem to be telling an increasingly coherent story; though Rabbits were known to the Romans in Spain, and taken to Italy from there, they did not bring them to Britain, the Anglo-Saxon settlers did not know them, and it was left to the Normans in the late 11th century to introduce them. Like so many other of their endeavours, they evidently did it very well.

Fitter (1959) remarks on a peculiarity of this story, that the Iberian race of the Rabbit, *Oryctolagus cuniculus huxleyi*, is distinguished taxonomically from the nominate race, *O. c. cuniculus*, elsewhere. They are not profoundly different; *huxleyi* is smaller (skull length 71–77.6 mm, compared with 78–82.4 mm in *cuniculus*; Miller 1912) but with proportionately longer ears. If the nominate race is everywhere a Norman/post-Norman introduction, it has probably acquired its distinction through being kept in captivity, and the 'typical' wild Rabbit is actually a feral domesticate. The genetic basis of the differences would be an interesting study.

A consequence of the introduction and exploitation of the Rabbit must have been the similar exploitation of the Ferret, for this was regularly used for obtaining Rabbits; there is a good illustration of a Ferret being put into a warren in the Luttrel Psalter of around 1343 (Backhouse 1989). The origin and time of domestication of the Ferret remain controversial; is it simply a domesticated Polecat *Mustela putorius*, or perhaps a domesticated Steppe Polecat *M. eversmanni*? It has regularly been claimed that it was tamed in North Africa and brought to Spain, but there is considerable doubt that Polecats ever occurred in North Africa. Given the early exploitation of Rabbits in Spain, it seems very probable that the Polecat was domesticated there into the Ferret, and then moved round Europe with its equally domesticated prey. There are bones of probable Ferrets in Mediaeval and 17th century Kingston Lacy (Locker 1994), as well as a record of a Ferret purchased there in 1391–92.

Miscellaneous Gleanings

The Domesday Book contains a few other snippets of information that merit notice. In the account for Norwich, the annual tribute in King Edward's time of a Bear for baiting has been mentioned (Chapter 4). In the account for Cheshire, it is recorded that the county town had to pay a tax of £45 and three timbers of marten skins – a timber was a bulk quantity of 40–60 skins (Yalden 1987). These could have been part of an import–export trade, not necessarily native Pine Martens, but it would fit the notion that the east of the county was still well-wooded at this time, and match the earlier record from Scandinavian York of bones bearing skinning marks (O'Connor 1991)

The only convincing descriptions of Beavers in Britain date to this period (Barrett-Hamilton and Hinton 1921).The Laws of Hywel Dda, supposedly dated to A.D. 940, state that 'the King is to have the worth of Beavers, Martens and Ermines in whatsoever spot they are killed, because from them the borders of the King's garments are made'. At that time, while an Otter was worth 12 pence and a Marten 24 pence, a Beaver was worth 120 pence. Clearly, they were already rare. Giraldus Cambrensis, alias Gerald of Barri, toured Wales in 1188 with the Archbishop of Canterbury, Baldwin. He gives a wonderfully confused account of Beaver natural history, some of which is apparently first-hand observation and some of which comes from the bestiaries. His most important observation is that 'The Teivi has another singular particularity, being the only river in Wales, or even in England, which has beavers; in Scotland they are said to be found in one river but are very scarce'. The detail with which he describes Beavers and their habits suggests that he either saw them himself, or at least spoke to local hunters who knew the habits of the animal well, for instance the description of them eluding hunters by disturbing the mud at the burrow entrance as they fled. On the other hand, he repeats the old bestiary tales of them castrating themselves to indicate to pursuing hunters that they are no longer worth pursuit, and the story of them grasping timber while being pulled along by a colleague. His account appears to have been picked up and recited by several later authors, who thereby imply that Beavers still occurred on the Teifi when they were writing. The export duties of David I, King of Scotland in 1124–53, mention skins of *Beveris*, but a similar Act of 1424 omits them, strongly suggesting that they had become extinct, at least commercially, between these dates. However, Boethius in 1526 wrote that Beavers abounded in Loch Ness, and the statement was repeated by his translator in 1536. There is supposed to be a traditional Gaelic name for the animal, *losleathan* (Broad-tail) or *dobrhan losleathan* (Broad-tailed Otter). The Welsh name in the Laws of Hywel Dda is *llost lydan*, clearly the same root. This suggests that *afanc*, discussed by Barrett-Hamilton and Hinton (1921) and more recently by Aybes and Yalden (1995) (see above) is not another name for Beaver, but in fact a mythical monster, a suggestion supported by a footnote in Barrett-Hamilton and Hinton. There seem to be no place-names, in Wales or Scotland, that are derived from *losleathan*.

Another rodent which had a profound effect on humans during Mediaeval times deserves notice. The Black Rat carries a flea *Xenopsylla cheopis* which is in turn the main vector for bubonic plague *Pasteurella pestis*. Though Black Rats apparently died out in early Anglo-Saxon times, they were back, for instance at York, by the end of the 9th century, and occur at most Mediaeval sites that have a small mammal fauna (Table 4.4). The Black Death which swept through Europe in 1347, and ravaged Britain in 1348–1349, is believed to have reduced the population of the British Isles from 5 millions to 3.5 millions, and it took 150 years for it to recover (McEvedy and Jones 1978). The better-known Great Plague of London, in 1665, was rather less severe, but still killed some 68 596 people in London alone (Twigg 1978).

Archaeological reports from Mediaeval sites do not add much to our general knowledge of the British mammal fauna in this period, except perhaps for a couple of interesting small mammal faunas reported from Mediaeval wells (Table 4.4). In Middleton Stoney, Oxfordshire, a latrine shaft in the castle dated prior to A.D. 1225 yielded 3283 bone fragments of small mammals and amphibians, though only 939 of these were identifiable to species. They included large numbers of House Mouse, Wood Mouse, Field Vole and Common Shrew, along with smaller numbers of Black Rat, Water Vole, Bank Vole, Pigmy Shrew and Water Shrew (Levitan 1984). Ecologically this is an odd mixture, and it is difficult to see how it came about. A commensal Barn Owl living in or around the top of the shaft seems to offer a more probable source than that such a strange mixture of commensal and wild rodents should have fallen down a virtual pitfall trap, for Barn Owls do take larger species such as Black Rats and Water Voles, and they will take large proportions of commensal rodents when they are available. Even then, the environment suggested for around the castle is an unusual one, for Wood Mice are rarely more available to Barn Owls than Field Voles. Perhaps stored grain attracted mice in large numbers. The site of Greyfriars in the City of London also provided a large sample of small mammals, from a well shaft dating to about 1480–1500. Among 2911 identified bones were representatives of 64 individuals of 12 species. The fauna was dominated by mice (17 *Mus* and 14 *Apodemus*, including one possible *A. flavicollis*); shrews were also numerous, all three species being present. There was only one Water Vole, along with three Field Voles and two Bank Voles. The fauna was completed by six Black Rats, a Stoat and a Hedgehog (Armitage and West 1985). Again, this is a mouse-dominated fauna, rather than a vole-dominated fauna as usually results from owl-pellets, and perhaps indicates the attraction of stored grain in the general area. This well does seem to have functioned as a pitfall trap.

Conclusion

The Anglo-Saxon immigrants found a farmed landscape, dominated by cattle and sheep, and though they cleared further woodland, they seem largely to have continued the farming activity of their predecessors. Wild mammals were not particularly important to them, though they hunted deer a little, and certainly noticed the appearance of Wolves in the countryside. The Norman invaders likewise continued farming, but were considerably more responsive to the game species, which they protected in forests, and enclosed in parks. It is doubtful that the wild species were any more important overall to the economy of the population in general; even in the elite establishments, domestic ungulates remained the main source of meat. However, the enactment of hunting legislation and the introduction of two important extra species, Fallow Deer and Rabbit, signals the changed perception of game, and portends the game legislation of the present day. It also perhaps marks the start of state-sponsored pest control, for the Wolf was clearly regarded as vermin, to be hunted and killed at any time. These are attitudes that continued to shape the ways that wild mammals (and birds) were treated, legally and socially, for most of the remainder of the millennium. Only in its last 75 years has a different, conservation-based, view become more widely espoused. The effects of these polarized views on the mammal fauna over the 400 years or so from Mediaeval time to the beginning of this century constitute the subjects for the next two chapters.

Chapter 6

Hunting and Harassment

Keeper's gibbet – Pine Marten, Polecat, Stoat and Weasel

The introduction of forest law by the Normans marked the introduction of the idea that some wild animals – game – might belong to someone, and the introduction of the crime of poaching, a concept that presumably could not have existed under Anglo-Saxon law. These concepts became increasingly important legal ideas during the following centuries. The forests were, however, increasingly unpopular with other landowners, and many were deforested during Tudor and Stuart times. With the loss of forest status, the deer were increasingly regarded not as valuable game, but as pests raiding the adjoining crops. They declined sharply. Woodland too was under increasing pressure, and with its decline, woodland species, including deer, became less common. Finally, commonland was seen as poorly used by agricultural improvers, so that common rights were increasingly bought out or legally extinguished, leading to further losses of open countryside. Privately owned land could be enclosed, but, more importantly, game animals became by enclosure the property of the new landowners, to be protected both against human poachers and wild predators. The period from Tudor times through to the early years of the 20th century thus saw wild places and wildlife under increasingly severe pressure. At least two more species became extinct during this period, and several species declined severely.

Extinction of the Natives

Of the 33 or so native terrestrial mammals inhabiting Britain at the start of the Post Glacial period, three, the Reindeer, Tarpan (Wild Horse) and Northern Vole seem to have died out within about 1000 years, presumably as a result of the climatic change and the change to more wooded conditions that accompanied it. Two more, the Elk and Lynx, were always assumed to have died out during Mesolithic times, that is before 5500 b.p., but radiocarbon dates suggest that they in fact survived to about 4000 and 1700 b.p. respectively, that is to Neolithic and Roman times (Chapter 3). The Aurochs and Brown Bear seem to have died out by Roman times, so far as the evidence shows at present, though the Bear might have survived a little later than this in Scotland (Chapter 4). The Beaver certainly survived into the Mediaeval period, but seems to have gone from Wales by 1200 and from Scotland by 1600 (Chapter 5). It is well known that two other species of large mammal were lost in this current millenium, Wild Boar and Wolf, but their fate is obscured by a mixture of legend and misinterpretation.

The Wild Boar (perhaps more accurately Wild Swine, but I find that term unfamiliar) is an inhabitant of deciduous woodland across a wide sweep of the Palaearctic, from France to Siberia and from the Mediterranean coasts of Spain and Italy northwards to southern Scandinavia. In much of Europe, it is still regarded as a major game animal, and from a European population of 0.5 million, a harvest of 0.4 million can be taken each year (Myrberget 1990). This substantial toll is, of course, only possible because of the large litters that sows produce, typically 4–6 young. The same productivity predisposed the species to domestication, along with its varied diet: acorns, beech mast and other seeds, certainly, but also roots and rhizomes, carrion, insects, and a great deal of grass in summer (Genov 1981, Groot Bruinderink *et al.* 1994). It has one definite limitation, a sensitivity to winter cold, especially snow cover. It had a sharp limit to its distribution in Russia at 60°N, though it has spread northwards into Finland during the 1980s as a consequence of a series of milder, less snowy, winters (Saez-Royuela and Telleria 1986). In Białowieża, though Wild Boar are relatively immune from predation by Wolves, they suffer from severe winters. Of 762 carcases found, 16% had been killed by Wolves, but 61% had died of disease and starvation, with snow cover and the acorn crop explaining most of the variation in death rate

from year to year (Okarma *et al.* 1995). In Norman England, the Wild Boar was clearly seen as a major game species, treated legally like a deer (Chapter 5), and was also enclosed in various parks. There were also numerous attempts to reintroduce it to parks that had formerly contained it, leading to a great deal of confusion about the date of its extinction. Rackham (1980, 1986) suggests that the Forests of Dean and Pickering were the last to house genuine native Wild Boar. There were records of them being supplied from Dean to the court of Henry III up to 1260, but none thereafter. A slightly later record of them being taken from Pendle Forest in 1295 has been noted (Chapter 5). However, there are records of Wild Boar in Chartley Park,

Wolves at a Sheep-fold

Stafford, in 1593 and 1683, in Savernake Forest in 1539 and 1543, and at Windsor in 1617 (Harting 1880). Contrarily, Rackham (1986) notes that there is no mention of Wild Boar in Savernake in Henry III's reign, and concludes that they must have been reintroduced. Rackham (1980) notes also that Chalkney Wood, Essex, was maintained as a park for hunting Wild Boar by the Earls of Oxford, the De Veres (who fancied that their family name related to the Latin *verres*, a boar), until Henry VIII's reign, when they were destroyed by the then Earl because of the damage they caused locally. This sounds very like the story related by Gilbert White (1788, Letter IX), that General Howe had turned Wild Boar out in the forests of Alice Holt and Woolmer, but that 'the country rose upon them and destroyed them'. The introduction of Wild Boar by James I from France in 1608, and from Germany to Windsor in 1611, is documented (MacGregor 1989); the latter were the ones recorded as being hunted there in 1617. His son Charles I similarly attempted to reintroduce Wild Boar from Germany to the New Forest, and again they caused noticeable damage to local agriculture; the intervention of the Civil War ended the experiment (Fitter 1959). In Scotland, Ritchie (1920) notes the Sheriff of Forfar witnessing the payment in 1263 for 4½ chalders of corn for the Wild Boar, which suggests that they were perhaps captive, or at least required support. Ritchie suggests that Wild Boar were extinct in Scotland by the early years of the 17th century, closely akin to Harting's suggestion that it died out in England also at the end of the 17th century. These dates ignore the likelihood that captive Wild Boar were maintained in several places, as status symbols and as meat on the hoof, long after they had in fact become extinct as wild animals, probably around the end of the 13th century.

The Wolf has an equally mythical end. Under Henry III (1216–72), the Engaine family held their lands at Laxton Pytchley in Northamptonshire by virtue of exterminating all the Wolves in the counties of Northampton, Huntingdon, Oxford and Buckingham, while in 1281 Peter Corbet was commissioned to destroy all the Wolves in the counties of Gloucester, Hereford, Shropshire and Staffordshire. The Wolf was vermin throughout, never a beast of the chase, and while under Henry II (1154–89) a head was worth only a few pence, by the time of John (1199–1216) the reward was 5 shillings; this implies that Wolves were becoming rare. There is reliable evidence of Wolves in northern Lancashire sometime later; in the Royal Forest of Lancaster, a man was paid 1s 2d to guard the calves against the Wolves in 1295–96, while seven calves were killed that year and a further eight cattle were killed by them in 1304–1305 (Tupling 1927, Shaw 1956). These may be the latest reliable records from England, but in Scotland the species certainly lasted much longer. Under James I, a law to enforce Wolf hunting, requiring attendance at three hunts during the cubbing season, was passed in 1427, and this continued at least until 1457. However, Wolves remained common into Mary's reign (1547–87) and the queen herself hunted Wolves in the Forest of Atholl in 1563 (Millais 1904). Deforestation seems to have expedited the destruction of Wolves (and their prey) during the 16th century (Ritchie 1920), and in 1684, 20 years after the last mention of them, Sir Robert Sibbald thought them extinct. Perhaps they survived a little later. One claimant for the last Wolf in Scotland was that killed by Sir Ewan Cameron of Lochiel at Killiekrankie, Perthshire, in 1680, and other plausible stories concern Wolves killed in Sutherland between 1690 and 1700. The even later date of 1743 attached to a large black Wolf killed in Findhorn by MacQueen of Pall-a-chrocain sounds like a fairy story; it had killed two children, who with their mother had been crossing the hills, not probable behaviour for a real Wolf (unless rabid). The story was not recounted until sometime in the following century.

A third extinction of a large mammal during this period has been largely overlooked. The hunting of large whales or at least the exploitation of stranded carcases has a long history. Clark (1947) reviewed the occurrence of cetacean remains in 50 Scottish archaeological sites, many of them fragmentary tools made from bone that cannot be identified to species. Of 16 whale skeletons found in the carse clays which extend far inland up the Firth of Forth, four had implements of deer antler associated with them, indicating that they had been exploited. However, one was a Blue Whale *Balaenoptera musculus* and another an unspecified rorqual, surely stranded rather than hunted whales. Clark remarks on the Basque hunting of Right Whales *Eubalaena glacialis,* which perhaps began as far back as the 10th century, and was at its peak in the 12th and 13th centuries. This fishery exploited three important features of these whales: they migrate in coastal waters, are relatively slow swimmers, and float when killed. They bred in the estuaries of the Bay of Biscay, and watch towers along the shore line were used to spot their arrival. Aguilar (1986) remarks that the calves were particularly targeted, for they were easier to kill, and their mother would attempt to protect her calf, making her also vulnerable. There is another species of whale that shows similar characteristics, the Californian Gray Whale. This is familiar as a migrant up the coast of California, breeding in winter in the warm shallow waters of Baja California and feeding in summer in the cold waters of the Bering Sea. It was originally named *Rhachianectes glaucus* in 1869, but fossil remains from the north-eastern Atlantic had already been named *Eschrichtius robustus* in 1866. It was another 15 years before the similarity of the Atlantic and Pacific species was remarked, and the middle of this century before their identity was accepted. Thus fragmentary sub-fossil remains of Atlantic specimens are the type material for a living Pacific species (Bryant 1995). Among these fragmentary Atlantic remains are one from Pentuan (Pentewan), Cornwall, with a ^{14}C date of 1329 b.p. and one from Babbacombe Bay, Devon, dated to 340 b.p. (A.D. 1610); these two are the most recent of the seven eastern Atlantic specimens (Fig. 6.1) which range back to 8330 b.p. (Table 6.1). A similar small population in the western Atlantic died out at about the same time (Mead and Mitchell 1984). Obviously, the species did survive into the 17th century, into the period of commercial whaling, and there are a few somewhat uncertain descriptions from early whalers also that seem to refer to this species. It is hard to escape the conclusion that, as nearly happened to the more numerous Right Whale, it was driven to extinction by human hunting. Targeting of calves and their mothers would be a certain way of exterminating a rare species, and it is evident from the descriptions of hunting in the Pacific that young Gray Whales were a particular target there.

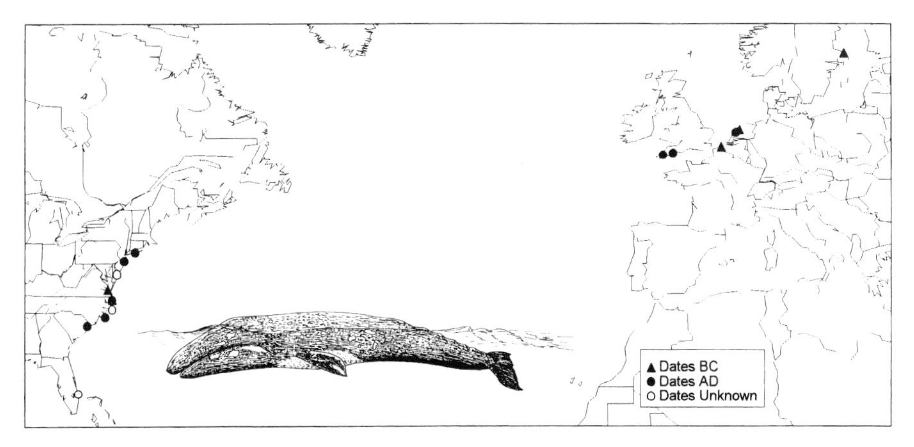

Figure 6.1 Distribution of sub-fossil remains of the Gray Whale *Eschrichtius robustus* in the North Atlantic (from Mead and Mitchell 1984, Bryant 1995).

Table 6.1. Occurrences of the locally extinct Gray Whale *Eschrichtius robustus* in the eastern and western North Atlantic (from Mead and Mitchell 1984, Bryant 1995).

Site	*Grid Ref.*	*Date*
Eastern Atlantic		
Ijmuiden, Netherlands	52°28′ N, 04°38′E	8330 b.p.
Graso, Sweden	60°22′ N, 18°30′E	4395 b.p.
Wieringermeer-polder, Netherlands	52°47′ N, 05°05′E	4195 b.p.
Oostduinkerke-Koksijde, Belgium	51°07′ N, 02°41′E	2024 b.p.
Ijmuiden, Netherlands	52°28′ N, 04°38′E	1730 b.p.
Pentewan, St. Austell, Cornwall	50°17′ N, 04°47′W	1329 b.p.
Babbacombe Bay, Devon	50°30′ N, 03°30′W	340 b.p.
Western Atlantic		
Chesapeake Bay, Virginia	37°10′ N, 76°20′W	10 140 b.p.
Corolla, North Carolina	36°15′ N, 75°45′W	2415 b.p.
Cape Lookout, North Carolina	34°30′ N, 76°40′W	1190 b.p.
Myrtle Beach, South Carolina	33°42′ N, 78°54′W	865 b.p.
Nags Head, North Carolina	36°10′ N, 75°45′W	865 b.p.
Toms River, New Jersey	39°57′ N, 74°12′W	455 b.p.
Southampton, New York	40°50′ N, 72°50′W	275 b.p.
Corolla, North Carolina	35°15′ N, 75°45′W	?
Rehoboth, Delaware	38°42′ N, 75°06′W	?
Jupiter Island, Florida	27°03′ N, 80°06′W	?

The status of seals also seems to have declined sharply during this period, although their record is poor. Hunting of seals has a similarly long archaeological history; there are records of both Grey and Common Seals in the Mesolithic sites on Oronsay (Grigson and Mellars 1987), and at Jarlshof on Shetland both those seals are present in Bronze Age to Viking age deposits; a few bones of Walrus were unearthed in the Late Bronze Age layers as well (Platt 1956). Grey Seals, which spend an extended period on land during the autumn breeding season, are particularly vulnerable to hunting, and their rookeries are all on remote cliff-bound beaches or uninhabited islands. More importantly, they colonized new islands when these were deserted by their human inhabitants. Thus North Rona, now home to the largest British colony, was deserted in 1844, and there were enough seals breeding by 1880 to invite exploitation (Boyd and Boyd 1990). The Outer Farne Isles which now host the breeding rookeries were abandoned by the lighthouse keepers in 1827; they were actively protected, as a nature reserve, from the 1860s (Hickling 1962). St Kilda was abandoned in 1930, and now produces about 100 pups a year. The Monach Isles were colonized after their abandonment in 1944, and now produce 3000 pups a year. Despite their preference for remote sites, Grey Seals were severely exploited throughout the historical period, and in 1914 it was argued that the total population was only 500 animals. It is very unlikely that the population was really quite so low, but it must indeed have been scarce, and legal protection was offered by the passing of The Grey Seals Protection Act of 1914, giving a close season from 1st October to 15th November. This was the first protection law for any British mammal.

DEFORESTATION AND THE DECLINE OF DEER

As the Crown lost interest in deer, so forests were deforested: sold to private land-owners or divided up by the Enclosure Acts. The ownership of Hatfield Forest was given away in 1238, and the forestal rights therein given to the Duke of Buckingham as early as 1446. Queen Elizabeth restored some interest in the forests as a source of timber, and Peak Forest saw the herd of surviving Red Deer build up from 30 in 1579 to over 120 in 1586 (Shimwell 1977). The Stuarts resumed some interest in hunting, both under James I and Charles I, the latter seeing a chance to create more income from enforcing or selling the forest rights. The attempt to reintroduce Wild Boar to the New Forest seems to have been part of this rekindled interest. However, increasingly the Crown's concern turned from deer to trees, oaks for naval shipping, and large areas of Dean, the New Forest, Alice Holt and Bere (Hampshire) were enclosed to be planted with oak. This was bound to reduce if not eliminate the deer. In the New Forest, the Deer Removal Act of 1851 stated its intention very clearly. Other forests were lost to the Enclosure Acts – Enfield Chase (Essex) in 1777, Needwood (Staffs) in 1801, Windsor in 1817, Neroche (Somerset) in 1830, Hainault (Essex) in 1851, and Wychwood (Oxford) in 1857 (Rackham 1980). In the New Forest, a census in 1670 suggested 7593 Fallow Deer and 375 Red Deer; these declined through the early years of the 19th century, and then were virtually exterminated during the three years following the 1851 Act (Tubbs 1986). The Fallow Deer did not regain 200 until around 1900, though they are now back to about 1000, a number maintained by culling. These details emphasize how vulnerable were deer to poaching, or, conversely, how much their survival depended on protection for organized hunting. Even in Henry III's time, Red and Roe Deer were scarce in southern England, and were culled mostly from his northern forests. By Tudor times, they were virtually extinct outside of protected areas. Gilbert White (1788) bemoaned the poaching of the Fallow Deer in Alice Holt and Red Deer in Woolmer Forest in the 1760s, despite the efforts of the keepers to protect them.

Roe Deer were apparently downgraded from being beast of chase to beasts of the warren in 1338, reputedly because they chased away other deer (Whitehead 1964). This meant that they no longer had the full protection of forest law, and could be hunted by private land-owners. Though there are numerous mentions of Roe in Mediaeval accounts, mostly of offences against the King's venison (for example, Pickering Forest Yorkshire in 1334; Cannock Chase, Staffordshire 1530; Forest of Dean 1282; Dorset in 1257), Roe Deer are believed to have become extinct in England by the end of the 18th century (Whitehead 1964, Taylor Page 1962). The last may have been that killed near Hexham, Northumberland, in George I's reign (1714–27). Although they may have survived in some of the wilder areas along the Scottish borders, as argued by Whitehead (1964), the evidence is weak, and the rarity, at least, of Roe is clear. The archaeological record of both Roe and Red Deer confirms their scarcity, at sites such as Okehampton (Maltby 1982) and Launceston (Albarella and Davis 1996) Castles, in the later Mediaeval period. As early as 1618, and again in 1633, Roe Deer were taken from Cumberland to London, which clearly indicates their absence from southern England. In Scotland, similarly, they declined to very low numbers by the same time; Ritchie (1920) suggests that they died out in the lowlands by the end of the 17th century, but planting of new woodlands in the latter half of the 18th century resulted in their recovery. They reappeared in the lower Tay valley toward the end of that century, in Clackmannanshire, Stirlingshire and Fife by 1828, and in 1840–45 penetrated to the south side of the Central Lowlands. This record of their return to southern Scotland makes it less likely that they could have survived in the English–Scottish border counties during the 17th and 18th centuries. Their return to England was the consequence of reintroductions: most importantly, from Perth to Dorset, at Milton Abbas, in 1800, and from Wurtenburg, Germany, to Norfolk in 1884 (Hewison 1995). They are still spreading out from these sites, and from the north into northern England.

The Red Deer fared little better. It is claimed that those of Exmoor represent survivors of the native Red Deer of the West Country, and more plausibly that those of the Martindale area of the Lake District are similarly descendants of native stock. Lowe and Gardiner (1974) analysed the skull shapes of British Red Deer, a good indication of relationships, and concluded that only the Lake District and Highland deer were native stock, the rest being various mixtures with introduced continental deer. Elsewhere, the species certainly became extinct in England, outside of deer parks, and died out in lowland Scotland too. Between 1750 and 1800, the human population of the Highlands increased rapidly, and realization of the ability of upland breeds of sheep (Blackface and Cheviot) to survive there led to the elimination of deer as competitors. It is thought that only six deer forests retained substantial populations of Red Deer in 1811 (Atholl, Black Mount, Glenartney, Glen Fidich, Invercauld and Mar). Paradoxically, they were saved by the rising interest in stalking, which made it worthwhile for some landowners to lease the sporting, rather than rear sheep, so that by the 1838, there were 45 deer forests. With the royal seal of approval to deer stalking given by Queen Victoria's purchase of the Balmoral Estate in 1852 and a fall in the price of wool in the 1870s, deer forests expanded further, from near 1 million hectares in 1883 to nearly 2 million by 1912 (Clutton-Brock and Albon 1989).

Deer suffered from both poaching and destruction of their woodland habitat. The Red Squirrel also suffered directly from the latter. The pollen record suggests that tree cover reached its minimum around 1700, a state which documentary evidence confirms. As Rackham (1990) puts it, the 18th century was an age of much tree-destruction. However, the modern type of forestry, planting large areas of trees, usually conifers, also began in the 18th century, with small-scale plantations of mostly coppice (deciduous) woodland. Thus woodland mammals were saved in time by the provision of new habitat. The Red Squirrel was the species most nearly lost. It appears to have been exterminated from most of Scotland. The last in Sutherland was recorded on 1630, in Moray in 1775, Ross and Cromarty in 1792 and Dumbarton in 1776. They were absent from Angus in 1813, Aberdeenshire in 1843, and Argyll in 1842 (Shorten 1954, Ritchie 1920). By this time, however, they were already being reintroduced, notably to Dalkeith, Midlothian in 1772 and Dunkeld, Perth in 1793 (Fig. 6.2). English squirrels were released at Dalkeith, but those at Dunkeld are believed to have come from Scandinavia. The present Scottish population derives largely from these and similar later reintroductions, though the possibility that they were reinforcing a few native survivors in Speyside and south-eastern Sutherland cannot be ruled out. The fact that landowners felt it worthwhile to reintroduce them at all emphasizes their scarcity.

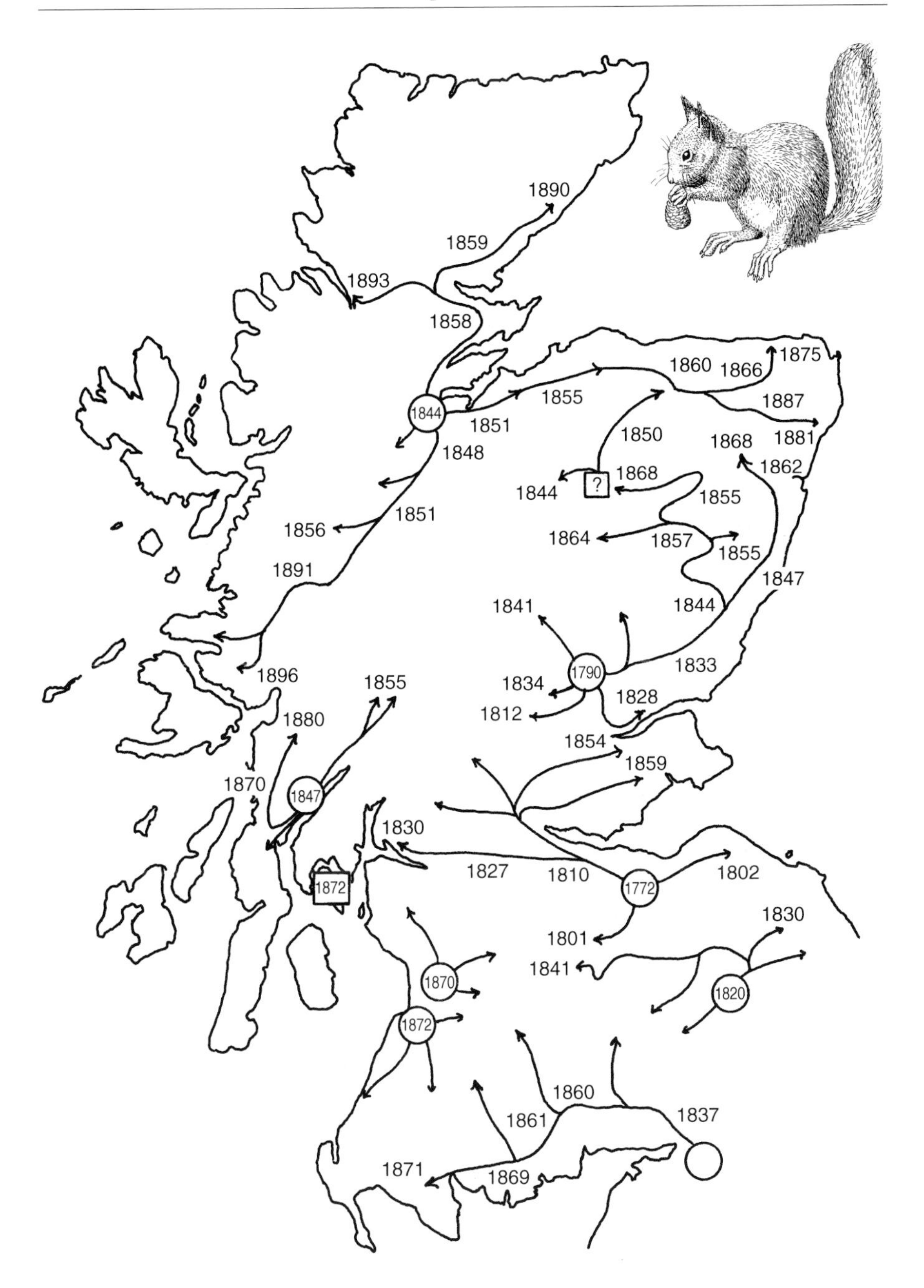

Figure 6.2 Reintroduction of the Red Squirrel *Sciurus vulgaris* into Scotland (after Ritchie 1920). The most important reintroductions were to Dalkeith in 1772 and Dunkeld in 1790, but there was also natural immigration from England into south-eastern Scotland in 1834. Dates of first appearance at various locations and suggested routes of spread are indicated.

Decline of the Predators

The protection offered to game species in hunting areas undoubtedly saved the deer from total extermination in Britain. The status of some of the carnivores may have saved them too. The Red Fox has always been an equivocal beast. For farmers it has always been a menace to poultry, and while its threat to lambs has always been greatly overestimated (Hewson 1984, Rowley 1970), it has certainly been persecuted with vigour in upland sheep-rearing areas. These prejudices were willingly adopted by the new gamekeeping profession in the 19th century, which saw any predator as a threat to the carefully husbanded Partridge and Pheasant. Nor were they mistaken. The Red Fox can have a demonstrable impact on the autumn harvest of gamebirds, because it tends to take the sitting hens and their eggs, just the stage of the life-cycle that needs to be protected (Tapper *et al.* 1996). So why was the Fox not wiped out in England, at least, as were several other predators? In some places it was; the Fox remained scarce in East Anglia until the 1960s, and has only recolonized within the last 30 years or so, having been exterminated last century (Tapper 1992). What saved it in many places is that it was itself the quarry of formalized hunting. In these hunt territories, coverts were created to give sanctuaries, and hedgerows were maintained to create interesting challenges to the huntsmen (as well as foraging areas for the Foxes). There are claims that Foxes were even imported from Europe, and were moved around the country from estates where they were 'vermin' to hunt territories.

The Otter similarly benefited from its status as the quarry of the Otter Hunts across most of England, Wales and lowland Scotland, balancing its pest status as perceived by game anglers, particularly in the Highlands. Prior to the development of pack-hunting, Otters were killed as part of the general war against predators, for their fur, and for the protection of fish ponds in particular (Howes 1976). At Arksey, between one and five Otters a year were killed through the 18th century, for a bounty of one shilling; in Doncaster, the payment in 1619 had been 6d, compared with 2d per Weasel and 4d per Polecat, so the Otter was always rarer and a more profitable target. However, the pelt was worth as much as one guinea, an even more tempting reason for killing one. Otter packs came into being by 1796. There were 23 Otter Hunts at their peak in the 1920s, and they killed 434 Otters in 1933, their highest toll (Jefferies 1989). Packs of otterhounds were extravagant status symbols, even more than fox-hounds, and depended on support not only by the members but also from spectators. Laidler (1982) remarks that a hunt required £600 a year to maintain the staff and the pack of hounds in the 1920s, equivalent to about £40 000 now. She deplores, understandably, the cruelty that was involved in hunting Otters, but the hunts did attempt a crude conservation policy; when Otters were scarce, they attempted to hunt but not kill their quarry, tried hunting in other catchments (even going to Ireland for a change of venue), and also maintained the habitat, including both cover and food, to ensure their next sport (Jefferies 1989). The fact that they were active undoubtedly reduced the other forms of persecution in their catchments.

The other predators, avian as well as mammalian, fared badly. Two important developments, one technical and the other social, combined to bring this about. The development of the shot gun, particularly the replacement of flint-locks by percussion caps in the 1830s and the replacement of muzzle-loading by breech-loading in the 1860s, provided a weapon which was sufficiently manageable to make driven gameshooting, an economical method of harvesting, realistically possible. The agricultural pressures which saw the open fields and common lands taken into ownership by the 2500 or so enclosure acts passed between 1750 and 1850 also gave the new landowners the rights to the game on their land. This in turn led to the development of the sporting estate, and the new profession of gamekeeper (Tapper 1992). There were about 17 000 gamekeepers by the 1871 census, rising to a maximum of 23 056 in 1911 (by 1981, there were only about 2500). They provided a coverage across the whole of England exceeding 0.4 keepers per 10 km^2, and over most of Scotland and Wales too (Fig. 6.3). In Norfolk, one of the most densely keepered counties, there was one keeper for every 444 ha (1100 acres); clearly the whole county was managed.

There were other technical and social factors involved, of course. The passing of the now somewhat infamous 'Act for the Preservation of Grayne' in 1566 declared as vermin not only the obvious pests of cereals, such as rats, mice and sparrows, but also various predators which might actually have helped to protect the grain from the vermin: Hedgehogs, Stoats, Weasels, Polecats, Badgers, Foxes, Wild Cats and Otters

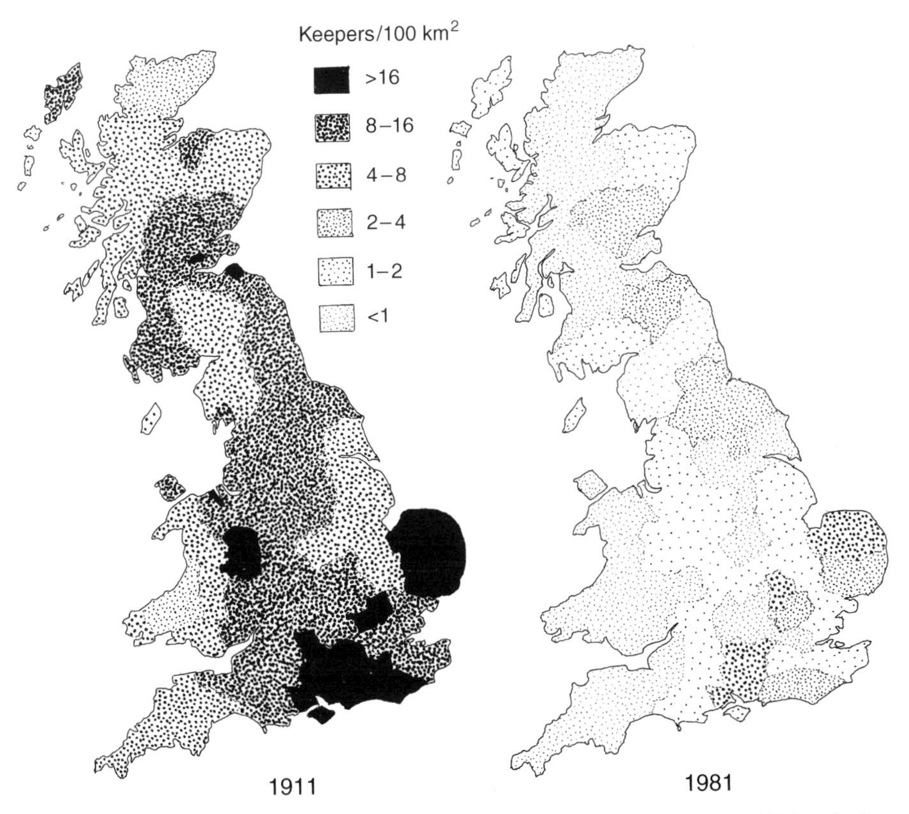

Figure 6.3 The density of gamekeepers in Great Britain (from Tapper 1992). In 1911, only four counties had fewer than four gamekeepers per 100 km^2, whereas by 1981, only four counties had more than this. (A 10-km square, as used on national distribution maps, has an area of 100 km^2.)

(Howes 1976). The Act encouraged the extermination of vermin by allowing churchwardens to pay a bounty on the corpses. Thus are detailed in many churchwarden's account books the level of cull which the vermin sustained. For Arksey in Yorkshire, Howes (1980) documents the take between 1720 and 1769 of 425 Polecats, a similar number of Weasels (which may have included Stoats) and smaller numbers (less than 50) of Otters and Foxes. This Act clearly set in train the perception of vermin which was pursued with vigour by the later gamekeepers and others. The addition of the gin trap to the gamekeeper's armoury in the mid-18th century made trapping of mustelids, in particular, much more effective. The Polecat seems to have been particularly vulnerable to this trap, and its use round rabbit warrens had severe consequences for this mustelid.

The three predatory mammals which suffered most were the Pine Marten, Polecat and Wild Cat. Their disappearance from successive counties was quite well documented by various naturalists at the end of the 19th century and the early years of the twentieth, particularly by writers of the relevant sections of the *Victoria History of the Counties of England.* These were summarized by Langley and Yalden (1977) in a series of maps which showed the distribution of the three at four critical times, 1800, 1850, 1880 and 1915. The Wild Cat declined first; it had already gone from southern England before 1800, and disappeared from northern Lancashire, Westmorland, Herefordshire and Shropshire early in the 19th century (Fig. 6.4). The latest county records from England are probably those from Yorkshire about 1840, from Loweswater, Cumberland and Castle Eden, Durham in 1843, and Eslington, Northumberland in 1853. In Wales, information is poor, but the species seems to have survived past 1850 only in the wilder counties of Merioneth, Caernarvon, Cardiganshire, Montgomeryshire and Breconshire. The latest record is probably that from Abermule, Montgomeryshire in 1862. In Scotland likewise the Wild Cat disappeared from the Central Lowlands before 1800, and was already rare then in the eastern counties. By 1850 it survived only in the Highlands, and by 1915 probably only north of the Great Glen.

The Pine Marten suffered from having a very desirable fur, as well as being a predator, and seems to have been rarer, that is had a lower population density, as well. It was already extinct in Anglesey and Wiltshire by 1800, though still present in other counties. It was however already scarce in south-eastern England and the central lowlands of Scotland, and had gone from these areas by 1850 (Fig. 6.5). It lingered into the 1870s in Devon, Cornwall, Somerset and Hereford, in Lincolnshire to the 1880s, and in various northern counties to about 1900, but by 1915 was thought to survive in England only in the Lake District, perhaps the Cheviot Hills, and parts of Yorkshire. In Wales, it had gone from five counties by 1870, from Flint as well by 1892, but probably survived in the other five counties. In Scotland, it was thought to be extinct in Angus and Kincardine by 1870, in Fife, Kinross and Ayrshire by 1875 and rare in the eastern parts of Sutherland, Ross-shire and Inverness-shire by 1881. However, it survived in the western parts of these counties to 1915, and beyond.

The Polecat was initially the most common and widespread of these three, perhaps because it was also the smallest. In 1800 it seems to have been still common in all counties, and it was still present in all of them in 1850, though becoming rare in some of the southern counties of England (Fig. 6.6). Between 1850 and 1880, however, it disappeared from Kent, Essex and Middlesex and from the south-eastern half of

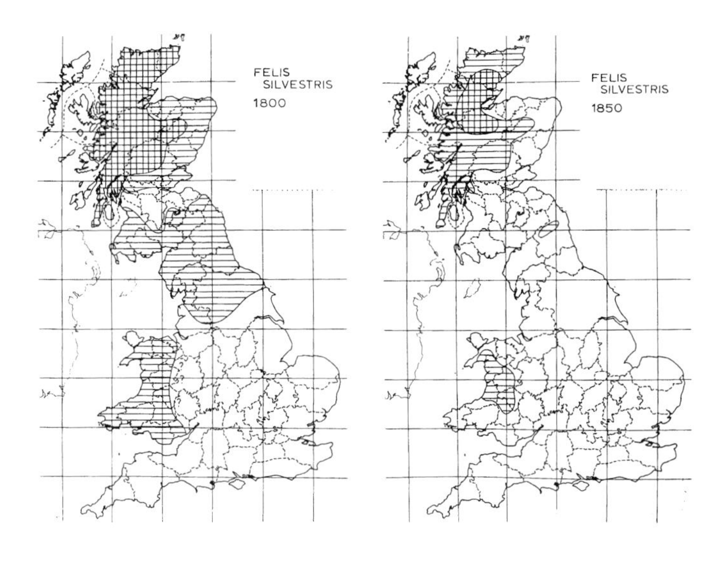

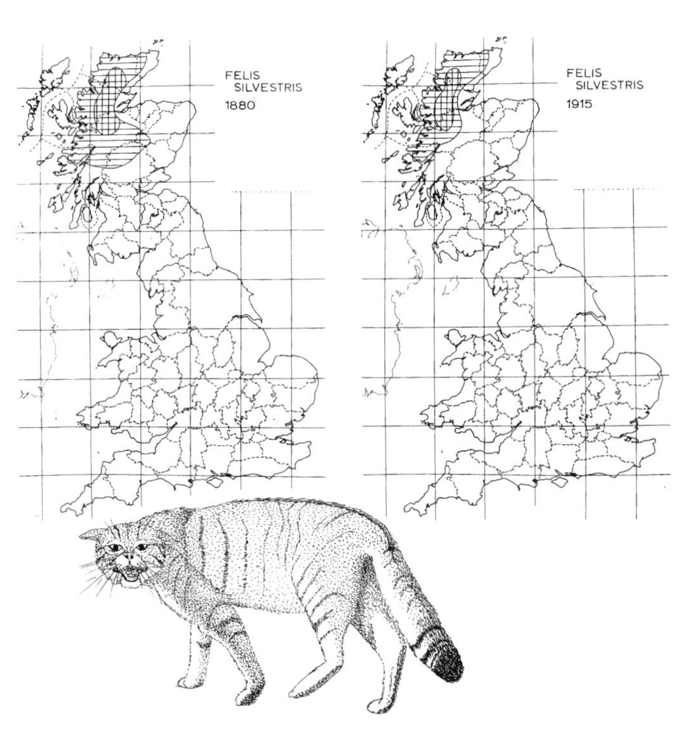

Figure 6.4 The decline of the Wild Cat *Felis silvestris* in Great Britain during the 19th century. The species survived only in extreme north-western Scotland. (From Langley and Yalden 1977 in *Mammal Review* **7**, by permission of Blackwell Science.)

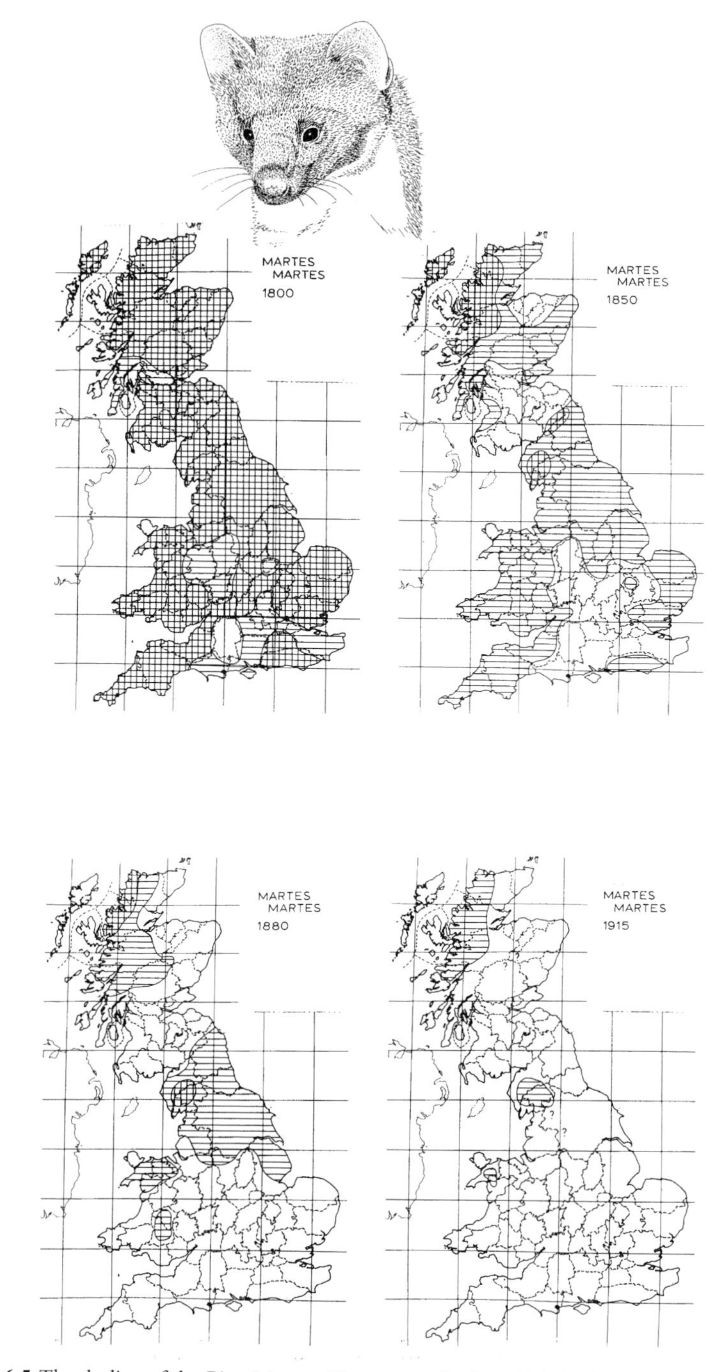

Figure 6.5 The decline of the Pine Marten *Martes martes* in Great Britain during the 19th century. The species seems to have survived in north-western Scotland, Cumbria and north Wales. (From Langley and Yalden 1977 in *Mammal Review* **7**, by permission of Blackwell Science.)

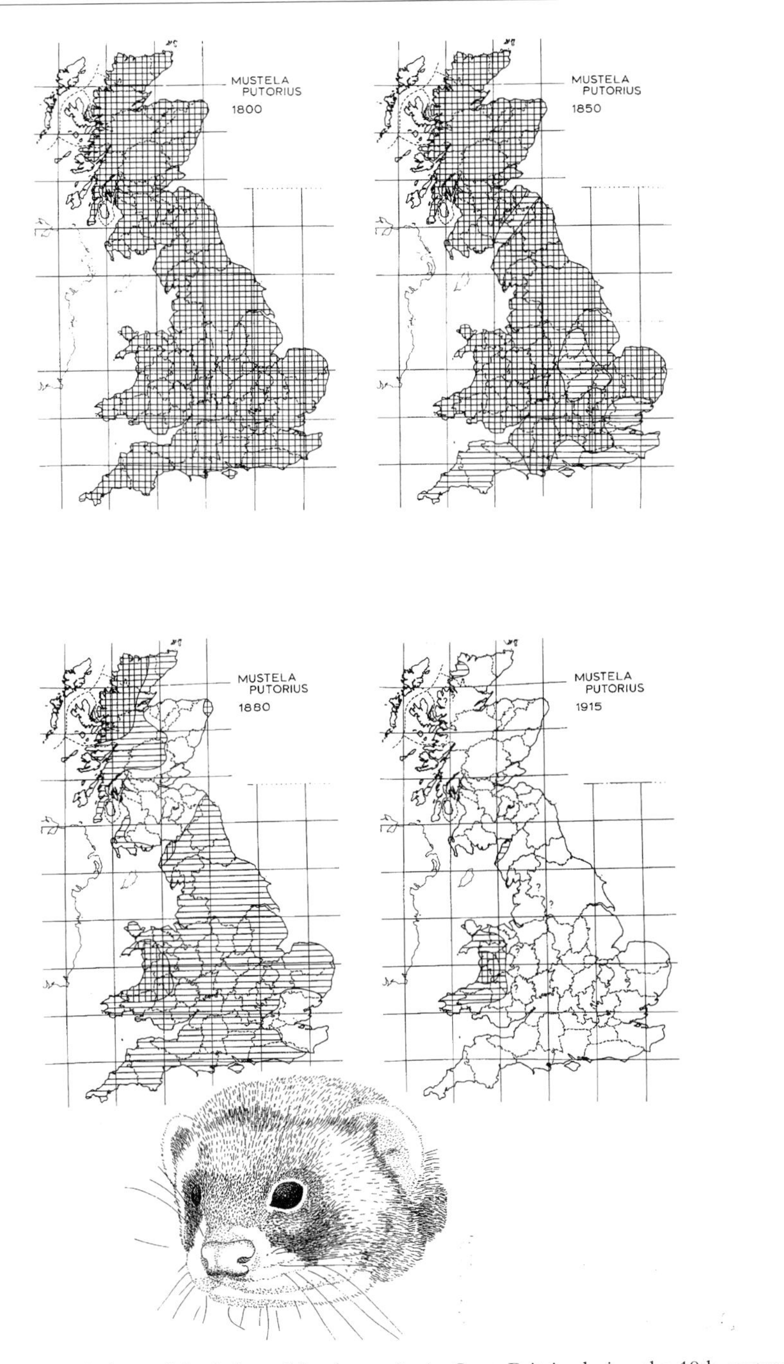

Figure 6.6 The decline of the Polecat *Mustela putorius* in Great Britain during the 19th century. Although there were late records from north-western Scotland and the Cumbrian coastal plain, it survived only in Wales and perhaps the Marches, where gamekeeping pressures (Fig. 6.3) were lowest. (From Langley and Yalden 1977 in *Mammal Review* **7**, by permission of Blackwell Science.)

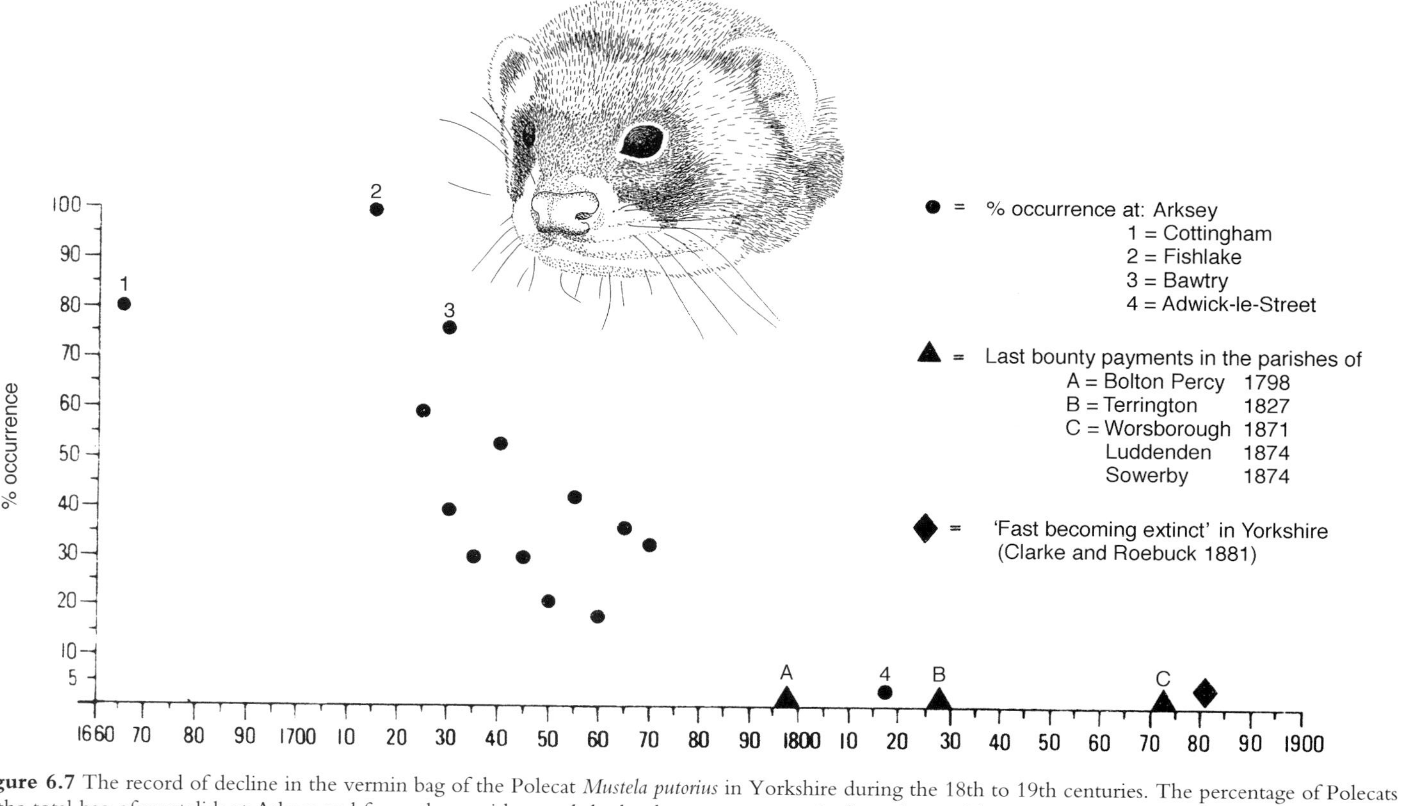

Figure 6.7 The record of decline in the vermin bag of the Polecat *Mustela putorius* in Yorkshire during the 18th to 19th centuries. The percentage of Polecats in the total bag of mustelids at Arksey and four other parishes, and the last bounty payments in five other parishes, suggest that Polecats were declining more quickly than implied by the last dates mapped in Fig. 6.6. (From Howes 1980 in *Naturalist* **105**, by permission of C.A. Howes.)

Scotland, and became rare throughout the rest of England. By 1915, it was believed to survive only in the extreme west of Sutherland, in a coastal strip of Cumberland, and in central and western Wales, being common only in an area of about 40 miles around Aberdovey. It may in fact have survived in Shropshire and Herefordshire, and perhaps in Yorkshire, as well, but like all extinctions, the last records are confused by misidentifications (of Ferrets as Polecats and Feral Cats as Wild Cats), by unrecorded releases or escapes of captive examples (particularly likely with Pine Martens, a popular menagerie animal), and by optimistic assessments of survival. In the case of Polecats, particularly, there is good evidence from Yorkshire that the pattern of decline recorded by Langley and Yalden (1977) is actually 'optimistic', that is, that the last dates as they compiled them stretch the survival of the species beyond the dates at which viable populations existed. An analysis of the churchwardens' accounts for south Yorkshire parishes shows that the major decline happened in the second half of the 18th century, rather than in the 19th century (Howes 1980; Fig. 6.7). If the same were true for the other species, as seems likely, the implication is that the rare predators were already becoming scarce before they were finally exterminated by gamekeepers and collectors.

CONCLUSION

By the beginning of the 20th century, all the larger native mammals were extinct or restricted to a small fragment of their natural range. The interplay of habitat loss and direct persecution which was responsible for the losses can rarely be teased apart satisfactorily, but the density of gamekeepers that existed in 1911, at their maximum, emphasizes something we have forgotten at the end of the 20th century: how many people lived and worked in the countryside then. The pressure that could be brought to bear on the larger and therefore more conspicuous mammals was very considerable. The converse point is made by the effectiveness of protection, usually for sporting purposes, of such species as Red and Roe Deer, which declined much earlier than the carnivores but also started to recover while they were still declining. The situation in nearby parts of Europe is also instructive. Wolves, for example, survived in France until the early part of the 20th century, and Brown Bears and Beavers were never quite exterminated. France did not have the same intensity of gamekeeping, and as a larger country with a similar sized population produced a lower human pressure on the other mammals. The extension of the conservation ethic to non-game species came rather late to Britain, and only just in time to save the Pine Marten. However, in the meantime, the introduction of various ornamental and game species did something to restore the British mammal fauna to the appropriate number of species. It is a moot point whether the exchange has been a beneficial one.

CHAPTER 7

RESTORING SOME BALANCE

Coypu and trap

IT is obvious that the stocking of parks with wild animals has a long tradition, back to Roman times at least, and that stocking of parks with deer and other animals has been practised in Britain since Norman times, particularly by royalty. Henry I enclosed the park at Woodstock, Oxford, with a wall in 1110, and housed there a menagerie including Lions, Leopards, Lynxes, Camels and a Porcupine; other game were hunted within the park (James 1990). The initial impetus for introducing species deliberately seems to have been an extension of this practice; species were released in the hope of providing extra or different hunting. Other species were released to be attractive additions to the fauna rather than for strictly utilitarian purposes. Yet others, of course, escaped accidentally from zoos, fur farms or parks. Together, the introductions represent a major source of interest and concern; some are major pests, others are of considerable conservation concern because of their effects either on native mammals or their habitats, and collectively they represent a massive ecological experiment, albeit unplanned. To the House Mouse, Black Rat, Fallow Deer and Rabbit, already long established by 1700, the additions in the 18th, 19th and 20th centuries include another four well-established and abundant mammals, now major elements of the mammal fauna, and at least a further four that have become locally established. Two other species that have come and gone again deserve mention as part of this unnatural experiment. Fortunately this topic has been well discussed in two major books, Richard Fitter's classic *Ark in our Midst* and Christopher Lever's *The Naturalized Animals of the British Isles*. Most of my information will come from them, and I will not cite separately all their sources.

The Major Newcomers

Probably the most important pest for most Britons, if only because most Britons are townies, is the Brown Rat *Rattus norvegicus*. No rats are native to Europe, and the arrival of the Brown Rat in Britain is moderately well documented. Its homeland appears to have been the steppes of central Asia, though it is now spread so far round the world that it is hard to be sure. It seems to have spread after multiplying to plague proportions in Central Asia and crossing the Volga River in 1727. The earliest in western Europe seems to have been recorded in Denmark in 1716, on a ship from Russia, and it reputedly reached Ireland in 1722. Pennant said in 1760 that it arrived in England 'about 40 years ago', so about the same time. They too are supposed to have arrived on a timber ship from Russia, although their alternative name Norway Rat indicates that at one time they were supposed to have come from there; Norway did not have them before 1762. They did not reach France until 1750, nor southern Europe (Spain, Italy) until the end of the 18th or middle of the 19th centuries.

Enemies of the Hanoverian kings claimed that they came to England with Brown Rats in their baggage in 1714, though this is surely a convenient 'political scapegoat' story. Within a very short time, Brown Rats had spread throughout Great Britain, reaching Scotland before 1754, and fast become a pest everywhere. In Anglesey in August 1762, they ate the standing corn while it was being reaped. They rapidly replaced the Black Rat as the major urban pest, and constitute an early, though unstudied, example of ecological replacement. Robert Smith in 1768 noted the ecological distinction which persists; Black Rats lived in the ceilings and wainscots, while Brown Rats lived in sewers and shores. Gilbert White commented on a Black Rat killed at Shalden, near Alton, in 1777 as a great rarity, adding 'the Norway rats destroy all the indigenous (sic) ones' (Johnson 1970). As early as 1783, Thomas Swaine said that in the 15 counties he worked as rat-catcher, he never met Black Rats except in Buckinghamshire – a few in High Wycombe – and Middlesex, but that in the City of London they remained abundant. Pennant in 1776 also noted that 'the Norway rat has also greatly lessened their numbers, and in many places almost extirpated them' (Barrett-Hamilton and Hinton 1916). Several surveys during this century have documented the scarcity of the Black Rat, which has become confined to a few ports and islands (Bentley 1959, 1964, Twigg 1992, McDonald *et al.* 1997). To colonies estimated at no more than 500 each on Lundy in the Bristol Channel and Garbh Eilean in the Shiants, Inner Hebrides, there may be no more than another 300 or so in the rest of Britain, and possibly no established breeding colonies (Harris *et al.* 1995). The Brown Rat, conversely, remains abundant despite being legally declared vermin, despite being the subject of a £50 million pest control industry, and despite the attentions of gamekeepers; it may total some 7 million animals. About 45% of agricultural premises and 3.5% of urban domestic premises in England are believed to be infested with them (Harris et al. 1995). It is intriguing to realize that this replacement of Black by Brown Rats has not occurred in southern Europe, where the Black Rat remains common, but the Brown Rat is scarce, and usually confined to ports. In New Zealand, though the Brown Rat was introduced first, the Black Rat has replaced it (King *et al.* 1996). Evidently both species remain numerous on, and to some extent dependent on, ships for their transport and support. Their geographical separation evidently reflects their original homelands and physiological adaptations; the longer tail and larger ears of the Black Rat must give it a better ability to lose heat in warm climates, but a worse ability to withstand cooler ones. So far as I know, however, no-one has investigated this topic.

The comparable replacement of the native Red Squirrel by the introduced American Grey Squirrel has been more thoroughly studied, yet we still do not understand in full how this has happened, either. The Grey Squirrel is a native of the deciduous woodlands of the eastern United States, extending northwards just into Canada, where it co-exists with at least five other squirrel species. There is another, larger, species of deciduous woodland, the Fox Squirrel *Sciurus niger*; a Red Squirrel *Tamiasciurus douglasi*, not closely related to ours, but looking remarkably like it, even to having tufted ears, and being a specialist on conifer seed; two ground squirrels, the large Woodchuck or Marmot *Marmota monax* and a little chipmunk *Tamias striatus*; and a flying squirrel *Glaucomys volans*. Given the competition back home, it is perhaps not surprising that the Grey Squirrel has done well in the deciduous woodlands of southern Britain.

The earliest certain introduction seems to have been the pair kept at Henbury Park near Macclesfield, Cheshire, brought from America in 1876, kept in a cage to show friends, and released when the attraction tired. They thrived, and shortly became a pest locally. Another introduction may have occurred about 1860 in Dunham Park, 15 miles away to the north-west, for there were early records there, and there are fairly convincing descriptions of what may be otherwise undocumented releases even earlier, for instance in Montgomeryshire in 1830, Denbighshire in 1828 and Kent in 1850 (Shorten 1954). The most important releases, as far as ensuring the establishment of Grey Squirrels in Britain is concerned, took place at Woburn Park, Bedford (ten released in 1890, rapidly increased), Richmond Park, Surrey (100 from America, released 1902) and Regent's Park, London (91 from Woburn, released 1905–1907). Important releases also took place in central Scotland (Loch Long, Dumbartonshire, 1892; Edinburgh 1913; Dunfermline, 1919), Yorkshire (Malton, 36 in 1906; Bedale, 1913; Bingley, 1914; Hebden Bridge, eight in 1921), Bournemouth and Exeter (four in 1915) (Table 7.1, Fig. 7.1). With such large numbers, reinforcements and geographical spread, the success of the species was assured. Fortunately, the spread of the species coincided with the beginning of ecological study in Britain, and its spread has been documented fairly thoroughly. An early survey by Middleton (1931) and later surveys by Shorten (1946, 1954) have been followed up by the Forestry Commission and others (Lloyd 1983, Gurnell and Pepper 1993); more detailed local studies (e.g. Reynolds 1985) have added to the story.

As early as the 1944 survey, it was evident that where Grey Squirrels had been established for about 20 years, Red Squirrels had disappeared. By that time, Grey Squirrels were well established across much of central and south-eastern England; the

Table 7.1 Sites where Grey Squirrels *Sciurus carolinensis* were introduced to Britain (from Shorten 1954).

Locality	*NGR*	*Date*	*Source*	*Number*	*Result*
Henbury, Cheshire	SJ8873	1876	USA	4	Increased
Bushy Park, Middlesex	TQ1569	1889	USA	5	Failed
Woburn, Bedfordshire	SP9632	1890	USA	10	Increased
Loch Long, Dumbarton	NS2292	1892	Canada	3	Increased
Benenden, Kent	TQ8033	?	?	?	Increased
Nuneham, Oxford	SU5599	?	Woburn	?	Increased
Richmond, Surrey	TQ1972	1902	USA	100	Increased
Wrexham, Denbigh	SJ3349	1903	Woburn	5	Increased
Lyme Park, Cheshire	SJ9682	1903–04	?	25	Increased
Regent's Park, London	TQ2882	1905–07	Woburn	91	Increased
Malton, Yorkshire	SE7871	1906	Woburn	36	Increased
Cliveden, Bucks	SU9185	?	?	?	Increased
Kew Gardens, London	TQ1877	1908	Woburn	4	Increased
Farnham Royal, Bucks	SU9682	1908	USA	?	Increased
Farnham Royal, Bucks	SU9682	1909	USA	5	Increased
Frimley, Surrey	SU8758	1910	USA	8	Increased
Dunham Park, Cheshire	SJ7488	1910	?	2	Increased
Sandling, Kent	TQ7558	1910	?	?	Increased
Bramhall, Cheshire	SJ8984	1911–12	Woburn	5	Uncertain
Birmingham	SP0787	1912	?	?	Increased
Castle Forbes, Ireland	N0981	1913	Woburn	8	Increased
Bedale, Yorkshire	SE2688	1913	?	?	Increased
Bingley, Yorkshire	SE1039	1914	London	14	Slight increase
Darlington	NZ2914	1914–15	?	?	Increased
Exeter	SX9292	1915	?	4	Increased
Stanwick, Northants	SP9871	1918	?	2	Increased
Dunfermline	NT0987	1919	?	?	Increased
Bournemouth	SZ0991	?	London	6	Increased
Hebden Bridge	SD9927	1921	?	8	Slight increase
Edinburgh	NT2674	?	Zoo	?	Occasional
Aberdare, Glamorgan	SO0002	1922	London	?	Slight increase
Needwood Forest, Staffs	SK1524	1929	Bournemouth	2	Uncertain

Possible additional releases took place at Northrepps, Norfolk; Ellington Castle, Ayrshire; in Nottinghamshire; and in Suffolk.

Hampshire and Home Counties populations had merged, but the Devon and northern populations were still separate (Fig. 7.2). Moreover, they had been slow to penetrate eastwards into East Anglia, and the Cheshire population had not crossed northwards into Lancashire, either. Over the next 40 years, Grey Squirrels have spread in all these directions, penetrated the Pennines into north Lancashire and the southern Lake District, spread throughout the south-west even to the tip of Cornwall, and throughout Wales. Only in Scotland does their progress seem to be slower, and even here they seem to be belatedly colonizing the Southern Uplands (Bryce 1997). The corollary, the decline of the Red Squirrel, also continues. For a long time, they

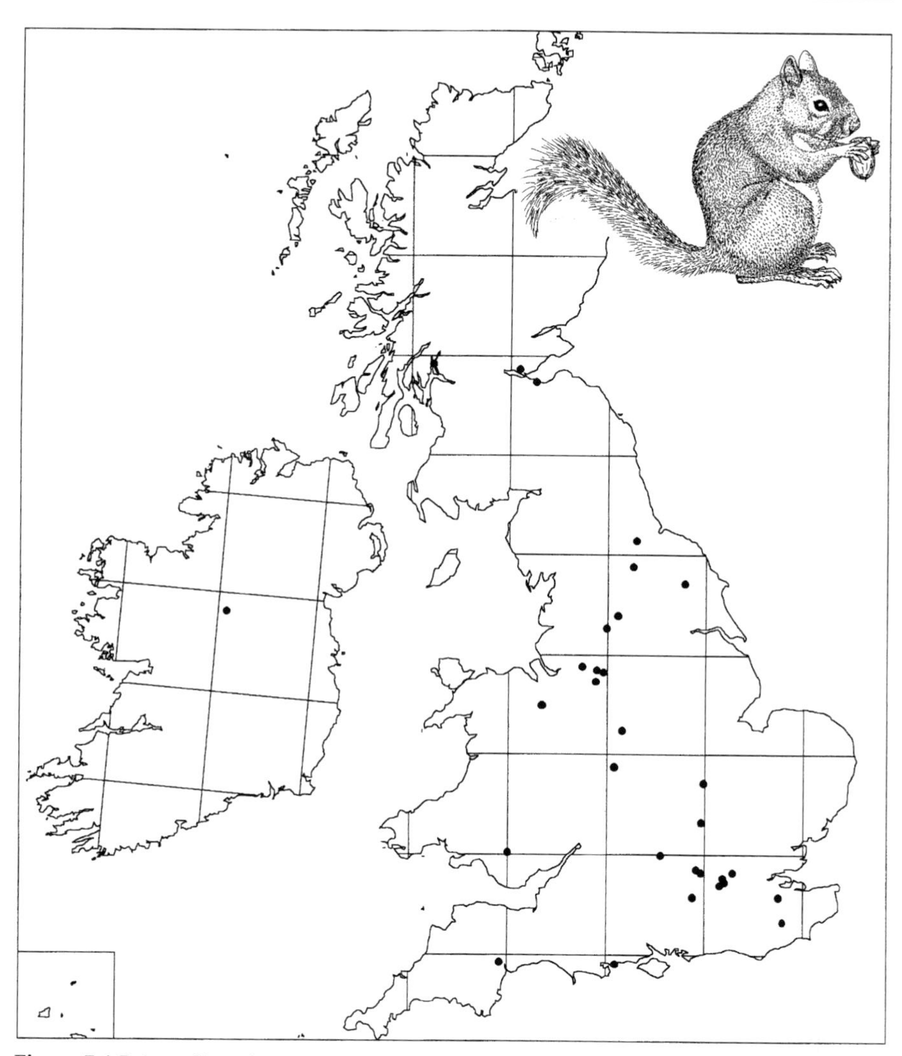

Figure 7.1 Points of introduction of the Grey Squirrel *Sciurus carolinensis* (after Shorten 1954).

held on in the conifer plantations of the Peak District, for example, but they seem to have died out in about 1994 (Grey Squirrels appeared in Sheffield itself by 1955, and spread into the valleys on the east side of the Peak District by 1974; Clinging and Whiteley 1980). In the Breckland conifer forests of Norfolk, numbers are very low, and an expensive experimental programme is underway to try to reverse the decline (Gurnell and Pepper 1993). A few remained in Cannock Chase, Staffordshire, until 1997, and they hang on in Anglesey and a few other Welsh forests. In the southern Lake District, the Grey Squirrel is spreading quickly, and Red Squirrels seem to disappear within a year or two of Greys appearing (Skelcher 1997). Only on the Isle of Wight, on Brownsea and Furzey Islands in Poole Harbour, and less certainly in the Caledonian pine forests of the Highlands, does the Red Squirrel seem reasonably safe.

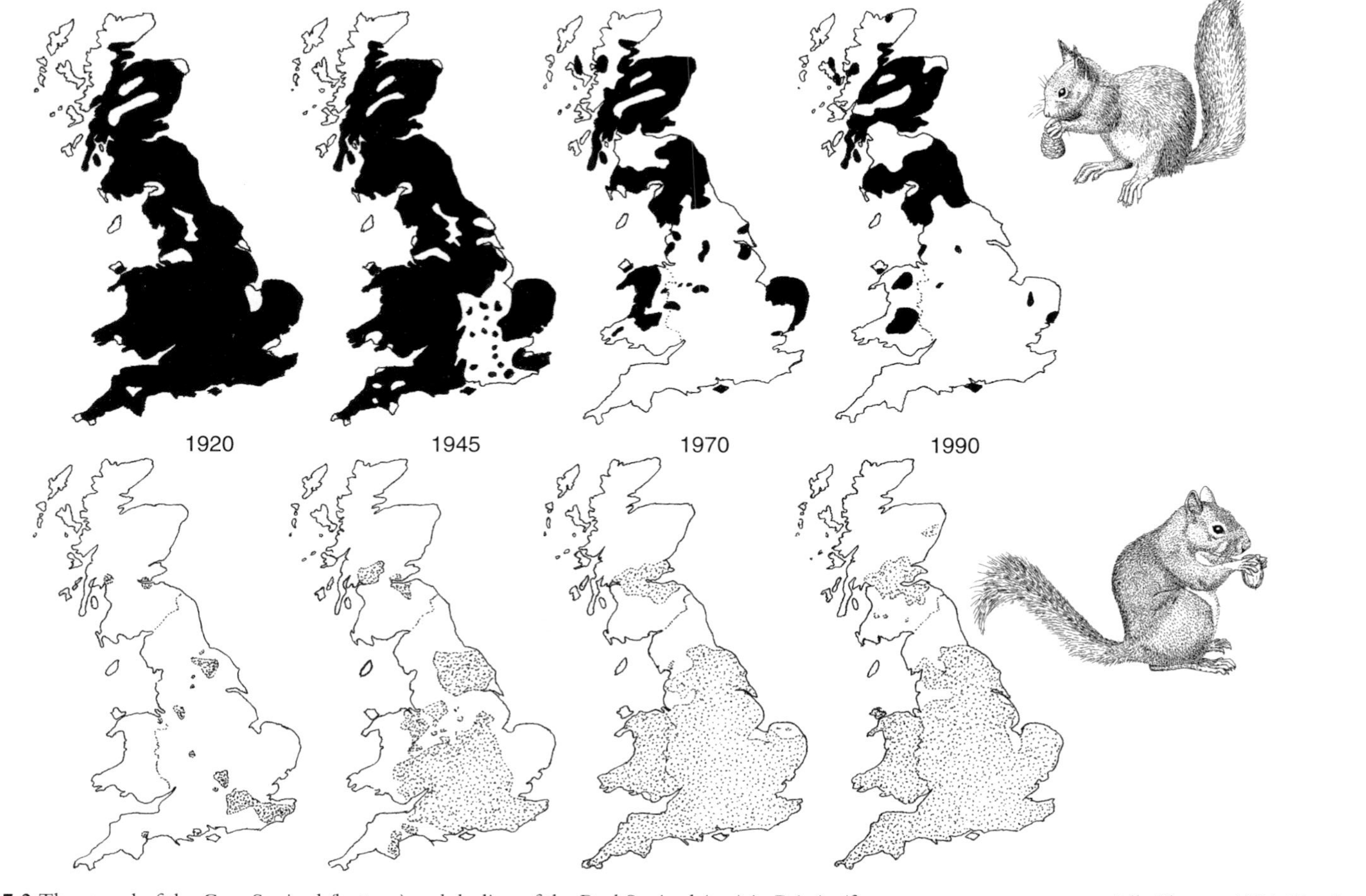

Figure 7.2 The spread of the Grey Squirrel (bottom) and decline of the Red Squirrel (top) in Britain (from numerous sources, especially Shorten 1954, Lloyd 1983, Gurnell and Pepper 1993).

Even in these refuges, careless or irresponsible introductions of Grey Squirrels cannot be entirely discounted; two individual Grey Squirrels have already been killed on the Isle of Wight, in 1978, but a population did not establish itself (Gurnell and Pepper 1993).

There has been much speculation about the causes of the displacement. Popular notions, that the Grey Squirrel outbreeds its smaller relative or fights and kills it, are not supported by much evidence. In appropriate conifer areas, densities and breeding success of Red Squirrels are just as good as those of Grey Squirrels in deciduous woodland (Shuttleworth 1997). Fights are recorded very rarely, not as frequently as sightings of the two species feeding amicably near each other. However, comparisons do show that Grey Squirrels have distinct advantages in deciduous woodland. For a start, they are about twice as large, lay down twice as much fat, and spend much more time on the ground; this is understandable, since foods like acorns and beech mast are shed in the autumn and have to be collected from and stored in the ground. Pine and spruce cones, by contrast, the main foods of Red Squirrels, are on the tree for two years while they ripen, and larch cones take a year to mature, hence cones are available for a more aboreal, lighter, squirrel. On this scenario, it is just unfortunate that the Red Squirrel in Britain lives in a country where most of the natural woodland is deciduous. In Oak–Hazel woodland, as on the Isle of Wight, Red Squirrels do eat Hazel nuts, as well as other seeds and fruits, but they are not very good at digesting acorns. Fed on them experimentally, they lose weight, and are found to have digested only 10% of the acorns, whereas Grey Squirrels maintain weight and digest 70% (Kenward and Holm 1989, 1993). It is uncertain, at present, whether Grey Squirrels have a similar disadvantage when presented with cones. It may be that they have trouble digesting turpenes, or have difficulty extracting the small seeds quickly enough to sustain themselves. It is a commonplace observation that they do eat pine cones, but they seem to die out in pure conifer woods during hard winters, whereas Red Squirrels obviously cope.

Lest this difference in dietary ability seem a complete answer to the question of why the Grey has displaced the Red Squirrel, it should be noted that there are complications. One is the parapoxvirus which certainly afflicts Red Squirrels, but whose origin and influence is uncertain (Sainsbury *et al.* 1997). It is possible, at least, that the virus has been introduced by Grey Squirrels, is relatively benign in them, but is more serious for Red Squirrels. Arguments against its importance include failure to notice any numbers of sick or dying Red Squirrels during their rapid displacement in for instance the southern Lake District (when people have been looking out for such signs) and, conversely, their long coexistence nearby in places like the Brecklands, where a virulent disease would surely have wiped out Red Squirrels 20 years ago. The slow displacement of Red Squirrels in Norfolk documented by Reynolds is particularly instructive. In some survey squares, Red Squirrels disappeared up to 18 years before Greys appeared, in others they co-existed for up to 16 years. This sounds like slow ecological replacement, rather than any dramatic effect like a virulent disease, or direct aggression. The possibility that the larger Grey Squirrels displace Red Squirrels passively (that Red Squirrels just keep out of the way) when food is short or times are hard cannot be ruled out as a factor, though. What is clear is that Red Squirrels, in southern Britain at least, are quite unable to cope with the presence of Grey Squirrels. The biggest problem for Red Squirrels is no longer anything to do with poor habitat,

A Red Squirrel and two Grey Squirrels

human persecution, or predation; it is simply their misfortune that we introduced their larger relative.

The third very successful introduction offers a similar example of unwitting effects, though at least in this case the introduction was accidental. The American Mink *Mustela vison* was imported to Britain for fur farming in the 1920s. After the First World War, surviving officers with pensions were looking for ways to establish new industries, and this seemed a possibility. The Mink is a widespread carnivore in North America, where it feeds on a variety of riparian prey, including Musk Rats *Ondatra zibethica* (see later). The first fur farms in England date to 1929, and in Scotland to 1938 (Dunstone 1993). Escapees were recorded in England in the 1930s, but the first recorded breeding in the wild was noted in Devon, on the River Teign, in 1956. Although Ian Linn and John Stevenson drew attention to the potential problem that this event signalled, there was at first scepticism that Mink could establish themselves here. Indeed, it took them 8–10 years to colonize a short stretch of the Teign, but then in three years from 1961 to 1963 they exploded over much of southern Devon. Another colony, on the River Ribble in Lancashire, behaved similarly. By the 1970s, Mink were established in every county in England, and most of Wales and Scotland too. Most seriously, from an ecological viewpoint, they have established themselves on Lewis, in the Outer Hebrides, an area that should have no ground predators, and is therefore a stronghold for ground-nesting birds such as waders.

The role of the Mink as a predator has created a great deal more heat than light. In

American Mink and Water Vole

a country which has Stoats, Weasels and Red Foxes, and which ought to have Wild Cats, Pine Martens and (similar-sized) Polecats, it is hard to believe that Mink should be a serious threat to native birds or mammals. Early research suggested indeed that it was a generalist predator, eating a lot of non-native Rabbits as well as a wide variety of other small mammals, common birds and fish. However, recent surveys of the distribution of the Water Vole *Arvicola terrestris* have revealed an alarming recent (post-1970s) decline in its distribution and abundance, as well as a worrying correlation between presence of Mink and absence of Water Vole (Strachan and Jefferies 1993). More detailed studies of sites where Mink have moved in to areas with Water Voles, resulting in the extermination of the voles, have also been documented; radio-tracked Water Voles have been killed in front of the biologist tracking them (Woodroffe *et al.* 1990, Strachan *et al.* 1998). Studies of some sea bird colonies in the west of Scotland have also suggested a more serious effect than previously realized. Common Terns *Sterna hirundo* declined from 1839 pairs to 1179 pairs over a 10-year study, a 36% reduction. Black-headed Gulls *Larus ridibundus* and Common Gulls *L. canus* similarly declined by 52% and 30%, though it is possible that the birds will redistribute themselves, in time, to less accessible islands (Craik 1997). It is the Mink's good fortune to have established itself at a time when its most likely competitors, Polecat and Otter, have been at low numbers. As the Otter recovers, there is some suggestion that it outcompetes the Mink; Mink numbers seem to be lower in river basins which now have Otters in them (Strachan and Jefferies 1996). Certainly Otters have recovered well in the rivers of south-west England, the area with the earliest and most extensive Mink populations, so Mink have in no way inhibited them, though this was at one time a popular fear. Studies in France suggest that Mink avoid Polecats, moving out of swampy areas in spring when Polecats move in to feed on frogs (Lodé 1993). While Polecats ate mostly voles, Brown Rats and amphibians, Mink ate fish and birds. Certainly Polecats seem to be spreading back into England from Wales at a rate which suggests that Mink are no serious rival (Birks 1993). Thus the concerns which have been expressed that Mink pose a threat to other carnivores have little foundation. On the other hand, they do pose a serious threat to Water Voles and other riparian or marine species. None of our native carnivores is a major predator of riparian life; Otters are fish specialists, and eat birds or mammals less frequently, while the other mustelids, though they can and do swim, are not truly aquatic. Perhaps it is the Water Vole's misfortune that it has only now encountered an aquatic carnivore. On the other hand, the situation has certainly been exacerbated by the appalling mismanagement (particularly overgrazing) of riverine vegetation. In areas of extensive swamp, Water Voles and Mink could probably co-exist, but in the tiny strip that is all that usually remains to them, Water Voles have no chance. This is the final stage of a decline that seems to have begun in Roman times (see Chapters 3 and 4).

The fourth of the well-established exotics is the most exotic of all, the Chinese Muntjac *Muntiacus reevesi*, a small deer (about as tall as a Fox and as heavy as a Badger) from subtropical south-east Asia. Given its homeland, it is surprising that it has done so well. It is not well adapted to the British climate, for, unlike all our other deer, it breeds throughout the year. However, though fawns may be born at any time of year, the young bucks vary the age (between 51 and 112 weeks old) when they shed their first antlers, so gain an antler cycle that is synchronized to the yearly weather cycle. Antlers are shed in May–June, grow through the summer, and shed the velvet in August–October. They remain in hard antler over winter. In China, it is a species of sub-montane forest, but in Britain it does as well in farmland and gardens as in woodland.

It was brought to Woburn Park in 1894, so far as can be established, but the story of its escape and establishment has been thoroughly confused, perhaps deliberately obfuscated. For a start, it was claimed that both Indian (*M. muntjak*) and Chinese Muntjac were established in the grounds of Woburn Park, had escaped and were established in the wild, and were even hybridizing. It is now evident that Indian Muntjac, which are appreciably larger, never did well, never established themselves in the wild, and never hybridized, at least not in the wild. (It is known from captive studies that the two can interbreed, even though one has a chromosome number of 46 and the other six (females) or seven (males); hybrids were thought to be fertile, but are now known to be sterile; Capanna 1973, Chapman and Chapman 1982.) It is furthermore now evident that, far from escaping into the wild and then spreading out, the previous (12th) Duke of Bedford had a policy of deliberately liberating small groups of Muntjac, typically five bucks and four does, at various likely sites around the country (Chapman *et al.* 1994). Some Muntjac, of both species, were turned out of Woburn Park as early as 1901, but there is little evidence that they established themselves until the 1920s, and they were still very local in the 1940s and 1950s. At about this time, however, there were surreptitious releases to various localities up to 100 miles away, to Elveden in Suffolk, Bix Bottom and Bicester in Oxfordshire, Corby in Northamptonshire and an unknown destination in Kent, at least. All except the last resulted in successful establishment of local populations (Fig. 7.3). The notion that

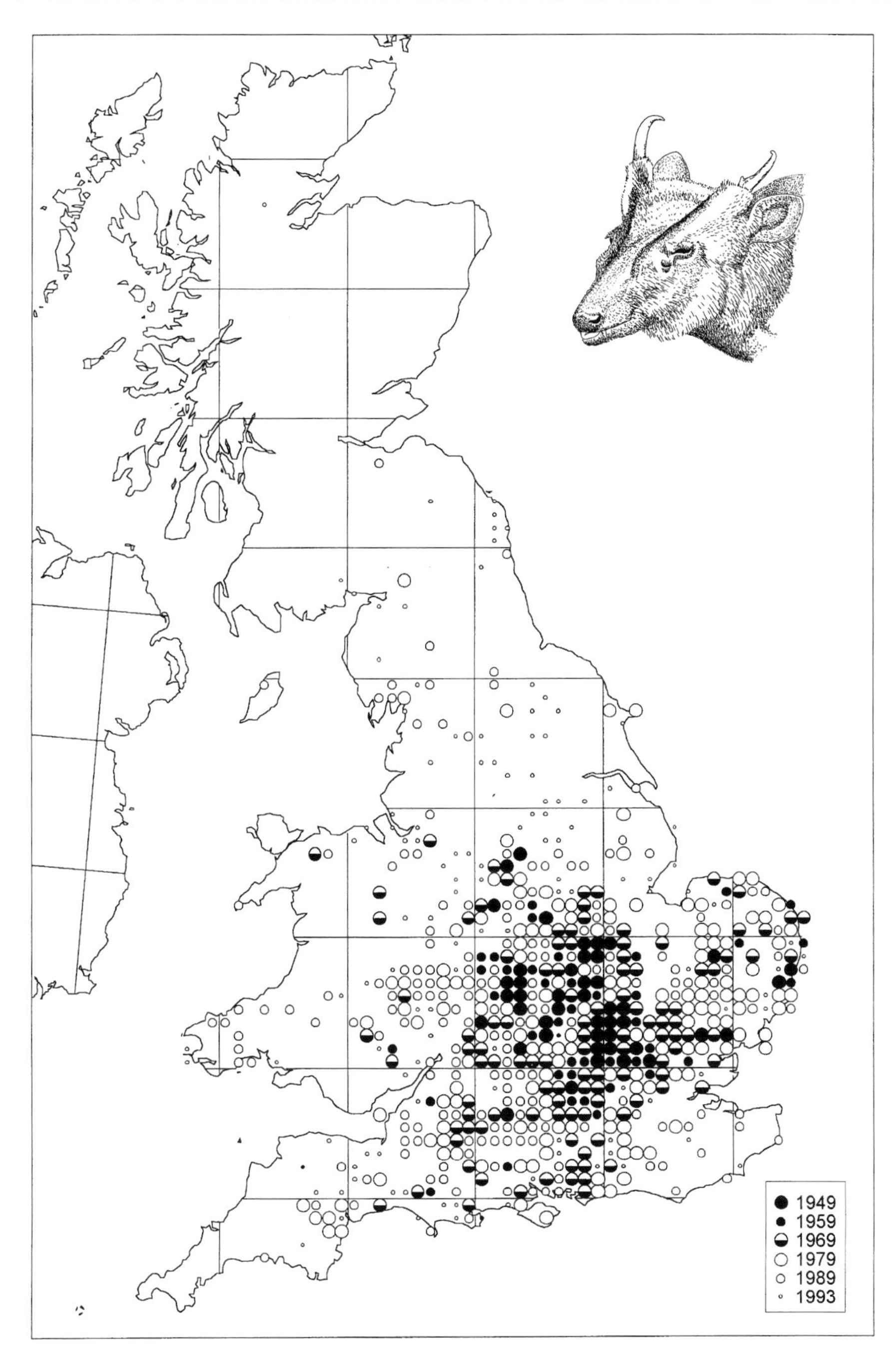

Figure 7.3 The spread of the Chinese Muntjac *Muntiacus reevesi* in Britain (after Chapman *et al.* 1994). The circles indicate 10-km squares in which the species had been recorded by various dates. The pattern of 'jumps', Muntjac appearing well beyond their previously known ranges, indicates that Humans have been spreading the animal. This is now illegal.

Muntjac showed a 'natural' rate of spread away from Woburn of around 2.5 km per year, much faster than other deer, is therefore totally misleading (Roe Deer have spread from Dorset at about 0.8 km per year, and Fallow Deer seem to have spread at similar rates in several countries; Chapman *et al.* 1994). It is not surprising that the species has established itself well.

The Chinese Muntjac seems to have only minor repercussions on the other mammal fauna, but more serious effects on the flora. Some birds are suffering loss of habitat, and it is eating the food plants of some butterflies. Where it has spread into the range of Roe Deer and Chinese Water Deer, it seems to have suppressed the numbers of both. It is now beginning to cause concern for conservationists managing woodlands for their flowers and those trying to retain or restore coppice management. Because they are so small, and because of the difficulty of managing a humane cull for a species that may be pregnant or lactating at any time of the year, Human culling has been modest, at best. Foxes kill some, especially fawns, and severe winters can cause severe mortality. However, there have been mostly mild winters in the last decade or more, and the population has increased substantially. In Monks Wood National Nature Reserve, the flowering of Bluebells, Primroses and Common Spotted Orchids has been greatly reduced (Cooke and Farrell 1995, Cooke and Lakhani 1996), while both there and in Hayley Wood coppice management has been disrupted because all the new growth is eaten. Fencing is rarely an efficient way of protecting woodland against deer for more than a few years (because snow, or falling branches, or animals, force breaches in it), but the idea that deer should be controlled by shooting is an anathema to many conservationists.

The Lesser Introductions

One of the more intiguing problems of the introduced mammal fauna is why the Edible Dormouse *Glis glis* has spread so little. Introduced to Tring Park, Hertfordshire, apparently from Hungary, by Walter, Lord Rothschild, in 1902, the species has been well established in a small area of the Chiltern Hills for 30 years or so. It has a range about 30 km north–south and 20 km east–west, between Aylesbury, High Wycombe and Whipsnade (Fig. 7.4). Although it can range up to 1700 m in its lifetime, over 85 years it seems to have spread at only 380 m per year. Thus apparently suitable habitat further north up the Chilterns, and further south into Oxfordshire, remains uninhabited. It is not clear whether this represents a species having difficulty coping with habitat or climatic factors outside its natural range, or the slow life style of dormice in general. Recent radiotracking studies show that the animals have very small home ranges (0.3 ha for a female and 2.4 ha for a male), so perhaps moving long distances is something foreign to them (Morris and Hoodless 1992).

Within this region, it can sometimes be a household pest, and there has been at least one serious outbreak of damage to commercial larch plantations (Platt and Rowe 1964). As a non-native mammal, it is illegal to release Edible Dormice back into the wild (without a licence), under Section 14 and Schedule 9 of the 1981 Wildlife and Countryside Act. However, under the Bern Convention, all Gliridae are protected, in Britain as in Europe. This makes it rather an intractable pest control problem, and many householders reputedly deal with the problem by surreptitiously releasing

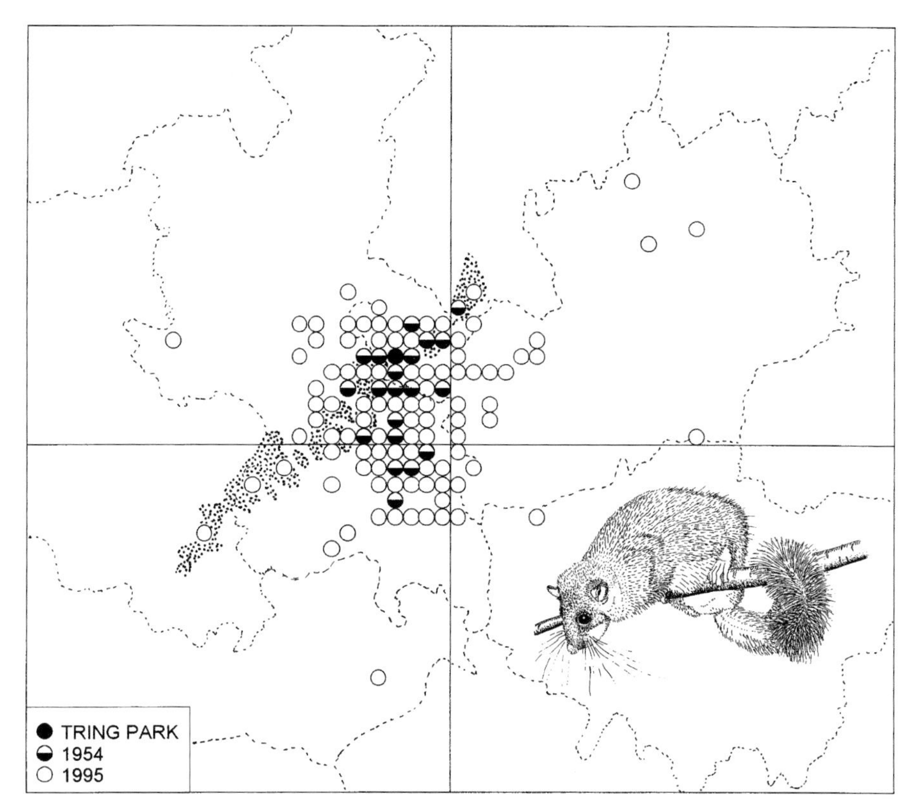

Figure 7.4 The spread of the Edible Dormouse *Glis glis* in the Chilterns (after Thompson 1953, Morris 1997). The spots indicate tetrads (2 × 2 km squares) in which the species has been recorded since its introduction at Tring in 1902. Stippled shading is land above 200 m.

Edible Dormice some distance from where they have been making a nuisance of themselves. As a way of ensuring that they do spread into Oxfordshire and elsewhere, this is hard to beat, but it is certainly illegal, and given the problems caused by the other introduced species, foolhardy.

The Chinese Water Deer *Hydropotes inermis* is a parallel case to both Edible Dormice and Chinese Muntjac. Initially introduced to the grounds of Whipsnade Zoo, where 32 were liberated in 1929–30, they multiplied, and can still be seen readily in the zoo grounds. Some were taken to Woburn, and supposedly escaped in the 1940s. Some were sent from Woburn to two estates in Hampshire, Leckford Abbas and Farleigh Wallop, from where they also escaped, and to Studley Royal in Yorkshire. At some time, some were also released or escaped in Norfolk, and it seems certain that, as with the Muntjac, there were in fact a number of deliberate but surreptitious releases (Chapman 1995). Though not much bigger, and no more conspicuous, than the Muntjac, the Chinese Water Deer has managed to establish only small populations. It is a native of flood plain grasslands in China, and there is precious little comparable habitat in Britain. One population has established itself in Woodwalton Fen, and others are in the Broads.

The species is trebly interesting as a deer. It is one of the few deer to lack antlers (the only one if musk deer *Moschus* are put in a separate family); instead, like Muntjac, the males bear large tusks. It is the only one that can have litters of 4–6 young (most deer have singletons, or twins at the most), though the average seems to be about 2.5 young per litter. Lastly, it is actually rather rare, perhaps endangered, in its homeland, because it is an animal of the lowlands, where most Chinese also live. Whether we should be hosting it in Britain is another matter.

The species that has done best after introduction while remaining as yet rather local in its distribution is another Asian deer, the Sika *Cervus nippon*. This is a much larger species, the size of a Fallow Deer, and a native of Japan and the adjacent parts of Korea and China. Initially it was introduced to Powerscourt in Ireland in 1860, but very successfully, for deer were passed on from there to many other sites in Britain. Other herds were founded at Fawley Court, Oxfordshire at an unknown but early date, and at Rosehall in Sutherland in 1900, but the sources of these two are uncertain. This is unfortunate, because two subspecies, the Japanese *C. n. nippon* and the Manchurian or Mainland *C. n. hortulorum* have been involved, and may have had different consequences.The most successful introductions were to the Mull of Kintyre in 1893 at Carradale (Fawley Court deer), which escaped in 1914–18, to Rosehall itself, and to Aldourie, Inverness (Rosehall deer) which were deliberately released in 1900 (Ratcliffe 1987). A herd on Brownsea Island in Poole Harbour swam to the mainland in 1896, and are the origin of the well-established population in the conifers of the surrounding basin, while two escaping from Beaulieu in 1904 and two more released deliberately the following year are the source of the population in the south-east sector of the New Forest. A population in the Bowland Forest/Ribblesdale area along the Lancashire/Yorkshire border originated in deliberate releases in 1906 to provide hunting for the buckhounds. Thus a complex mixture of deliberate and accidental establishment occurred (Ratcliffe 1987).

Until about 1972, none of these populations had extended very far from its source, and the total range of the species in Britain extended into only 54 10-km squares. Since then, largely taking advantage of newly maturing conifer plantations throughout its range, it has spread extensively. Ratcliffe (1987) suggests a range extending to 166 10-km squares. Thus its range doubled in a decade, and this has been especially evident in Scotland (Fig. 7.5).

The main consequences of these have been twofold: a new forestry pest, and a serious threat to the native Red Deer. Perhaps because they resemble their native habitats, Sika have taken to the commercial conifer plantations, especially in Scotland, far better than either of the native deer, or the other introductions. Much of Kintyre is now covered in such plantations, and Sika have spread accordingly. They are adept at stripping bark from young spruce, and can cause serious commercial damage, but they are very difficult to cull in such dense woodlands, most of which were not designed with deer culling in mind. Though they are the same size as Fallow Deer, competition between these two species seems not to have occurred, largely because the two do not meet; in the New Forest, Fallow Deer live north-west of the London to Bournemouth railway, and Sika to the south-east. The big problem is of hybridization with Red Deer. The two species are, somewhat unexpectedly, totally interfertile, with not even any evidence that one parent or the other does better under some particular environmental conditions. Initially hybridization seemed unlikely; the large

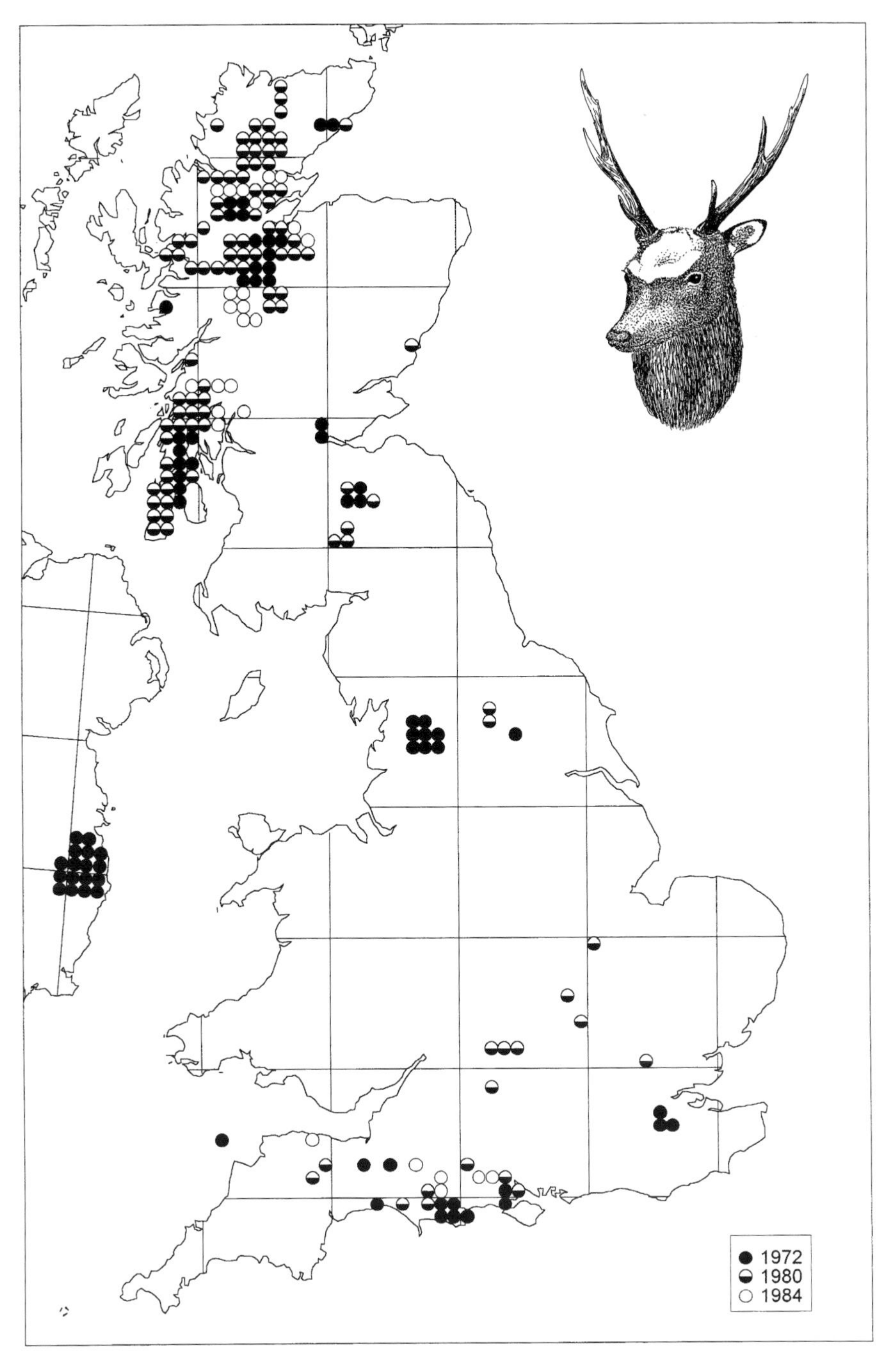

Figure 7.5 The spread of the Sika *Cervus nippon* in Britain (after Clarke 1974, Ratcliffe 1987). The rapid expansion in the conifer forests of Scotland is very obvious.

Red stags could not mate successfully with small Sika hinds, and the small Sika stags failed to impress the Red hinds. However, when young male Red deer have met Sika hinds in parks, they have hybridized readily, and are then fully interfertile. It is possible that hybridization has occurred more readily with the larger Mainland race than with the smaller Japanese, but since the source of several of the Sika populations is uncertain, so is this hypothesis. Hybrids were reported as early as 1940, in the southern Lake District, and it was established some time ago that the deer living in the Wicklow Mountains in Ireland, not too far from Powerscourt, were neither Red Deer nor Sika, but a completely hybridized population (Harrington 1973). The same happened in the southern Lake District (Lowe and Gardiner 1975). It is now evident that the same thing is happening in Scotland, and even apparently pure-looking Red Deer living some distance from the nearest Sika turn out to have Sika genes in them. It is considered that the process has probably already gone too far to be reversed – many of the hybrids are unrecognizable as such, so culling them out is not an option (Pemberton *et al.* 1998). This seems a very sad way to lose our largest native land mammal. In the southern Lake District, the genuine native Red Deer of the Martindale area are already threatened by Sika spreading north from the Bowland population, or rather by the hybrid swarm that has been formed between them and escaped park Red Deer.

At least the last, and most unlikely, of these introductions seems to pose no threat, and may well die out despite the best efforts of various owners to establish new populations around the place. The Red-necked Wallaby *Macropus rufogriseus* is of course a marsupial, a native of Australia. Strictly it is the Tasmanian race, also known as Bennett's Wallaby, which has been established here. This race, unlike its mainland cousin, has a breeding season, making it better adapted to the British climate, and, since Tasmania is about 42° south, and has mountains reaching 1617 m, it is reasonably well preadapted to our climate. The earliest established colony was perhaps on Herm, in the Channel Isles, where a small number lived from about 1890 to 1910. There is a thriving colony, which has reached 400–600, within the grounds of Whipsnade Zoo; this has provided the source for other collections. A park collection in Sussex seems to have leaked a few escapees that reputedly survived in the Weald from 1940 to 1972, but seem to have died out. The best studied population started with Whipsnade stock introduced to a private zoo in Staffordshire. Five were released to take their chance at the start of the war, when private menageries and their fencing became unsupportable. They increased to about 50 animals, until the 1962–63 winter killed many of them, but they survived in small numbers through to the 1990s. Another colony of about 26 lives on Inconnachan Island in Loch Lomond, the result of releases in 1975. A large number of casual sightings of this species in the wild has also been reported (Baker 1990), for this is a popular menagerie/park/pets corner species, relatively inexpensive, and inexpensive to feed, yet sufficiently 'different' to attract interest. Fox predation of young can be serious, and so can severe winters, but the main problems, at least for the Peak District colony, have been road casualties and general disturbance from people and their dogs. They live rather secretive lives, like overgrown hares; moving little, sitting tight if they think they can escape notice, but panicking if startled. They have been vulnerable to jumping over cliffs and in front of vehicles. Like deer, they only have one young a year, so replacement of losses is slow, and in the last 10 years it has been the young potential recruits that have died (Yalden 1988).

Here Today and Gone Tomorrow

Two more fur bearers, introduced like American Mink in the inter-war period, provide very instructive examples of what can be achieved by way of control when the financial means are provided as a consequence of the political will being summoned and a reasonable possibility of success being perceived. The Muskrat *Ondatra zibethica*, a large North American vole, and the Coypu *Myocastor coypus*, an even larger South American caviomorph (guinea-pig relative), are the two species concerned.

Muskrats were introduced to Europe, specifically to what is now the Czech Republic, in 1905, as a new source of fur. They did extremely well, and were believed to number 100 million by 1926. They had spread to neighbouring countries, and are now well established across most of western Europe. In the Netherlands, they do considerable damage to dykes, and 289 116 were killed in 1987, at a cost of about £7 million. In Britain, there were 87 fur farms by 1930, and escapes were noticed in 1927. Martin Hinton, Keeper of Mammals in the British Museum (Natural History), South Kensington, recognized the problem that they were already causing in Europe, and when escapes were reported in Britain started to lobby the government (Sheail 1988). In fact, a large area of the Severn valley in Shropshire was found to be infested, and smaller colonies in Sussex, Surrey and on the Stirlingshire/Perthshire border area were subsequently discovered (Warwick 1934; Fig. 7.6). Independently, the Irish authorities found a colony established around Lough Derg on the River Shannon. An eradication campaign was mounted, using the then standard fur-trappers' leg-hold traps, and nearly 4500 Muskrat were caught in the official campaigns (Table 7.2). So, unfortunately, were a large number of non-target species; in two years' trapping in Scotland, 2305 Water Voles and 2178 Moorhens were killed, among a number of other species, and figures for England were never published. Private operators, often working ahead of the official campaigns, killed about a further 350 Muskrat. The effectiveness of these campaigns was a triumph for the combination of clear advocacy, a well managed and scientifically led programme, and the professionalism of the staff involved. The deployment of teams of trappers, 23 in England and 13 in Scotland, working inwards, was adequate first to prevent the further spread of the species, and then to eliminate it within four years. As a measure of the professionalism of the programme, some trappers were retained until March 1937 in England and August 1937

Table 7.2 The progress of the campaign to trap out the Muskrat *Ondatra zibethica* in Britain (from Warwick 1934, 1941). These are the numbers trapped in the official campaigns; private individuals trapped around 350 further animals. Warwick gives the total catch in Ireland as 487, but the annual figures do not match the total.

Site	*1932*	*1933*	*1934*	*1935*	*1936*	*1937*	*Totals*
Scotland	51	744	151	10	1	1	958
Shropshire	1468	1064	181	7			2720
Surrey		42	12	1			55
Sussex		144	18	1			162
Lough Derg, Eire		297	59				487
							4382

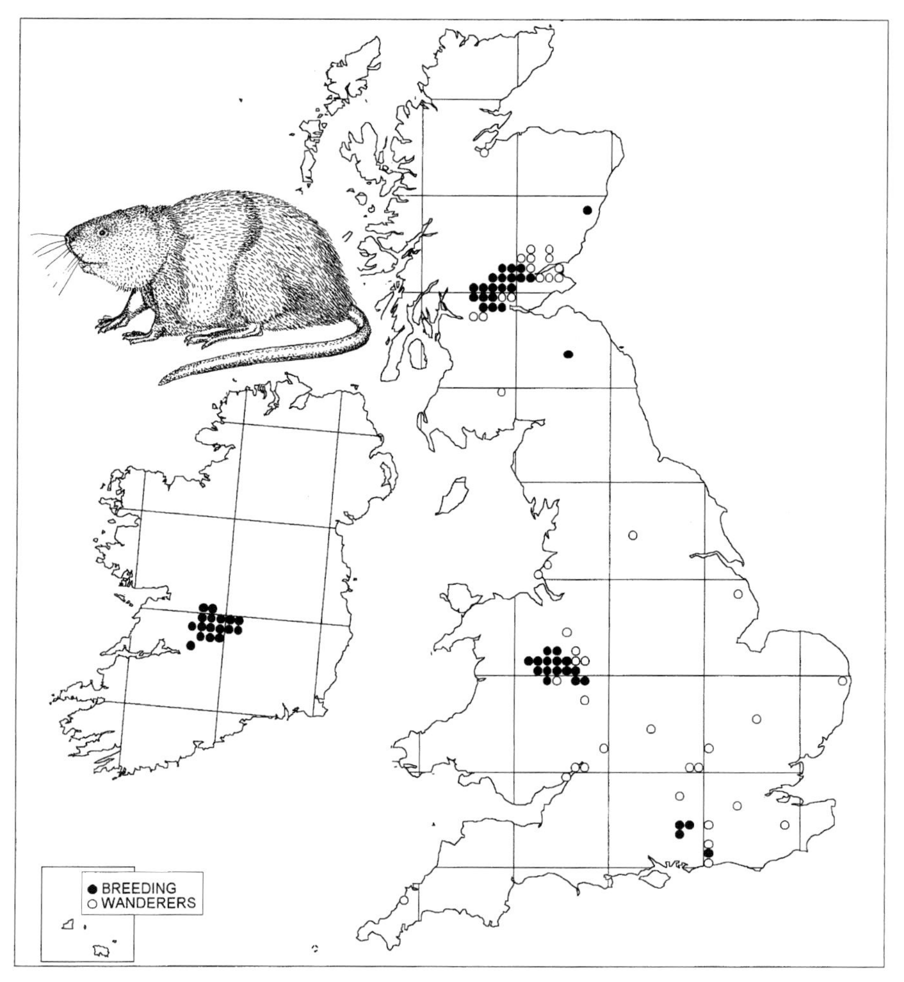

Figure 7.6 The distribution of the Muskrat *Ondatra zibethica* in the British Isles in 1929–35 (after Warwick 1934, 1941).

in Scotland, up to two years after the last had been killed in the wild, to ensure that none had been overlooked (Gosling and Baker 1989).

Meanwhile, the much larger Coypu had also been introduced, to around 50 fur farms by 1939, but the advice from Hinton, based on discussions with German scientists who had been dealing with both species, was that Coypu were such slow breeders (only two young per litter, usually) and so poorly equipped to cope with winter weather, that they would provide no threat to agriculture (Sheail 1988). As a consequence, little concern was shown when Coypu first escaped in the 1930s, and although 193 were killed in Norfolk in 1943–45, the severe winter of 1946–47 killed most of them (much as German experience had suggested). It was not until the mid–late 1950s that the animal was perceived to be a serious agricultural pest, both through its burrowing activities in low-lying, artificially drained, country, and because of its raids on root crops, especially sugar beet. In 1962, MAFF (the Ministry of Agriculture, Fisheries and Food) started a campaign with the modest aim of limiting it to the area of the Broads. It had spread well beyond this across East Anglia, and it was felt at the time that eradicating it from Broadland would be impossible (Fig. 7.7). The severest winter of the century, 1962–63, offered an immediate providential

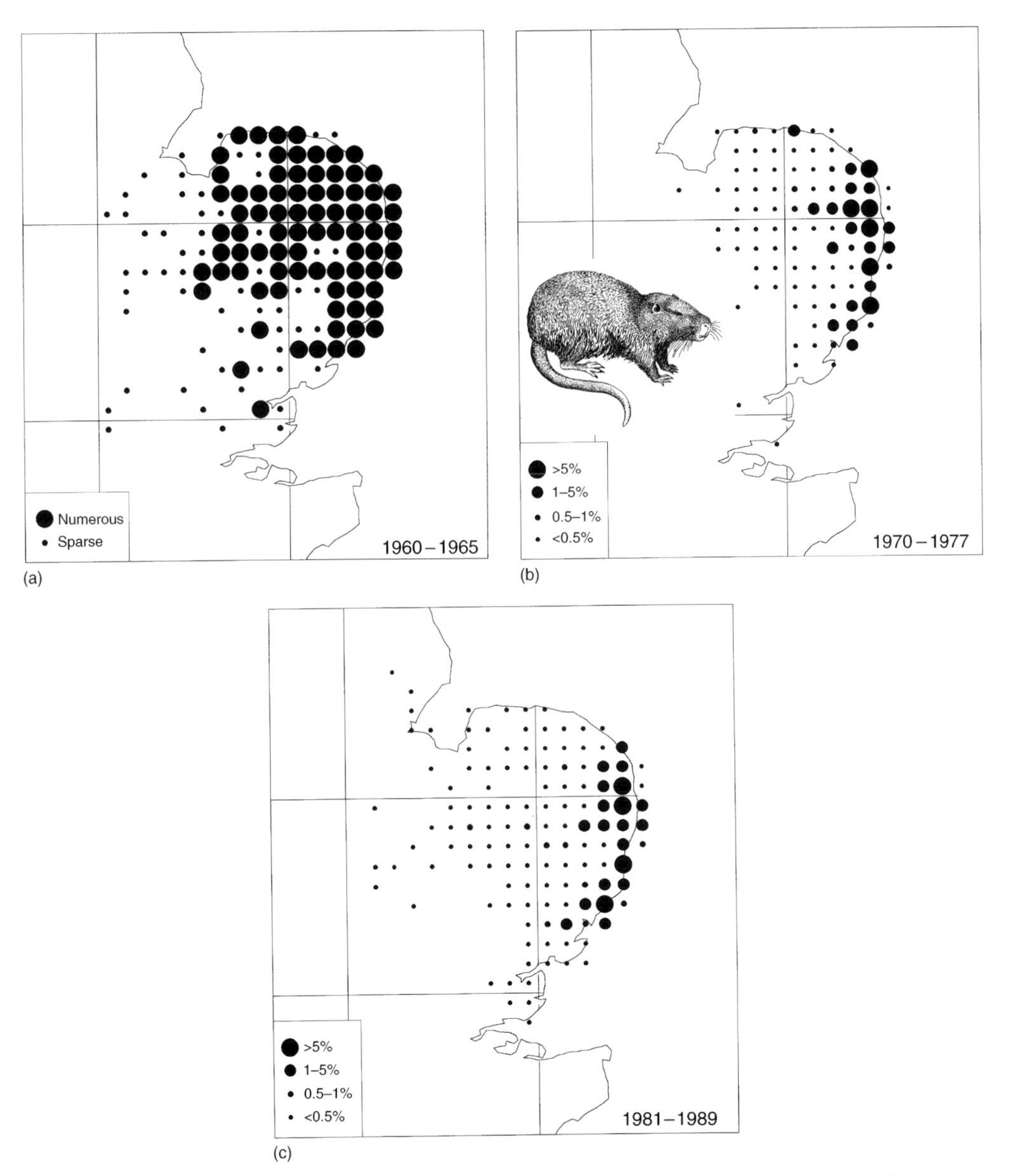

Figure 7.7 The distribution of the Coypu *Myocastor coypus* in England, (a) in 1961–65 (40 461 killed), (b) 1970–77 (59 290 killed) and (c) 1981–89 (34 838 killed) (after Norris 1967a, 1967b; Gosling and Baker 1989).

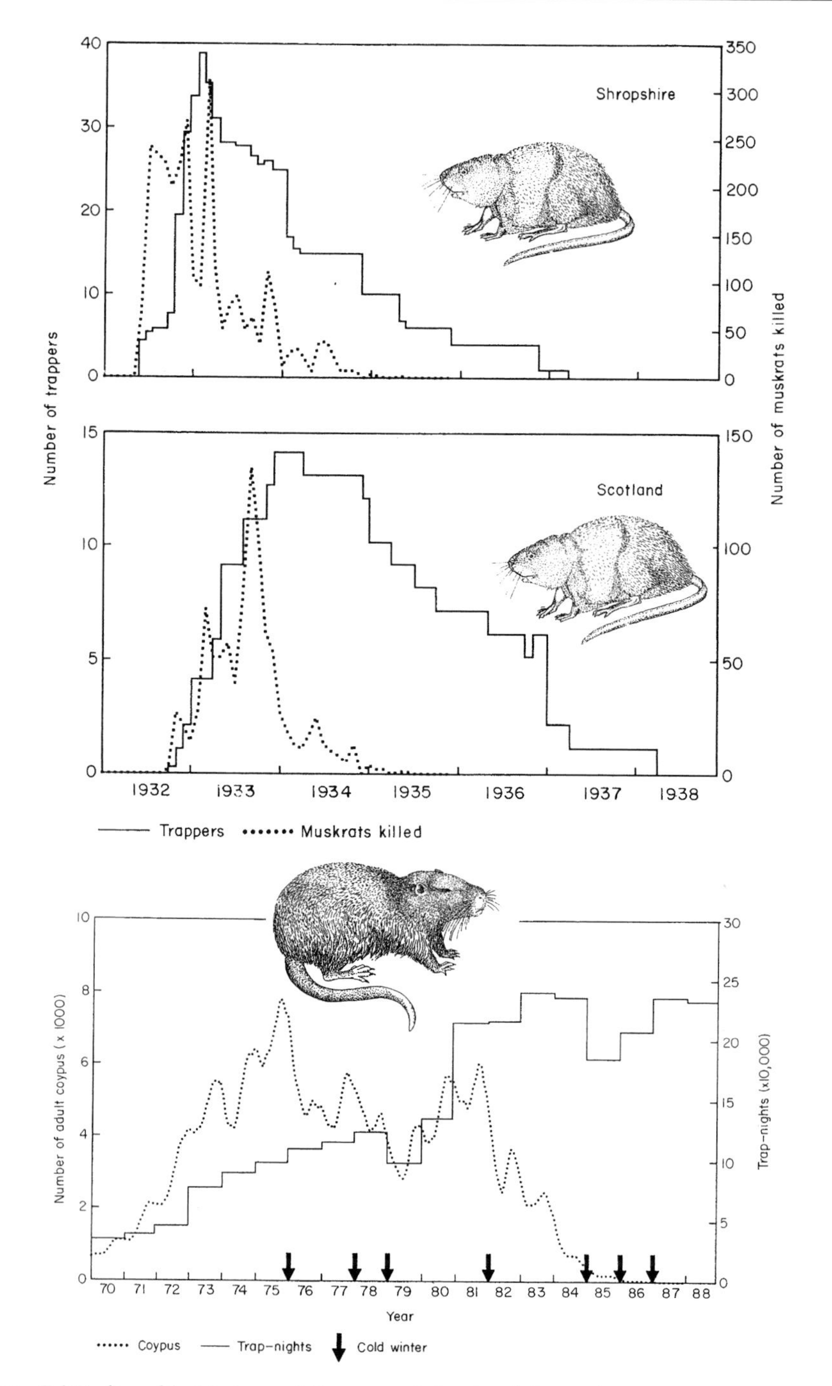

Figure 7.8 Decline of the Muskrat and Coypu populations in relation to the trapping campaigns against them. The numbers of trappers (Muskrat) and trap-nights (Coypu) were maintained well after the apparent decline of these rodents, ensuring their extermination. (After Gosling and Baker 1989 in *Biological Journal of the Linnean Society* **39**, by permission of Dr L.M. Gosling and the Linnean Society of London.)

hand, and numbers and range were severely reduced (Norris 1967a, 1967b). There were believed to be 200 000 loose in 1962, but 90% of them were killed. Between March 1963 and the end of 1965, the Coypu was largely eliminated outside of the Broads, and 18 000 had been killed. Unlike the Muskrat campaign, the Coypu campaign used cage traps, and caused very little mortality; a few animals were killed by drowning when water levels rose after heavy rain, but most captures were alive. Coypu were shot, and other species, which included Moorhens and Water Voles most commonly, but also Brown Hares, Rabbits, Brown Rats, Otters, Hedgehogs, Mink (but only five), Stoats and one Lamb, were released. The campaign used 12 trappers, 3000 traps, four boats, 13 vans, and caught 40 461 Coypu at 1.6 per trap night. It cost the tax-payers £70 500. The official campaign then finished, but limited local trapping (by 4–6 trappers) was thought sufficient to restrain the population in its strongholds. Indeed numbers remained low, and its range restricted, until 1971, when a series of mild winters once more allowed a rapid expansion first in numbers and then in range (Gosling *et al.* 1981). Numbers of trappers were increased, to 15 from 1973 to 1979, and then to 20. Meanwhile, an intensive research programme was underway, and indicated that with appropriate management and persistance, total eradication might be possible. Thus the final campaign was undertaken, and by 1987, the Coypu was extinct in Britain. It required 24 trappers, setting 48 traps per day, or 216 000 trap-nights per year, and cost £2.5 million. Trapping efficiency was increased by spreading the traps out to match the density of Coypu (fewer traps in the wider range, and a greater density in the Broads). Traps sited on rafts were found to be 50% more effective than land-based ones, and 600 were used, despite the added difficulties of servicing them. The population declined from an estimated 6000 in 1981 to 0 in 1987 (Fig. 7.8); the payment of a large bonus to the trappers if they completed their task by 1987, in advance of the termination of their contracts in 1991, ensured their persistance (Gosling and Baker 1988, 1989).

Discussion

This account of the recent introductions makes a rather dismal read. One of our most popular mammals, the Red Squirrel, has been exterminated from most of southern Britain by one introduction, another, the Water Vole, is certainly at risk, and has been recently added to the list of protected species (Schedule 5 of the Wildlife and Countryside Act). Our largest native land mammal, the Red Deer, looks as though it is doomed to be lost into a hybrid that is neither Red Deer nor Sika. The successful eradications of the Muskrat and Coypu are perhaps the two bright lights in this gloom, but it is quite clear that their eradication was the consequence of a number of fortunate characteristics that do not apply to most introductions. Riparian habitats are of limited extent, and constrain the distribution of amphibious animals to a relatively small and readily targeted part of the countryside. The species were both attacked while still restricted to relatively small geographic ranges, though in both cases it was a close run thing. Commercial trapping techniques and expertise were already available. Most importantly, scientific advice was available and heeded. By contrast, the species that have established themselves widely were not perceived to be problems until far too late. Although concern was expressed as soon as Mink were reported breeding in the wild, it was not taken seriously. Grey Squirrels were given 50 years to establish themselves thoroughly in the countryside before they were declared vermin in 1933, and Sika had at least 70 years before the serious threat they posed to either Red Deer or forestry was acknowledged. The parallel with the Ruddy Duck *Oxyura jamaicensis* or Canada Goose *Branta canadensis* is rather obvious. The sentimental response in some quarters is that any animal has the 'right' to live here. I cannot agree. The native species certainly have the right to be here, more right than we have, but the exotic introductions have no right to be here, and no right to eliminate our native species. The whole point of biodiversity is that different parts of the World have different faunas and floras. I do not want to go to Australia to see Red Foxes and Rabbits, I would expect to see Quolls and Wombats. Likewise, I expect to see Deer Mice and White-tailed Deer in America, not House Mice and Fallow Deer. Conversely, I want to see Red Squirrels and Polecats in the British countryside, not Grey Squirrels and American Mink. Increasingly, I am disappointed. Should I be sad that my local wallabies are doomed? No, not really. They have provided me with much interest, and several research papers, over the years, but they should be safely behind park walls, or in Tasmania. I should be seeing Roe Deer in their place.

CHAPTER 8

ISLAND RACES

Scilly Shrews

So far, this account of the history of British mammals has concentrated on the history of mammals in Great Britain. Even Ireland has received only a few passing mentions, and the smaller islands have been similarly ignored. However, there is an interesting story to be told, and an interesting small mammal fauna to discuss. While the connection of Great Britain to the rest of Europe during the low sea-levels of glacial times is clear enough, and the severance of the land connection in post-glacial times a critical event in the fauna's history, the connection of other islands to Great Britain is much less certain, and the times of severance perhaps therefore irrelevant. The presence of white-toothed shrews on the Scilly Isles and the Channel Isles, of continental field voles in Orkney and Guernsey, and of various sub-species or races of Bank Voles (e.g. on Skomer and Jersey), Wood Mice (e.g. on St Kilda, various Shetland Isles) and Field Voles (e.g. in the Hebrides) have played a large, and misleading, part in the accounts of our fauna that have been written. The poverty of the island faunas – the number of species they are missing – is also an important part of the story.

IRELAND: THE BIG ONE

It will undoubtedly upset my various Irish friends to discuss Ireland as though it is just the largest of the offshore islands. Yet the Irish mammal fauna encapsulates the problem of island faunas and island races in the most dramatic, but also increasingly enlightening, way. To start with, Ireland is large enough and has a wide enough range of habitats; there seems no reason to explain species' absences purely because the island is too small. Yet there are no Moles, Common Shrews, Water Shrews, Weasels, Polecats, Roe Deer, Field Voles, Water Voles or Dormice. In the fossil record for the island, neither Beaver nor Elk appear.

Of the mammals that do occur, the Mountain Hare is present in the form of a well marked race, the Irish Hare *Lepus timidus hibernicus*, which is somewhat larger and browner than the Scottish Mountain Hare. The Stoat *Mustela erminea hibernica* is convincingly smaller, at least in Northern Ireland, and has a wriggly flank line between the dorsal brown and ventral white (unlike the straight line of English Stoats, but like Weasels). The Otter, more surprisingly, also seems to be distinctively darker, almost black, and with a smaller pale throat patch – *Lutra lutra roensis* (Dadd 1970). Less certainly, the Wood Mice may also be a distinct race (*Apodemus sylvaticus celticus*). Other species that are reckoned to be native include the Red Fox, Badger, Pine Marten, Pigmy Shrew, Red Squirrel, Red Deer and, of course, several bats. Pipistrelle, Leisler's, Brown Long-eared, Daubenton's, Natterer's, Whiskered and Lesser Horseshoe Bats are certainly present (O'Sullivan 1994). Leisler's Bats are more widespread in Ireland than in Great Britain, and although confined to the west, the Lesser Horseshoe Bat numbers about 12 000, comparable with its British population of 14 000. Notice that there are several conspicuous absentees even among the bats: Noctule, Serotine, Barbastelle, Bechstein's and Greater Horseshoe Bats, at least, in that 1994 list. Several others are probably missing too, but are rare enough in Great Britain that they could have been overlooked so far (Grey Long-eared and Brandt's, perhaps also the recently discovered Nathusius' Pipistrelle). To emphasize the last point, I am told that Noctule, Barbastelle and Nathusius' Pipistrelle have been reported during 1997, though these survey results are not yet published (Ian Montgomery, pers. comm.). The absence of competition from the Noctule may explain the relative abundance of Leisler's Bat in Ireland, since their diets are very similar (Vaughan 1997).

A number of other species are now present in Ireland, but there are strong reasons for believing that they have been introduced fairly recently. Hedgehogs are known to have been in Ireland since about 1700, but are supposed to have been introduced, perhaps for food (Fairley 1984). There are supposed sub-fossil remains from some of the Irish caves (Edenvale, Newhall and Barntick Caves, Co. Clare; Barrett-Hamilton and Hinton 1911), but the contemporaneity of the Hedgehogs with the other cave fauna is doubtful. Bank Voles were certainly absent from Ireland until very recently. In 1964, a student looking for fleas on Irish mammals trapped a 'mouse' that he did not recognize, and fortunately took a specimen to show his supervisor (Claassens and O'Gorman 1965). The distribution of the Bank Vole in Ireland has since been mapped several times; it is confined to the south-west, but is spreading outwards at about 3 km per year (Smal and Fairley 1984; Fig. 8.1). Extrapolating backwards from its periphery in 1970, and assuming the same rate of spread, it can be suggested that it arrived in Ireland

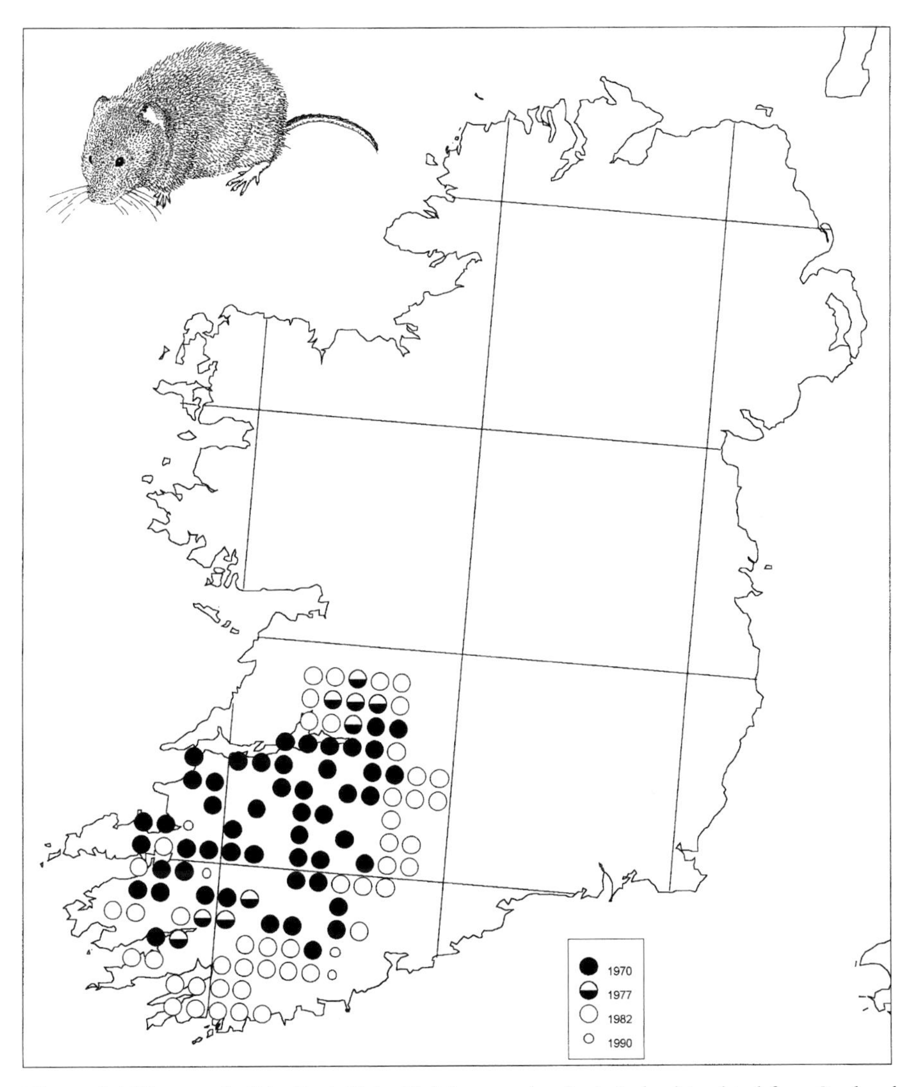

Figure 8.1 The spread of the Bank Vole *Clethrionomys glareolus* in Ireland (updated from Smal and Fairley 1984).

in about 1950, at a site on the south shore of the Shannon estuary in Co. Limerick. The Brown Hare has been introduced to Ireland for sporting purposes, but seems not to have spread very far; its farmland habitat seems to be occupied by the Irish Hare far more strongly than in Scotland or England, where it seems able to restrict the Mountain Hare to moorlands. Several exotic species that have been very successfully introduced to Great Britain also appear in Ireland: Rabbit, Fallow Deer, Grey Squirrel, Mink and Sika, though not the less widespread ones (Muntjac, Chinese Water Deer, Edible Dormouse). The 'old commensals', House Mouse, Black Rat and Brown Rat, also reached Ireland, and the later rat displaced the earlier one in the same way.

Thus the Irish mammal fauna, just like that of Great Britain, is a mixture: some ancient and presumably native species, some certain late introductions, and some of dubious status, probably early introductions. The prime puzzle is to explain how the native species reached Ireland when so many British species did not; a secondary issue is to disentangle the genuine native species from the early but undocumented introductions. The species which are sufficiently distinct to be recognizable subspecies have presumably been isolated in Ireland for a long period, but how long? When, indeed, did they get to Ireland?

Geological Evidence

The Pleistocene history of Ireland's flora and fauna is a long and extensive record, rivalling that of Great Britain in its scope and early study. Unfortunately, many of the mammal faunas came from caves that were excavated at the turn of this century, well before the complexities of Pleistocene climatic changes were appreciated, and with only passing regard for the stratigraphy. The pollen record of the Irish peat deposits is an excellent one, discerned in parallel with studies elsewhere in the British Isles and in Europe, but it does little to illuminate the cave faunas. Until very recently, therefore, the viscissitudes of the Irish climate and vegetation, which parallel those elsewhere in Europe, have been well understood, but their relationship to the mammal faunas has been obscure (except for Irish Elk from under the self-same peat bogs; see Chapter 2). An important paper giving ^{14}C dates for many of the significant cave fossils (Woodman *et al.* 1997) has now solved many of the most obvious problems, though not without raising a few more.

The pollen record shows at least two interglacials in Ireland prior to the present one; the earlier, named the Gortian, is well represented and is broadly contemporary with the Hoxnian in England, while the shadowy existence of a later one might be equivalent to the Ipswichian (Coxon and Waldren 1995). The Devensian Glaciation of England is matched by the Midlandian of Ireland. This was a period when ice covered most of northern Ireland down to about the latitude of Tipperary, and also the mountains of the south-west, but left a cold tundra plain across from Kerry to Waterford (Figure 8.2). As in England, south of the contemporary ice sheet, a fauna of lemmings and reindeer seems likely, but it is hard to imagine that most of the current fauna or flora could have survived. As the ice started to melt, around 15 000 years ago, a better tundra vegetation of willows and dwarf birches developed, and two mammals were certainly present then, Reindeer and Irish Elk (Stuart and Wijngaarden-Bakker 1985, Stuart 1986; see Chapter 2); in Ireland, this is called the Woodgrange Interstadial, clearly the equivalent of the Windermere Interstadial in Britain and the Allerød in Scandinavia. However, the fossil record from the peat bogs contains no other mammals (though rodent tooth marks on an Irish Elk specimen at Ballybetagh indicate that others were present). The peat deposits at such sites as Ballybetagh are terminated by sterile sands and gravels that indicate a return to colder conditions, the Nahanagan Stadial, equivalent toYounger Dryas or Pollen Zone III times. These wiped out the Irish Elk (Chapter 2), and it is a plausible argument that the conditions were too severe for any other, now native, species to persist. July temperatures are considered to have been around 8°C, compared with 18°C in London

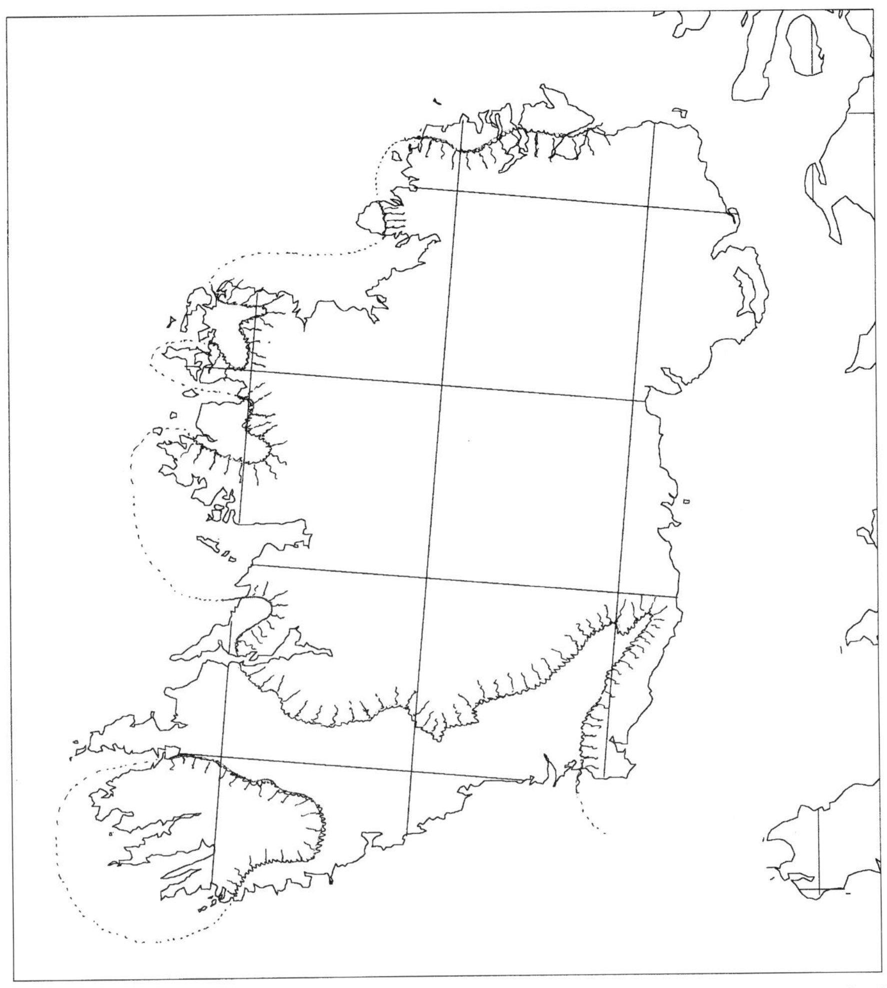

Figure 8.2 The extent of ice cover at the maximum of the Last Glaciation (Midlandian) in Ireland (after Woodman *et al.* 1997).

at present (Yalden 1982). In that case, all the present native species had to get to Ireland in the Post Glacial, and presumably from Great Britain. If they did, where was the land bridge, and how long did it last? On the basis of the evidence for lowered sea-levels and the timing of the rising sea-levels in the Post Glacial (Chapter 3), there was only a short period of about 2000 years available for mammals to cross the southern North Sea into England. However, there is an 80–90 m deep channel, scoured out by melting glaciers, up the whole length of the Irish Sea, and in the north the weight of remaining ice caps depressed northern Ireland and south-western Scotland even lower relative to sea-level. Although I tried to argue for a short period of relatively low sea-level that might have allowed a short-lived and soggy land-bridge from Scotland across to Ireland at about 8500 b.p. (Yalden 1981), this was surely wrong (Devoy

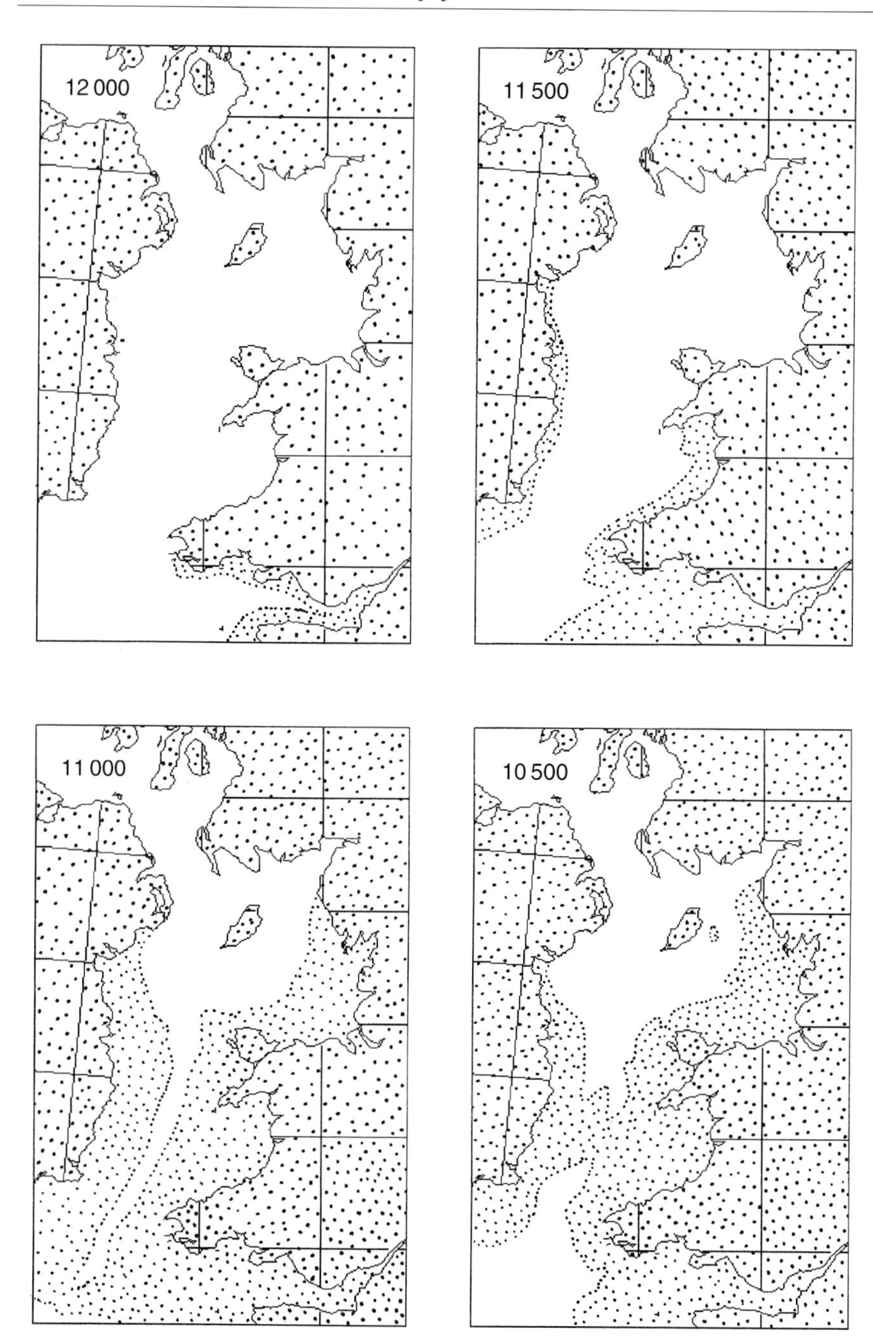
12 000
11 500
11 000
10 500

Figure 8.3 The fore-bulge theory of land-bridge formation to Ireland (after Wingfield 1995). The weight of ice on the Scottish Mountains would have pushed the land down relative to sea-level, creating a bulge around the periphery of the depression. This bulge would have produced a low-lying and transitory land-bridge that migrated northwards up the Irish Sea as the weight of ice melted from the Scottish Mountains.

1985, 1986, 1995). A 90 m deep channel would surely have been flooded by rising sea-levels by about 12 000 b.p. (cf. Fig. 3.5), in the Late Glacial rather than in the Post Glacial, and there is direct evidence of marine sediments off-shore in Dublin Bay as early as 12 000–11 000 b.p. (Devoy 1995). Further north, between Islay off Scotland and north-east Ireland, where the current sea-floor is only about –55 m at present, the land was in full glacial times depressed by the weight of the ice-cap over the Scottish mountains to as much as 80–90 m below present sea-level. While it started to 'bounce back' as the ice started to melt about 16 000 b.p., it is inconceivable that it produced a land-bridge in this region.

Thus it is clear that the hypothesis developed by Yalden (1982) was wrong in some of its details. Two modifications are indicated. One is that the south of Ireland was deglaciated somewhat earlier, and provided habitat for some mammals, in the Late Glacial. It has been well established that Reindeer and Irish Elk reached Ireland in the Late Glacial, and other species must have accompanied them. While conditions in the Nahanagan/Younger Dryas were evidently too severe for the Irish Elk to survive, Reindeer did survive through to Post Glacial times, as they did in England, and some of the more tolerant of the native species, notably the Stoat and Mountain Hare, perhaps also the Pigmy Shrew (which currently range in Europe well north of the Arctic Circle), could well have entered Ireland with the Reindeer, and survived with them to become part of the native fauna. The problem remains to determine when the land-bridge existed, and why various other species (notably Field Vole, Water Vole and Common Shrew, which also range far to the north in Europe) did not also cross. A very intriguing mechanism is suggested by Wingfield (1995). He develops the theory of a fore-bulge land-bridge, low lying and transitory in nature, passing up the Irish Sea during the period of deglaciation (Fig. 8.3). The fore-bulge is a ridge in the earth's crust caused as a balance to the depression caused by the weight of the ice-cap further north, and as the ice melted, so such a ridge would migrate northwards, leaving little direct evidence of its existence. On his suggestion, a land-bridge to south-west Ireland might have existed around 11 000–10 000 b.p. It would have allowed larger mammals, and those not needing to burrow, to cross to Ireland, while excluding species such as the Mole and Common Shrew. One problematic species is the Wood Mouse, a species which has been regarded as a human introduction, perhaps by the Vikings (Berry 1969). However, it is a species which does travel long distances, and could well have traversed a low-lying land-bridge. On sand-dunes, average nightly movements can be 1200 m, ten times larger than movements in woodlands (Corp *et al.* 1997). A few teeth were found in Mesolithic deposits near Dublin, immediately below a soil dated to 7600 b.p. (Preece *et al.* 1986), indicating that it was certainly present in Ireland much earlier than Viking times, and there are also Iron Age records from Dublin (Wijngaarden-Bakker 1986). This matches the considerable biochemical variation in Irish Wood Mice, which argues for an extended presence, whereas the uniformity of Irish Bank Voles matches their known recent introduction (Byrne *et al.* 1990, Ryan *et al.* 1996).

The Lusitanian Element

Discussion of the Irish fauna has in the past been complicated by consideration of the 'Lusitanian Element' in the fauna and flora. There are a number of species – no

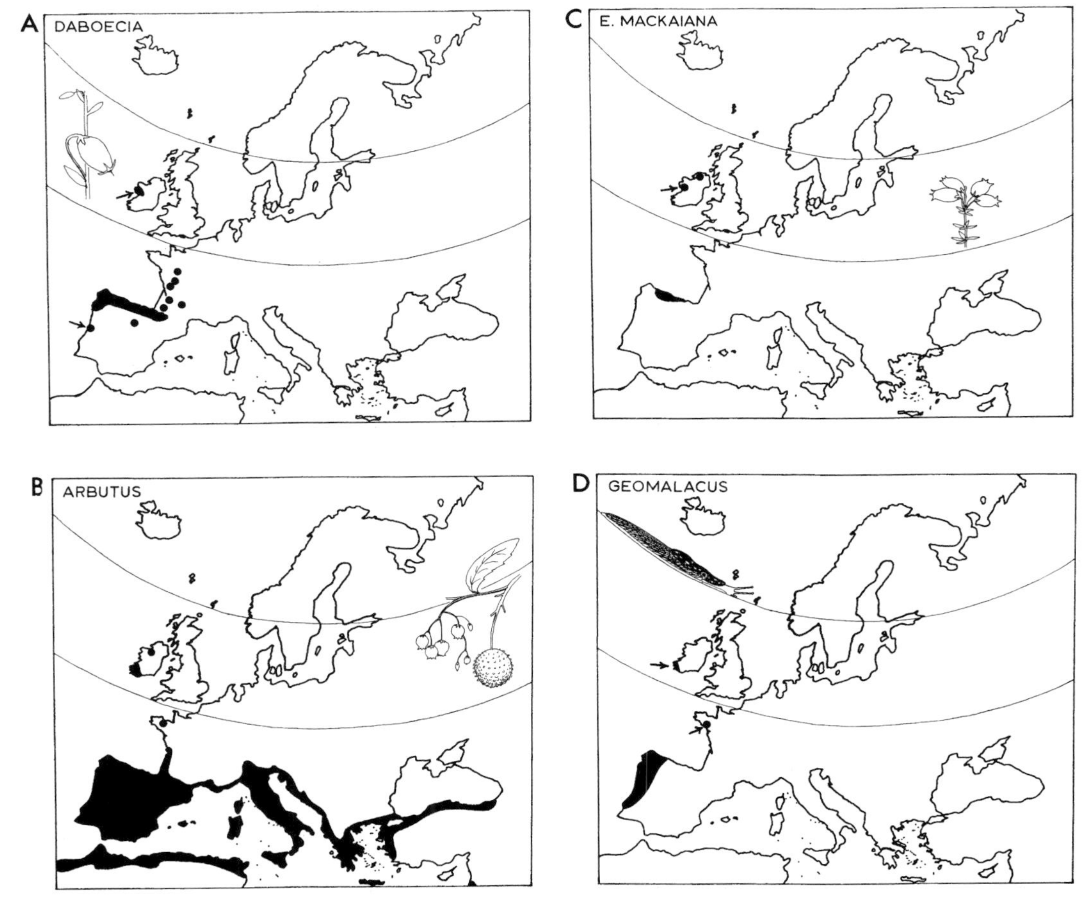

Figure 8.4 Distribution maps for some of the Lusitanian elements of the Irish biota (after Yalden 1982, and sources therein, in *Mammal Review* **12**, by permission of Blackwell Science).

mammals – which occur in Ireland, particularly in the south-west, but not in England. They do occur further south, in northern Spain, Portugal or western France. Among these species are the Strawberry Tree *Arbutus unedo*, two heaths, *Erica mackiana* and *Erica mediterranea*, the Kerry Slug *Geomalacus maculosus*, a small moth *Calamia tridens* and a couple of woodlice (Fig. 8.4). Geographically allied to them are a few other species that occur also in south-west Britain, such as the Cornish Heath *Erica vagans*. The Natterjack Toad *Bufo calamita*, which occurs in Kerry, and in Britain more extensively up the Irish Sea coast as far as the Solway Firth, could also be regarded as a member. The argument has been that these characteristically south-western species might have survived in a now-submerged land area, 'Lusitania', off the south west of Britain in the Celtic Sea during the cold stages of a glacial period (or even periods, since several of them seem to have been present in Ireland during the Gortian Interglacial). Implicit in this theory is the idea that Ireland, then as now, was bathed by the Gulf Stream, and was relatively mild. This is simply not true. Ireland was as cold as the rest of Britain, during the Devensian as a whole and during the Nahanagan Late Glacial cold phase in particular. The polar front, where cold Arctic water met the milder Atlantic water, was off south-west Ireland in Nahanagan times (cf. Fig. 1.7). These Lusitanian elements are noted for their intolerance of cold conditions, especially frost, and could not have survived there then. On the other hand, since the climate improved so rapidly in the Post Glacial, it is possible that these coastal species took early advantage of the improving conditions, and even made use of a fore-bulge land-bridge. More probably, they, or some of them, took assisted passage, either with birds, or with early traders; trade between Spain, Ireland and Cornwall (for tin) was established in pre-Roman times.

Palaeontological Evidence

The dates recently published by Woodman *et al.* (1997) add much more certainty to this discussion. There was a mammal fauna in Ireland in the Mid-Devensian, including Spotted Hyaena, Brown Bear, Reindeer, Red Deer, Giant Deer and Mammoth, with dates between 39 000 and 30 000 b.p. (Table 8.1). It is represented by fossil faunas from the caves of Castlepook, Co. Cork, Ballynamintra and Shandon, Co. Waterford (Fig. 8.5). This is roughly contemporary with, perhaps slightly later than, the faunas of the Upton Warren Interstadial in Britain (see Chapter 2). Slightly later mammals from the same caves, dated 28 000 to 24 000 b.p., include Norway Lemming, Irish Hare, Wolf, Brown Bear, Spotted Hyaena, Horse, Red Deer (surprisingly), Reindeer and Mammoth. At about the Glacial maximum, with dates of 20 300–19 950 b.p., only Arctic Fox, Arctic Lemming and Mammoth were certainly present, all from Castlepook Cave.

The most interesting records for a discussion of the present-day fauna are those dated to the Woodgrange (= Allerød) Interstadial. In addition to the well-known presence of Reindeer and Irish Elk, Red Deer, Wolf, Brown Bear and Irish Hare were present, with dates of 12 190–11 790 b.p. During the Nahanagan Cold Stage, there are three records of Arctic Lemming and two of Stoat to accompany the various dates for Reindeer. It is clear that Irish Elk did not survive this cold phase (Chapter 2), and Red Deer seem to have died out as well, but the other species present in the Woodgrange could well have survived through to the Post Glacial. Direct evidence of mammals early in the Post Glacial from this project include two Brown Bears (from Derrykeel Bog, 8880 b.p. and Donore Bog, 8930 b.p.), a Wild Boar (Kilgreany Cave, 8340 b.p.) and, very remarkably, Lynx (also Kilgreany Cave, 8875 b.p.). Wild Cat bones were not dated in this programme, and Wild Cats are not now found in Ireland, but bones from both cave (Edenvale Cave, Co. Clare) and archaeological sites (Newgrange) indicate that it was once a member of the native fauna (Savage 1966, Wijngaarden-Bakker 1974).

Negative information should be noted, as well. As in Great Britain, it has been suggested that the Horse might have been wild in Ireland, before the Neolithic introduction of domestic horses. Seven likely specimens, from caves, gravels and peats, were dated for this programme, but only one (from the Mid-Devensian at

Figure 8.5 The location of cave and other archaeological sites in Ireland.

Table 8.1 Radiocarbon-dated records of Irish mammals (from Woodman *et al.* 1997)

Site	*IGR*	*Specimen*	*Date*
Arctic Lemming *Dicrostonyx torquatus*			
Castlepook Cave	R603009	Mandible	20 300 b.p.
Kilgreany Cave	X172944	Mandible	10 360 b.p.
Keshcorran Cave	G710130	Mandible	10 060 b.p.
Edenvale Caves	R322747	Mandible	10 000 b.p.
Norway Lemming *Lemmus lemmus*			
Castlepook Cave	R603009	Mandible	27 930 b.p.
Irish Hare *Lepus timidus*			
Shandon Cave	X292950	Humerus	28 240 b.p.
Plunkett Cave	G710130	Foot bones	12 190 b.p.
Ballynamintra Cave	X108955	Tibia	1 500 b.p.
Brown Bear *Ursus arctos*			
Castlepook Cave	R603009	Femur	37 870 b.p.
Ballynamintra Cave	X108955	Calcaneum	35 570 b.p.
Castlepook Cave	R603009	Calcaneum	33 310 b.p.
Shandon Cave	X292950	Calcaneum	32 430 b.p.
Foley Cave	R686099	Teeth	26 340 b.p.
Plunkett Cave	G710130	Humerus	11 920 b.p.
Red Cellar Cave	R645417	Tibia	10 650 b.p.
Donore Bog	S370877	Teeth	8930 b.p.
Derrykeel Bog	N169032	Tooth	8880 b.p.
Spotted Hyaena *Crocuta crocuta*			
Castlepook Cave	R603009	Metapodial	>45 000 b.p.
Castlepook Cave	R603009	Bone	34 300 b.p.
Castlepook Cave	R603009	Scapula	24 000 b.p.
Wolf *Canis lupus*			
Shandon Cave	X292590	Mandible	27 500 b.p.
Castlepook Cave	R603009	Mandible	23 470 b.p.
Plunkett Cave	G710130	Mandible	11 150 b.p.
Elderbush Cave	R310740	Radius	1730 b.p.
Arctic Fox *Alopex lagopus*			
Shandon Cave	X292950	Tibia	29 980 b.p.
Castlepook Cave	R603009	Ulna/Tibia	19 950 b.p.
Lynx *Felis lynx*			
Kilgreany Cave	X172944	Femur	8875 b.p.
Stoat *Mustela erminea*			
Killavullen Cave	W655998	Tibia	10 680 b.p.
Kilgreany Cave	X172944	Skull	9980 b.p.
Coffey Cave	G710130	Mandible	7650 b.p.
Foley Cave	R686099	Humerus	305 b.p
Pine Marten *Martes martes*			
Foley Cave	R686099	Femur	2555 b.p
Kilgreany Cave	X172944	Tooth	2780 b.p.

Site	*IGR*	*Specimen*	*Date*
Mammoth *Mammuthus primigenius*			
Castlepook Cave	R603009	Rib	34 100 b.p.
Castlepook Cave	R603009	Bone	33 500 b.p.
Shandon Cave	X292950	Molar	27 150 b.p.
Castlepook Cave	R603009	Rib	20 360 b.p.
Horse *Equus caballus*			
Shandon Cave	X292950	Scapula	27 630 b.p.
Newhall caves	R322747	Metatarsal	1675 b.p.
Plunkett Cave	G710130	Tooth	1580 b.p.
Drumquin	H310720	Tooth	635 b.p.
Sydenham Station	J370750	Tooth	120 b.p.
Wild Boar *Sus scrofa*			
Kilgreany Cave	X172944	Metapodial	8340 b.p.
Sutton	O260390	Ulna	7140 b.p.
Dalkey Island	O270260	Scapula	6870 b.p.
Irish Elk *Megaloceros giganteus*			
Castlepook Cave	R603009	Phalanx	37 200 b.p.
Castlepook Cave	R603009	Phalanx	32 200 b.p.
Castlepook Cave	R603009	Astragalus	32 060 b.p.
Ballybetagh	O2122	Antler	15 170 b.p.
Shortalstown	T1130	Ass. Plants	12 160 b.p.
Garransdarragh Bog	W645785	Skull	11 820 b.p.
Edenvale Caves	R322747	Radius	11 750 b.p.
Killuragh Cave	R782488	Phalanx	11 510 b.p.
Unknown	?	Antler	11 380 b.p.
Ballynamintra Cave	X108955	Ulna	11 110 b.p.
Kilgreany Cave	X172944	Tooth	10 960 b.p.
Unknown	?	Ribs	10 920 b.p.
Ballybetagh	O2122	Mandible	10 610 b.p.
Red Deer *Cervus elaphus*			
Shandon Cave	X292950	Humerus	26 090 b.p.
Plunkett Cave	G710130	Tibia	11 790 b.p
Stonestown	N472729	Femur	4190 b.p.
Ventry Bay	S380990	Antler	3985 b.p.
Magheralin	D140540	Antler	3760 b.p.
Newhall Cave	R322747	Maxilla	2270 b.p.
Sydenham Station	J370750	Mandible	2020 b.p.
Killuragh Cave	R782488	Metatarsus	955 b.p.
Reindeer *Rangifer tarandus*			
Castlepook Cave	R603009	Radius	38 650 b.p.
Castlepook Cave	R603009	Metacarpal	35 200 b.p.
Ballynamintra Cave	X108955	Metatarsal	33 630 b.p.
Shandon Cave	X292950	Radius	30 840 b.p.
Foley Cave	R686099	Teeth	28 000 b.p.
Castlepook Cave	R603009	Antler	12 480 b.p.
Kilgreany Cave	X172944	Tooth	10 990 b.p.
Edenvale Caves	R322747	Radius	10 850 b.p.
Kilgreany Cave	X172944	Bone	10 700 b.p.
Roddans Port	J6373	Antler	10 250 b.p.

27 000 b.p.) was older than 1580 b.p.; horses were not native wild mammals in Post Glacial Ireland, and their earliest appearance is the example from Newgrange, dated to about 4000 b.p. (Wijngaarden-Bakker 1974). A similar result was obtained for Elk (*Alces alces*). Though generally regarded as absent from Ireland, four specimens have been recovered from peat bogs (Monaghan 1989). Two turn out to have dates of 10 b.p. (i.e. about 1940) and must represent discarded souvenirs of hunting trips to Scandinavia or even Canada. Even more remarkably, there are no remains of Red Deer dated earlier than the one, 4190 b.p., from a peat bog at Stonestown, Co. Longford. The supposed record from the Mesolithic site of Mount Sandel is now considered to have been misidentified (Woodman *et al.* 1997). The Red Deer has always been confidently listed as a native Irish mammal, but it seems that either human introduction must be invoked, or Red Deer swam from Scotland at some fairly late date in the Post Glacial.

In summary, it seems that the present mammal fauna of Ireland includes, as native mammals, Stoat and Irish Hare which survived through the Nahanagan cold period after they colonized Ireland in the Woodgrange Interstadial, while sea-levels were still low. The now-extinct Wolf, Brown Bear and Lynx must also have done so. Three other native species which occur, now, well north in Europe, Pigmy Shrew, Red Fox and Pine Marten, may be presumed to have been also part of this early fauna, but direct evidence is lacking. A further suite of supposed native mammals are inhabitants of deciduous woodland, and surely cannot have survived the Nahanagan cold period in Ireland. The evidence of one of them, Wild Boar, implies that they reached Ireland very early in the Post Glacial, yet the evidence for a land-bridge is sketchy, at best. Wingfield's hypothesis of a fore-bulge land-bridge offers the best possibility, but implies a very narrow latest land-bridge at about 9600 b.p. This is contemporary with such sites as Star Carr and Thatcham, where there is evidence of Wild Boar and Badger in Britain; other Irish members of this group would be Wood Mouse, Red Squirrel and Wild Cat, and perhaps Red Deer if, despite the lack of current evidence, it was in fact native to Ireland. Among the species that did not get to Ireland, the most surprising absentees on this scenario are Beaver and Elk, both capable swimmers, both distributed well north in Europe at present, both present early in Post Glacial England, and both capable, one would think, of negotiating a narrow land-bridge. On the other hand, they and the Water Vole, another absentee, are fond of freshwater habitats, and water weeds are part of their diet. A maritime land-bridge might not have been attractive. Sand-dunes are not attractive to Bank Voles and Field Voles, either, so perhaps they and their main predator, the Weasel, are absent for that reason. King and Moors (1979) have argued that Weasels must have reached Ireland, when there were lemmings for them to eat, and then died out when their prey disappeared. Perhaps, but there is no evidence that they did, despite the evidence for Stoats in four caves (Kilgreany, Foley, Coffey and Killavullen Caves) and one Iron Age site. Sleeman (1986) remarks on the scarcity of prey species in this fauna, compared with the wealth of carnivores; it is hard to believe that Mountain Hares, Wood Mice and Wild Boar really sustained such a large suite of carnivores, but that is what the present evidence suggests.

A couple of anomalous details deserve a mention. An apparent ox bone from Sutton Shell Midden, Co. Dublin, gave dates of 6660 and 6560 b.p., but Aurochs has never been reported from Ireland, yet domestic cattle had not been introduced to

Western Europe, let alone Ireland, so early as this. The bone in question is the shaft of a long bone, not very diagnostic, and could belong to some other species (perhaps Brown Bear). Despite the absence of Field Voles from Ireland, there is a toothless skull from Kilgreany Cave of one individual, which seems morphologically close to specimens from the island of Muck (Savage 1966). Savage argues that it was dropped in the cave by a passing owl. Perhaps a Short-eared Owl, migrating from Scotland, called by, though the species is not known for roosting in caves. There is at least one record of a Mole being recovered from an owl pellet in Ireland, so the argument is plausible (Fairley 1984).

Extinctions and Introductions

Just as in Great Britain, the current fauna of Ireland has been impoverished by extinctions and enriched by introductions. The Wild Boar seems not to have survived into historical times, but the timing of its disappearance is obscured by the appearance of domestic pigs, which might have been domesticated from it or interbred with it. The archaeological records, of Wild Boar at Neolithic Newgrange and Dalkey Island, indicate that it survived to at least 5600 b.p., but its later history is lost. The Brown Bear likewise survived to Neolithic times, being recorded from Carrowkeel, Co. Sligo and Lough Gur, Co. Limerick. The Wild Cat is recorded from Neolithic Lough Gur, Bronze Age Ballinderry, Co. Offaly, and Iron Age Larrybane, Co. Antrim and Uisneach, Co. Westmeath. These three species certainly seem to have become extinct before the beginning of the Christian era (Wijngaarden-Bakker 1974, 1986). The Wolf has a better documented history, and a much later survival. Historical mention of the Wolf as a major pest in Ireland by Augustin in 655 A.D. and the passing mention by Shakespeare of 'Irish wolves howling against the moon' indicate that they were well known. Measures to eliminate Wolves in Ireland were passed in 1614, 1652 and 1653, when they were clearly still numerous. However, the active extermination of Wolves which these measures engendered had perceptible success by 1683, when it was remarked that Wolves were scarce in Co. Leitrim. Fairley (1984) collates numerous records of the 'last' Wolf, in various counties, starting with Antrim in 1692 and 1712. The absolute last is about 1786 from Mount Leinster, on the Carlow/Wexford border. The Red Squirrel is confidently regarded as native to Ireland, but it certainly became extinct by the 18th century; Fairley (1983) draws attention to its surprising absence from the fur returns which document the export between 1697 and 1819 of an estimated 114 million Rabbits, as well as 10 000 deer, 42 053 Otters, 49 302 Foxes and 14 294 Mountain Hares. The current population of Red Squirrels is the consequence of well-documented reintroductions, starting from 1815 (Shorten 1954). In the absence of an archaeological record (but squirrels are always rare in archaeological sites), it has been suggested that they might have been imported at an early (prehistoric) date for their fur, and the suggestion has been repeated for Red Fox and Pine Marten (Fairley 1984). A dearth of evidence forces us to leave this as speculation, but the earliest Pine Marten bones so far dated (Table 8.1) are consistent with this possibility.

The influence of humans in importing mammals is most clearly seen in the case of Sheep and Cattle. Radiocarbon dates of 5050 b.p. for a Sheep humerus from Dalkey

Island and for Cattle of 5510 b.p. from Ferriter's Cove and of 4820 b.p. from Dalkey Island show that domesticated ungulates were introduced to Ireland as early as to anywhere in the British Isles. Domestic dog was recorded even earlier, from the Mesolithic of Mount Sandel. The earliest mention of Fallow Deer in Ireland is from 1296, and of Rabbits from 1282 (Fairley 1984). Sika were released into the Wicklow Hills from Powerscourt in 1860, and Grey Squirrels were released at Castle Forbes in 1913 (see Chapter 7). As in Great Britain, Sika have hybridized with Red Deer, and Grey Squirrels are displacing the Red Squirrels, though their spread is slower because of the scattered nature of woodland in Ireland.

Smaller Islands

If the evidence for land-bridges to Ireland in the Post Glacial is so doubtful, the situation for most of the other islands is consequently but paradoxically clearer; most of them, particularly the northern ones, cannot possibly have been connected to mainland Britain. The notion that the Outer Hebrides, St Kilda, Shetlands or Orkneys got their sparse faunas across a land-bridge can be discounted. The situation for some of the Inner Hebrides is less certain, and the Isle of Man may have been connected to England for a short time in the Post Glacial if Wingfield (1995) is correct about the place and timing of a forebulge land-bridge. The situation of the Isles of Scilly and some of the Channel Isles is also less certain. What, in detail, is the problem, and how much evidence do we have to solve it?

Idiosyncratic Faunas and Obsolete Theories

When Millais was returning one evening in August 1886 from fishing on Orkney, he noticed what appeared to be a Water Vole running along a sheep track (Berry and Rose 1975). This large, almost black, animal turned out to be something entirely new to Britain, duly described as *Microtus orcadensis*. Later collecting revealed similar but slightly different voles on several of the other Orkney Islands (Sanday, Westray, South Ronaldsay, Rousay), all in due course made subspecies of *orcadensis*. It was pointed out at the time that these voles more closely resembled the Continental Common Vole *Microtus arvalis* than the usual British Field Vole, *M. agrestis*, but it was only in 1959 that cross-breeding experiments, confirming that Orkney and German *M. arvalis* would readily interbreed, convinced everyone to regard the Orkney Vole as a subspecies, albeit well-marked, of the Continental vole. Meantime, similar large, though not dark, voles from Guernsey had also been described as a new species, *M. sarnius*; that too is now regarded as a subspecies of *M. arvalis*. Even more remarkable than the discovery of the Orkney Vole were the white-toothed shrews found on the Scilly Isles in 1923. These too were originally named as a new species, *Crocidura cassiteridum*, but are now regarded as a subspecies of the more widespread Continental *C. suaveolens*. The Channel Isles also have white-toothed shrews, but of two species. Jersey and Sark have the Lesser White-toothed Shrew *C. suaveolens*, as have the Scilly Isles, but on Guernsey, Alderney and Herm it is the Greater White-toothed Shrew *C. russula* which is present (Delany and Healy 1966). Other islands have races of the more familiar small mammal species. Most diverse, at least in the taxonomists' lists, is the Wood Mouse. Subspecies have been named for St Kilda, for Foula, Fair Isle and Yell in the Shetlands, for Mingulay and Lewis in the Outer Hebrides, Rhum, Tiree, Gigha and Islay in the Inner Hebrides, and for Bute, Great Cumbrae and Arran in the Firth of Clyde (Table 8.2). Subspecies of Bank Vole are named only for Raasay, off Skye, Mull, Skomer off Pembrokeshire and Jersey. Field Vole subspecies are described from North Uist, Eigg, Muck, Islay and Gigha; a Highland race, *M. a. neglectus* has also been distinguished from a race in the rest of Britain, *M. a. hirtus*. The House Mice of St Kilda were also a distinct subspecies, *Mus musculus muralis,* but they became extinct after the human population evacuated the island in 1930. These subspecies have been distinguished on a variety of supposed discriminatory features, particularly large size, but also variations in coat colour and tooth morphology. Surprisingly, the shrews have not been subjected to much taxonomic discrimination.

The patterns of supposed subspecies are closely related to the idiosyncratic distribution of the different species on the different islands (Table 8.3). The Scilly Isles have Wood Mice and House Mice as well as white-toothed shrews, but no other small mammals. The islands with large forms of Bank Vole, Skomer, Raasay and Jersey, do not have Field Voles, though Mull has both. The Field Vole *Microtus agrestis* is present on most of the Inner Hebrides, and on North and South Uist, but is absent from Lewis and Barra, as well as Orkney and Shetland, Isle of Man and all the Channel Isles.

Table 8.2 The subspecies of small mammal named from the British Isles and Channel Isles. Very few of these would be regarded now as taxonomically valid (identifying recognizably distinct forms), or useful. Those marked ⋆ might be useful.

Microtus agrestis		*Apodemus sylvaticus*	
M. a. exsul	North Uist	*A. s. celticus*	Ireland
M. a. mial	Eigg	⋆*A. s. fridariensis*	Fair Isle
M. a. luch	Muck	*A. s. granti*	Yell, Shetland
⋆*M. a. macgillivrayi*	Islay	*A. s. thuleo*	Foula
M. a. fiona	Gigha	⋆*A. s. hirtensis*	St Kilda
M. a. neglectus	Highland Scotland	*A. s. hebridensis*	Lewis
M. a. hirtus	Southern Britain	*A. s. nesiticus*	Mingulay
		A. s. hamiltoni	Rhum
Microtus arvalis		*A. s. tirae*	Tiree
⋆*M. a. orcadensis*	Mainland Orkney	*A. s. maclean*	Mull
M. a. ronaldshaiensis	South Ronaldsay	*A. s. larus*	Jura
M. a. rousaiensis	Rousay	*A. s. tural*	Islay
⋆*M. a. sandayensis*	Sanday	*A. s. ghia*	Gigha
M. a. westrae	Westray	*A. s. butei*	Bute
⋆*M. a. sarnius*	Guernsey	*A. s. cumbrae*	Great Cumbrae
		A. s. fiolagan	Arran
Clethrionomys glareolus		*Mus domesticus*	
⋆*C. g. skomerensis*	Skomer	⋆*M. d. muralis*	St Kilda
⋆*C. g. alstoni*	Mull		
⋆*C. g. erica*	Raasay	*Sorex araneus*	
⋆*C. g. caesarius*	Jersey	*S. a. granti*	Islay, Jura

Table 8.3 The distribution of small mammals on the various British Isles, and the Channel Islands (after Corbet 1961, Yalden 1982). OH = Outer Hebrides, IH = Inner Hebrides. [1]The provisional distribution atlas (Arnold 1978) included single records of Bank Vole from Islay and Common Shrew from Lewis, but these have not been confirmed. [2]The Common Shrew on Jersey is, strictly, the French Shrew *Sorex coronatus*. Note that the Wood Mouse is found most widely on the islands, and the Pigmy Shrew is found on more of the outer islands than the Common Shrew.

Islands	*Common Vole* Microtus arvalis	*Field Vole* Microtus agrestis	*Bank Vole* Clethrio. glareolus	*Wood Mouse* Apodemus sylvaticus	*Common Shrew* Sorex araneus	*Pigmy Shrew* Sorex minutus	*White-toothed Shrews* Crocidura suaveolens	Crocidura russula
Shetland, St Kilda	–	–	–	+	–	–	–	–
Orkney	+	–	–	+	–	–	–	–
Ireland, Man	–	–	–	+	–	+	–	–
Lewis, Barra (OH)	–	–	–	+	–[1]	+	–	–
N. & S. Uist (OH)	–	+	–	+	–	+	–	–
Eigg, Muck (IH)	–	+	–	+	–	+	–	–
Raasay (IH)	–	–	+	+	+	+	–	–
Mull, Bute	–	+	+	+	+	+	–	–
Skye, Islay, Jura, Gigha, Arran, (IH)	–	+	–[1]	+	+	+	–	–
Skomer	–	–	+	+	+	+	–	–
Scilly Isles	–	–	–	+	–	–	+	–
Jersey	–	–	+	+	+[2]	–	+	–
Sark	–	–	–	+	–	–	+	–
Alderney, Herm	–	–	–	+	–	–	–	+
Guernsey	+	–	–	+	–	–	–	+

Both Common and Pigmy Shrews are present on most of the Inner Hebrides, but only the Pigmy Shrew is present on the Outer Hebrides, as it is also on the Isle of Man. On the other hand, the Pigmy Shrew is missing from the Channel Isles and Scilly Isles. Jersey, alone of the Channel Isles, has a larger, red-toothed, shrew, but it is the French Shrew *Sorex coronatus*.

Glacial Relict Hypotheses

The explanations originally conceived to account for the distribution and subspecific variation of the island small mammals invoked the interplay of varying glaciations and repeated invasions of Britain by these small mammals. Martin Hinton, who was a rodent taxonomist at the British Museum (Natural History) when he was not arguing for the control of Muskrat, described many of these races himself, and also described the small mammal faunas from many Pleistocene cave excavations. However, he thought that there had been only one glaciation, rather than the three or more that are now recognized, and did not appreciate their full climatic severity. He argued that Orkney Voles, for example, were a relic of an earlier, pre-glacial, immigration to

Britain, which survived on Orkney, in isolation, through the glacial period, but were replaced in mainland Britain by the later-invading Field Vole. He thought that he had discerned, in *Microtus corneri*, the fossil evidence for the earlier presence of the ancestors of Orkney Voles in British cave faunas. There is evidence of white-toothed shrews (though not the small *Crocidura suaveolens*) in a few cave faunas (Chapter 3), which could be taken as evidence of their prior existence in Britain, from which the island populations would be regarded as survivors. When the evidence of several glaciations became incontrovertible, more complex theories of peripheral survival of the relicts of earlier interglacial faunas were provided, notably by Beirne (1952). As a student, I spent a long time trying to construct a coherent story based on his account that fitted all the various races and species, to no avail. The patterns of the different species did not make sense.

That is the certain message from these faunas: they do not conform to a coherent story. If any species were to survive on, say, Orkney or the Scilly Isles through the Last Glaciation, or through the Younger Dryas, it would surely not be the more southerly *Microtus arvalis* and *Crocidura suaveolens*, but the more northerly distributed *Microtus agrestis* and *Sorex araneus*. Similarly, if any species were going to spread back quickly after a glacial period and reach more islands than any other, it would be the northerly Field Vole, not the much more southerly Wood Mouse (Fig. 8.6). Moreover, as already argued, the north of Britain was depressed by the weight of ice, and lower relative to sea-level than it is now (put the other way round, Late Glacial raised beaches, up to 150 m above present sea-level, indicate how much higher apparent sea-levels were then around western Scotland). The possibilities of a Post Glacial land-bridge to allow Wood Mice to reach St Kilda, the Outer Hebrides or Shetland are nil. There is some possibility that there was a connection to the Isle of Man, but in that case, as for Ireland, it must have been a very selective land-bridge, probably soggy and low-lying, to filter out the voles, Mole and Common Shrew. The situation of the Channel Isles is particularly intriguing. They were far enough south to escape any depression due to ice, and the seas between Jersey and France, in particular, are shallow enough that a reduced sea-level of only 10 m would create a land-bridge (Lister 1995, Fig. 8.12). That should allow most small mammals to immigrate, so why is Jersey missing Water Voles, Field Voles, Pigmy Shrews, Water Shrews and Weasels? It is large enough to host Stoats, Moles and Red Squirrels, though the latter are supposed to have been introduced about 1885 (Le Sueur 1976). Why should Guernsey, further from France than Jersey, have Common Voles but not Bank Voles? And why should the two white-toothed shrews occur, on different islands, when *C. suaveolens* does not even occur on the nearby French coast?

One element in the solution to this puzzle is surely the role of human introduction, probably accidental, advocated strongly by Corbet (1961). He points out that small islands in particular have impoverished economies, as well as faunas, and depend heavily on the importation of livestock, fodder and bedding, as well as building materials and other goods, all of which would give small mammals ample opportunity to stow away. The species most closely associated with human habitation, including the rats, House Mouse and Wood Mouse, would be the most probable hitch-hikers. Voles are less likely, and red-toothed shrews least likely, to be involved, just because of their habits and habitats, but white-toothed shrews do tend to be more commensal (Yalden *et al.* 1973, Genoud 1988). A history of importation would certainly explain the

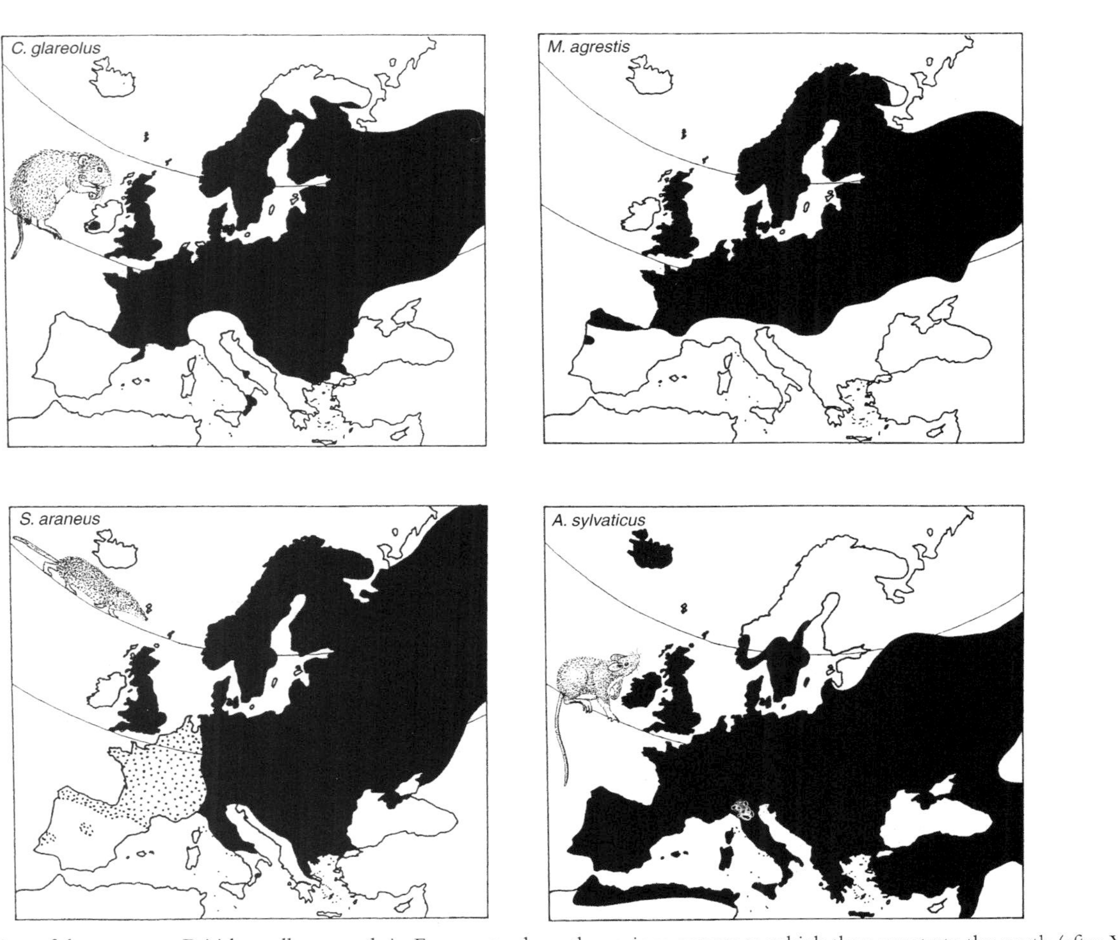

Figure 8.6 Distribution of the common British small mammals in Europe, to show the various extents to which they penetrate the north (after Yalden 1982). Field Voles and Common Shrews reach to 70° N, Bank Voles only to about 67° N, and Wood Mice only to 63° N.

irregularities of their distributions. What about the evidence, implicit in their subspecific status, that they had been around for long enough to diverge recognizably from their mainland relatives? Here more recent collections and taxonomic analyses have been invaluable. For a start, most of the island races differ far less than appreciated when the subspecies were first described. Many of the differences turned out to be exaggerated by comparing small samples of specimens, perhaps from opposite ends of the country if that is how museum collections had been assembled. One major feature of the island races is that they tend to be larger. However, as an example, the Bank Voles from Skomer, Mull, Raasay and Jersey, while much larger than the voles from south-east Britain with which they were compared, are not so strikingly large when compared with Scottish Bank Voles (Corbet 1964; Fig. 8.7). Similarly, while the Wood Mice from St Kilda seem enormous by comparison with those from south-east England, those from various other islands and from Scotland provide an almost complete size series. When Mike Delany (Delany 1964, 1965; Delany and Healy 1964, 1967a, 1967b) reanalysed variation in *Apodemus*, he found for instance that those from the Outer Hebridean island of Lewis, Raasay and Mull grouped with those from Mainland Scotland, those from the islands of North Uist, South Uist and Barra, also in the Outer Hebrides, grouped together, while those from Colonsay and Rhum were distinct from all others. This is not the pattern of geographical differentiation (recognizing a *hebridensis* group on the Outer Hebrides) that classical taxonomy had suggested. The importance of other characters has also been exaggerated. Bank Voles on Skomer, Jersey and Raasay have more complex molars than those from mainland Britain, but the difference is due to only one or two genes, judging from breeding experiments, and the variation seen in all populations overlaps (Fig. 8.8). The voles with complex molars come from islands that lack Field Voles, and it is likely that the Bank Voles have adapted to a harder diet. The pattern of Field Voles on the Scottish islands and adjoining mainland is particularly intriguing, because it does offer some evidence for a relict distribution. In the Highlands, north of the Great Glen, and on many of the islands, Field Voles have an extra loop on the inner side of their upper first molars (Fig. 8.8); this is the 'complex' condition, and diagnostic of the form *Microtus agrestis neglectus*. South of the Great Glen, and on the islands of Muck, Mull, Colonsay, Jura, Lismore, Arran and Bute, this loop is missing, the 'simplex' condition. Interestingly, Islay, next beyond Jura, and Luing, geographically part of the southern group of islands, have voles with 'complex' molars. The presumption is that the 'complex' condition occurs in an older population of voles, which invaded Scotland first, and reached many of the islands, but was later replaced by form with simpler molars, which reached most of the southern islands, but not the northern ones. On this basis, the voles with complex teeth on Luing and Islay are genuine relicts of an earlier population, but must still be a Post Glacial colonization.

In summary, the island races tend to be at the extreme end of a range of variation seen in mainland populations, distinctive certainly, in some cases, but not so different as to require thousands of years of evolution in isolation. Given the likelihood of only a few founding animals, and the powerful selective forces operating on island populations (severe winter food shortages, perhaps, no ground predators, exposure to wind and rain; Berry and Jakobson 1975), the changes could have arisen in a few generations. They do not provide any evidence for long isolation and survival from preglacial times.

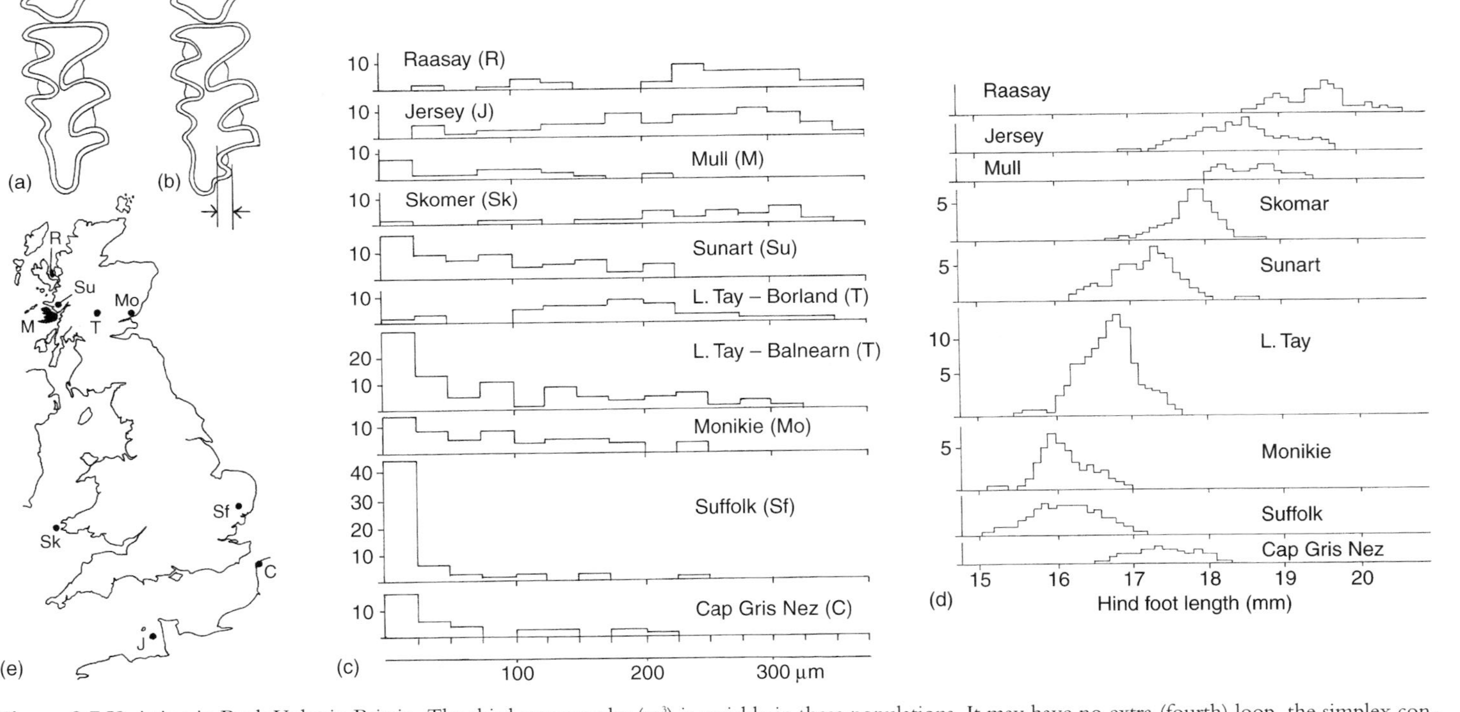

Figure 8.7 Variation in Bank Voles in Britain. The third upper molar (m^3) is variable in these populations. It may have no extra (fourth) loop, the simplex condition (a) or a well-developed loop (b), the complex condition. The size of the loop is usually small, but in island populations from Raasay, Jersey and Skomer it is large (c), and one forestry plantation (Borland) at Loch Tay also had a complex population in 1955–1957, though by 1975 this had reverted to simplex. The size of these Bank Voles, best measured in museum specimens from the length of the hind foot (d), is also variable, much larger on average in those from the islands (Raasay, Jersey, Mull, Skomer) than in those from southern England. However, those from Scotland occupy an intermediate position, and show that the island populations are at the extreme of the usual range of sizes rather than totally distinctive. The inset map (e) identifies the location of the various collection sites by their initials, expanded on (c). (From Yalden 1982 in *Mammal Review* **12**, by permission of Blackwell Science.)

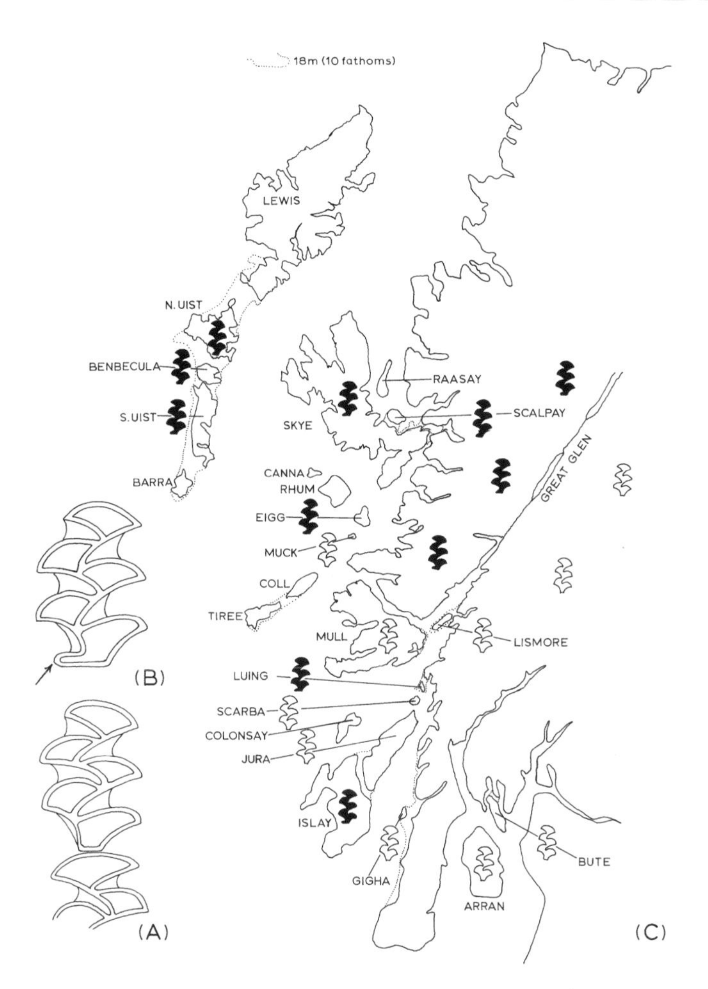

Figure 8.8 Variation in Field Voles in Britain. The first upper molar (m^1) may be simple (a), or may have an extra loop (b), like the one that characterizes the second upper molar in this species. (c) Populations north of the Great Glen typically have this loop (are complex), while those south of it have simple molars. On the islands, the pattern is confused. Most of the more northerly islands have a complex m^1 (black silhouettes), but Muck, and also the more southerly islands of Mull and Lismore have simple m^1 (white silhouettes). Conversely, most of the southern islands have simple m^1, but Islay and Luing have them complex.

Those islands lying near to the mainland, separated by channels no more than 10 m deep, were probably colonized naturally by voles from the nearest mainland (Skye, Mull, Lismore, Luing, Scarba, Jura, Islay and Bute?). In that case, the Islay population, with its complex molars, may be a relict of an early colonization of 'complex' voles, later displaced on Jura by 'simplex' ones. The outer islands, such as North and South Uist, Benbecula, Muck, Eigg, Colonsay and Arran, separated by channels deeper than 50 m, probably owe their vole populations to accidental human introduction. Other islands, no more or less isolated, seem to lack Field Voles. (From Yalden 1982 in *Mammal Review* **12**, by permission of Blackwell Science.)

The role of founder effect has been explored, and exploited, further in studies by Berry (1969, 1970) on the epigenetic variants in the skulls (which are the usual museum relics of collecting trips to these islands). Epigenetic variants are small variations, such as single or double foramina, extra small bones present or absent, which are believed to be genetically determined (Fig. 8.9). Their frequency varies from population to population, and the difference in average scores of these variants in nearby populations indicates how closely related the populations are to each other. On this basis, Berry (1969) argued that Wood Mice in the Shetlands, and on St Kilda and on Iceland too, came from Norway, rather than from Scotland. Those on some of the Hebridean islands also came from Norwegian stock, rather than from the Scottish mainland, though those on Mull and Colonsay came from the mainland. The Wood Mice on the Scilly Isles and Channel Isles, also, seem to have come from their nearest mainlands (Berry 1973; Fig. 8.10). The distinctive House Mice of St Kilda, and those of the Faroes, show similarities to those of Norway, strengthening the argument for Viking involvement. For the Orkney Voles, the nearest relationships seem to be with the Common Voles of the Balkans, among the material available, not the geographically nearest voles of northern Germany (Berry and Rose 1975). Common Voles do not occur as far north as Norway, so they could not have been brought from there.

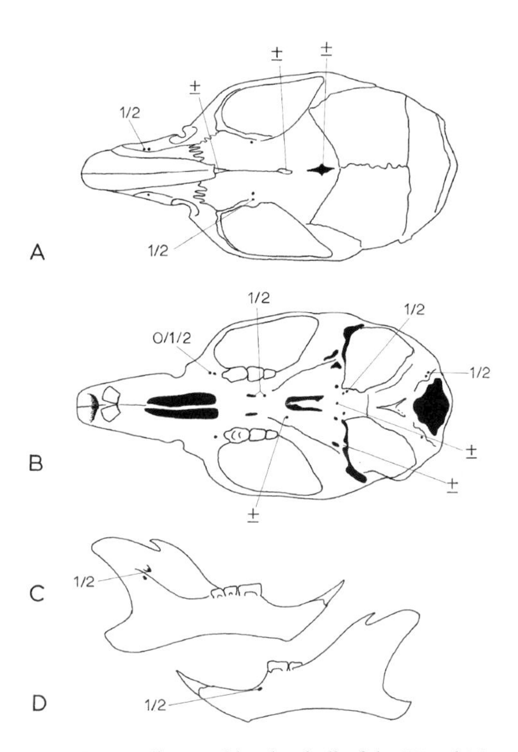

Figure 8.9 Epigenetic variation, as illustrated by the skull of the Wood Mouse. Some small foramina may be single or double (1/2); other foramina and small bones may be present or absent (±). These variants occur in different proportions in different populations, and reflect how closely related each population is to its neighbours. (From Yalden 1982 in *Mammal Review* **12**, by permission of Blackwell Science.)

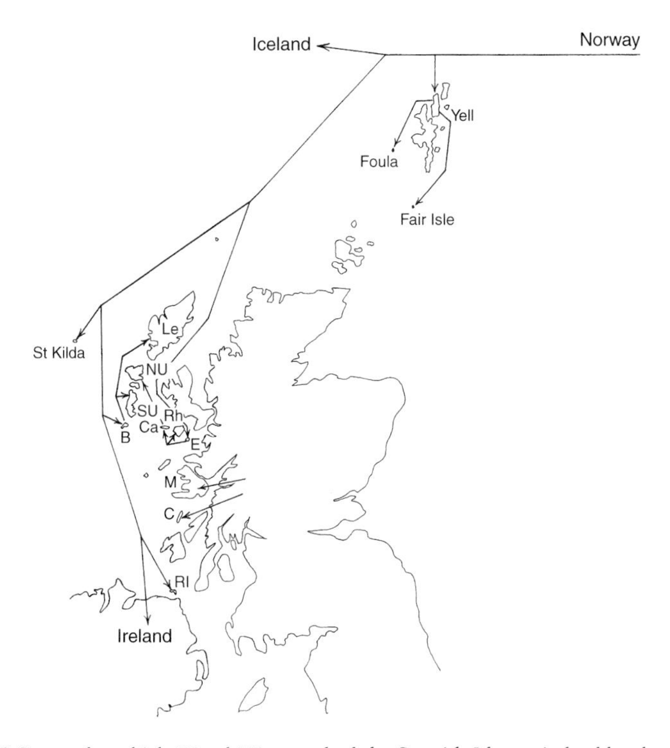

Figure 8.10 Routes by which Wood Mice reached the Scottish Isles, as judged by the similarity of epigenetic variation between the different island populations (after Berry 1969, 1973).

Mice from Norway are more closely related to many of the island populations than mice from Scotland. Berry suggests that seven separate introductions of Norwegian mice may have occurred, (1) to Iceland, (2) to St Kilda, (3) to Yell and thence to Foula and Fair Isle, (4) to Barra, thence to South Uist and Lewis, (5) to Eigg, thence to Rhum and Canna, and from Canna to North Uist, (6) to Rathlin Island and (7) to Ireland. The mice of Mull and Colonsay came from the Scottish Mainland. (From Yalden 1982 in *Mammal Review* **12**, by permission of Blackwell Science.)

Archaeological Evidence

It would be nice to have sub-fossil specimens to support some of these suppositions, but there is very little. Several islands do have archaeological sites, and some have even produced small mammal faunas as well. They tell a rather incoherent story, not least because there are too few of the islands with good faunas, and too few faunas across the range of time involved.

Shetland

Jarlshof on Mainland Shetland has a continuous record of human habitation from Late Bronze Age onward (Platt 1956). In the earliest levels sheep and cattle dominate the mammalian remains, along with fewer pigs, cats and ponies. A single dog skeleton, the lower jaw, rather improbably, of a Wolf, and a few bones of Common and Grey Seals plus two vertebrae of a Walrus and fragmentary remains of a large whale complete the fauna. In the 1st to 2nd century A.D., sheep, cattle, pigs and Grey Seals remain common, while Common Seal, whale and dog are less frequent. In the 3rd to 8th century, Red Deer and Wild Cat are recorded, as well as the domestic species, and in the early 9th century (Viking) levels, sheep, cattle, pigs, pony, and dog plus Common and Grey Seal and whales are all numerous. The Wolf, Red Deer and Wild Cat must all have been imported from either Scotland or Norway. Unfortunately no small mammals are recorded to help understand their early history. As Berry and Johnston (1980) remark, only the Otter and Wood Mouse have been on Shetland longer than written history, the other terrestrial mammals are moderately well documented introductions. The Brown Hare was introduced around 1830, and became quite common, but died out about 1937. The Mountain Hare was brought from Scotland about 1907, and remains well established on Mainland. Hedgehogs were imported about 1860, and Rabbits were present by 1654. Stoats were reputedly brought in sometime before the 17th century, in revenge for failure to pay a tithe of poultry, but perhaps were brought to control Rabbits.

Orkney

Several sites in Orkney have produced a valuable record. At Quanterness on Mainland Orkney, as well as at Links of Notland on Westray and Holm of Papa Westray, Neolithic sites have yielded large samples of bones including rodent teeth (Corbet 1979, 1986). All three Neolithic samples, with dates as early as 5400 b.p. at Quanterness, have *Microtus* molars that are, surprisingly, as large as the modern voles; in fact, those on Westray seem to have become appreciably smaller over the last 4000 years, and those on Mainland are also a little smaller (Fig. 8.11). Buckquoy, too, has large *Microtus* though at later (Pictish and Viking) levels (Bramwell 1977b). A converse sequence, of small immigrants getting larger with time, would have been more probable. It is also very intriguing to find the voles present in Orkney so early in the Neolithic; it seems that the earliest Neolithic farmers must have brought these voles along with their sheep and other livestock. Wood Mice are also present at Neolithic Quanterness (Corbet 1979, 1986) and at Isbister, slightly later (Sutherland 1983), so even if the Vikings were responsible for introducing Wood Mice to Shetland and the Outer Hebrides, they were several thousand years too late to bring them to Orkney. There is a burial of 13 articulated Red Deer skeletons at Links of Noltland, and both antlers and butchery remains suggest that Red Deer were important quarry. Red Deer bones are also present, though few in number, at Neolithic Quanterness and pre-Norse and Norse age Buckquoy. It seems possible that they had been brought over to establish feral populations by early farmers, for it is barely conceivable that they were native. There are also a few bones of Roe Deer at Buckquoy and Knap of Howar, and Fox at Quanterness. Otters occur, less surprisingly, at Knap of Howar, Isbister, Quanterness and Calf of Eday. However, at all these sites, the bones of domestic ungulates predominate, and wild mammals were clearly no more than a small supplement to the diet (Noddle 1977, 1983, Barker 1983, Clutton-Brock 1979).

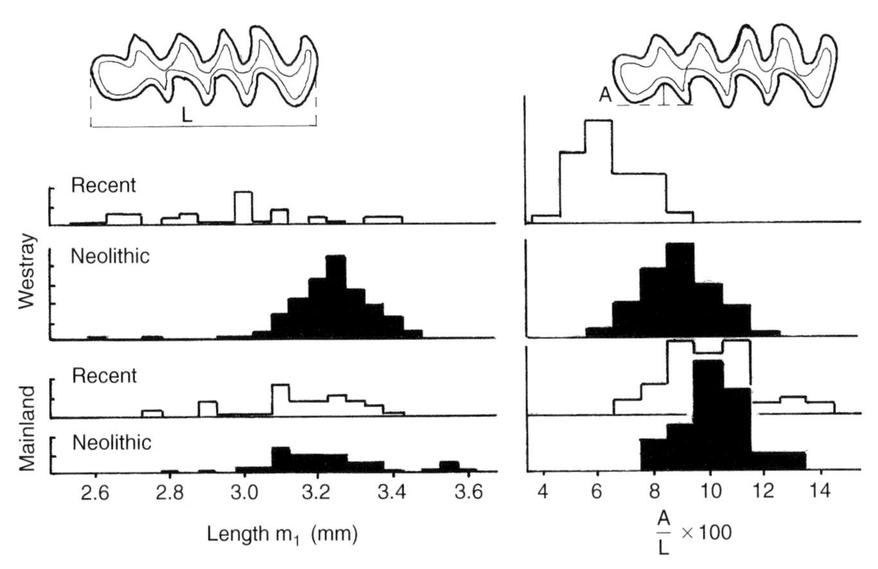

Figure 8.11 Changes in the size and shape of Orkney Vole teeth over time since the Neolithic (after Corbet 1986). Recent Westray voles are smaller, and the entrant A is smaller, than in the early Neolithic specimens. On Mainland Orkney, they have changed little in this time.

Hebrides

Oronsay has yielded a Mesolithic fauna (see Chapter 3) that includes Red Deer and seals, but no small mammals. The poverty of the mammal fauna suggests that the island had no native mammal fauna, but the vagaries of sampling could be blamed instead for the poor record.

On Iona, both the Iron Age mound of Dun Bhuirg and the Columban-Mediaeval Monastery have yielded mammalian bones, though mostly of domestic mammals, mostly sheep, cattle and pigs with some horse. However, Red Deer were also numerous at all levels, and small numbers of Roe Deer and seal bones also figured, while Badger and Fox appear in the later Monastery deposits (Noddle 1981). It is suggested that the latter were scavenging in the midden, but since neither now occur on the neighbouring much larger island of Mull, it is likely that all the wild mammals, like the domestics, were imported to Iona. It seems a particularly small site for Red Deer, and a very unlikely place for Roe Deer, even though pollen analysis did provide evidence of birch–hazel scrub at some time. Unfortunately, as on so many archaeological sites, no small mammal bones were reported.

Jura has produced one of the few small mammal archaeological samples that helps to confirm part of the historical hypothesis. *Microtus* molars from there show a complex condition in the Neolithic, fitting the notion (above) that this was an early condition, now relict on Islay further south-west. They have become simpler with time on Jura, either by arrival of new genetic forms, or by gradual evolution over 4000 years of the originally 'complex' population (Corbet 1975).

Isle of Man

The Isle of Man lacks many of the species which are missing from Ireland (Field Vole, Water Vole, Common Shrew, Water Shrew, Mole, etc.), and a few others, notably Red Deer, Red Fox and Badger. The sea-floor to the east is relatively shallow, and one would expect most mammals that got into Mesolithic England to have been able to cross to the Isle of Man. It being a smallish island, however, some of them were probably hunted out. Interestingly, one place-name, Cronkshynnagh Farm, has the Gaelic root *sionnach*, suggesting the former presence of Foxes (Taylor 1911). Red Foxes have recently reappeared on the island, apparently by surreptitious reintroductions (S. Fargher, pers. comm.). Red Deer, Roe Deer, Wolf, Fox and Wild Cat are reported from cave deposits, and the Giant Deer *Megaloceros giganteus* is a well known sub-fossil inhabitant (Garrad 1972, Mitchell 1958).

Isles of Scilly

An interesting small mammal fauna, perhaps dated to Bronze Age or possibly somewhat later, Iron Age, has been excavated from the small island of Nornour (Pernetta and Handford 1970, Turk 1978). Wood Mice are now present only on the two largest islands, Tresco and St Mary's, but were present then on Nornour. However, a lowering of sea-level of 10 m would see most of the Isles of Scilly joined as one larger island, and it seems likely that Nornour was then part of such a larger land mass. The Scilly Shrew was also already present. Most remarkably, however, the Root Vole *Microtus oeconomus* was the most abundant species. It is not clear whether this represents a true glacial relict, or an early accidental introduction. Whereas Wood Mice and Scilly Shrews could easily have been introduced by traders from the south, the Root Vole is a northern species, and certainly could not have come from that direction, so perhaps it was a Late Glacial relict.

Jersey

The situation of the Channel Isles suggests that small mammals would have survived there, if anywhere, from the milder Windermere Interstadial through the Younger Dryas into Post Glacial times. The very shallow sea-floor to nearby France, which would form a land-bridge to Jersey with a drop of only 20 m in sea-level, and to the rest of the group with a drop of 40 m, suggests that most small mammals should have reached them in the Post Glacial (Le Sueur 1976; Fig. 8.12). Yet they have as idiosyncratic a fauna as islands further north. Jersey also has a rich history of archaeological, including cave, sites, which should tell something of the faunal history. Indeed, there is a distinctive small Red Deer present in the Eemian (equivalent to Ipswichian) Interglacial of Jersey, discovered at Belle Hougue on the north coast (Lister 1995). 'Full-sized' Red Deer reached Jersey during the preceding glacial (Saalian, equivalent to Wolstonian), being recorded at La Cote de St Brelade on the south side of the island; Chamois *Rupicapra rupicapra* and Birch Mouse *Sicista betulina* also reached Jersey then, the nearest they ever got to Britain (Callow and Cornford 1986). During about 6000 years, the Red Deer shrank to about 56% of the size of their contemporaries in Britain and France (linear size, from length of limb bones). Lister estimates that they

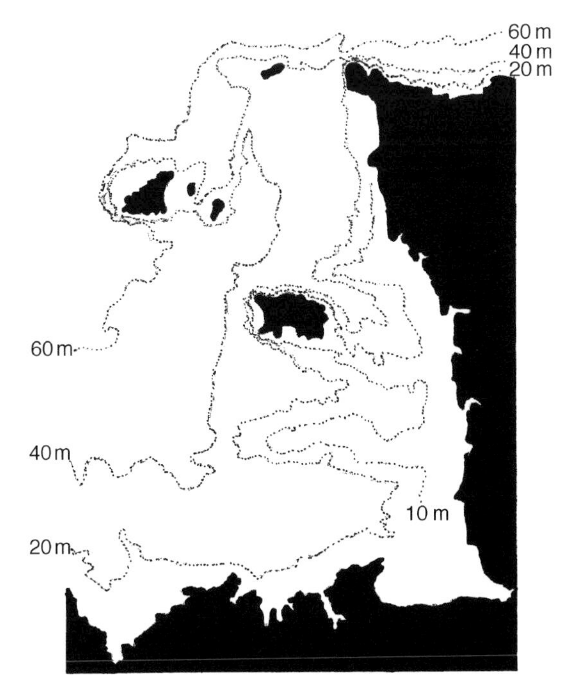

Figure 8.12 The submarine contours around the Channel Isles, and potential land-bridges to France (after Le Sueur 1976, Lister 1995). On present topography, a drop in sea-level of only 10 m would connect Jersey to France, but a drop of 40 m would be required to connect the other Channel Isles.

weighed about 36 kg, whereas their contemporaries weighed about 200 kg (Fig. 8.13). Then in the Devensian, when lowered sea-levels renewed the land connection to France, large Red Deer reappeared on the island, along with Woolly Mammoth, Woolly Rhinoceros, Horse, Bison, Reindeer, Spottted Hyaena, Brown Bear, lemmings *Dicrostonyx torquatus*, Root Vole *Microtus oeconomus* and sousliks (Marett 1916). This is probably an early Weichsellian (Devensian) fauna, and might have been reduced by the severity of the full glacial conditions about 20 000 b.p., but no more detailed record is available, and nothing to document the Post Glacial colonization of the island.

Discussion

The faunas of the small islands have provided a great deal of scope for zoological exploration over the last century. Initially, when that was the fashion, new forms were discovered and described, and specimens duly deposited in museums. With the fuller appreciation of evolutionary taxonomy, which recognized the importance of variation in populations, rather than the distinctiveness of individual specimens, larger samples were collected, to provide evidence that the differences between the island and mainland populations were not so great as once thought, but were instead just as interesting for the light they shed on the origins of the populations, and for the

Figure 8.13 The dwarf Red Deer of Jersey in the Ipswichian Interglacial, compared with its contemporary relative in Britain (after Lister 1995). It weighed about 36 kg, and had a shoulder height of 70 cm, compared with 200 kg and 125 cm for the mainland form. It also had much simpler antlers.

dramatic examples they provided of microevolution in progress. Theoretical explanations for the ways in which their founders arrived have been developed alongside these more recent studies, but the nature of the archaeological record on small islands is so patchy that proof is unlikely ever to be uncovered. The value of the Irish record is that archaeological confirmation may well be possible, and the recent renewal of interest in this puzzle by Irish colleagues holds great promise of further revelations.

Irish Hare

CHAPTER 9

THE TWENTIETH CENTURY AND BEYOND

Polecat and Brown Rats

THE current century has seen an enormous increase of interest in, research on and knowledge of our mammal fauna. Classics such as Millais (1904) and Barrett-Hamilton and Hinton (1910–1922) have been joined by two New Naturalist volumes on mammals as a whole (Matthews 1952, 1982) and a specialist volume on seals (Hewer 1974). The *Handbook of British Mammals* has already gone through three editions (Southern 1968, Corbet and Southern 1977, Corbet and Harris 1991). Three distribution atlases have been published (Corbet 1971, Arnold 1978, 1993), and an attempt has been made to enumerate the British mammals in a J.N.C.C. report (Harris *et al.* 1995). Professional surveyors and an increasing army of volunteers have carried out specific distribution surveys of, amongst others, Badgers, Brown Hares, Otters and Water Voles. A national bat survey, looking at feeding habitats used by bats, and another looking at their use of churches, have extended our knowledge substantially. Surveys of deer, particularly Red Deer, are carried out by the Deer Commission in Scotland, and of seals, especially Grey Seals, by the Sea Mammals Research Unit. Gamekeepers continue to record both their vermin and their game bags, and the Game Conservancy collates increasingly valuable reports of their work (Tapper 1992). Some 1700 Mammal Society members and 2000 Bat Conservation Trust members are involved in supporting the professional surveyors. What does this greatly increased body of knowledge and expertise tell us of the recent changes and current state of our mammal fauna?

Recent Changes and Current Status

Insectivores are generally difficult to survey, and changes in their numbers not readily apparent. The Hedgehog remains widely distributed, but the evidence for that comes largely from road casualties. This raises legitimate concerns about the importance of this cause of mortality. Does a regular toll on the roads indicate a healthy population of them, or does it indicate a drain that could eventually be catastrophic? The only direct evidence of a decline comes from the gamekeepers' toll (Tapper 1992), suggesting a decline of 80% over 30 years from the 1960s. It is not certain, though, if this represents a real decline in the number of Hedgehogs; it could instead be that gamekeepers are concentrating on the more serious predators of game (corvids and Foxes), or no longer reporting Hedgehog kills so assiduously. There has been no adequate study of the level of road mortality on a study population in Britain, but in southern Sweden, about 10% of adults were killed on the roads, compared with 28% dying in hibernation (Kristiansson 1990). Deaths due to predation by Badgers may also be more serious than road casualties, on present evidence (Doncaster 1992). Moles and the three mainland shrews remain widely distributed and relatively numerous. Agricultural pressures, including removal of hedgerows (removing their habitat) and application of pesticides (removing their food supply), must have had a profound effect, but it does not show in the relative scales of abundance which are all that are available. Moles suffer especially from ploughing, which greatly reduces the supply of earthworms, but direct persecution seems to have slackened; Moles are less frequently strung up on fences as a token of the molecatcher's efficiency and the farmer's antipathy. In most circumstances, molehills are a minor nuisance that can be eliminated by harrowing or raking the tilth over the ground. The campaign to prohibit strychnine poisoning, a particularly inhumane method of control, may offer further help (Atkinson *et al.* 1994). The Scilly Shrew is scheduled for particular protection, on account of its limited range in Britain, but also on account of rather limited knowledge of its status. Harris *et al.* (1995) thought, rather diffidently, that it might number only 14 000. This prompted a more thorough survey in 1996, which suggested instead that there might be as many as 99 000 (Temple and Morris 1998). The absence of other shrews and the abundance of the (also introduced) terrestrial sandhopper *Architalitrus dorrieni* on which it feeds may explain this unsuspected abundance.

Bats remain among the most difficult of mammals to survey adequately, despite the best endeavours of Bat Conservation Trust members. All were added to the list of protected mammals in Schedule 5 of the Wildlife and Countryside Act 1981, moreover their roost sites were also given protection. This was, and is, controversial, for many bats depend on buildings for their roosts, and the notion that bats might have precedence of legal status when they roost in someone's house, or church, has certainly annoyed some writers. Actually, the legislation does not say that bats must be tolerated in someone's house, but it does require that advice be sought from the nature conservation agencies (English Nature, Countryside Commission for Wales, Scottish Natural Heritage) before any disturbance takes place. This piece of legislation has had a profound effect on our knowledge of bats, because it puts bat colonies and surveyors in touch with each other. As a result, a number of Bat Groups, now banded into the Bat Conservation Trust, have been formed, to provide the necessary network of surveyors, and some 2500–3000 colonies are visited each year (Mitchell-Jones *et al.* 1986, 1993). Most are Pipistrelles *Pipistrellus pipistrellus* which, as the commonest bat in Britain, are not themselves of immediate conservation concern, but a number of rarer species that do merit concern are discovered as well. In the first two years, the 628 colonies visited included 367 (58%) Pipistrelles and 191 (30%) Long-eared, but also nine other species, including such rarities as both horseshoe bats, Barbastelle and Leisler's Bats.

This increased interest in bats has led to various more specialist surveys that give us some hints about their changing status in Britain. A survey of a sample of 538 English churches and chapels in 1992 found that 142 of them were used by bats as roosts (Sargent 1995). This is 26%, and extrapolating to the total of over 30 000 churches in England suggests that they might contain over 6000 bat roosts. Pipistrelle and Long-eared Bats were the most frequent, but Greater and Lesser Horseshoe, Serotine, Natterer's and Daubenton's Bats were also reported.

One obvious follow-up to the roost visits has been to count the bats coming out from a number of roosts repeatedly each summer. These counts suggest that the number of bats emerging fell sharply over the first two years of the survey, especially in southern England. By 1992, colonies were only 27% of the size they had been in 1978 (Fig. 9.1). They also indicate that bat colonies (which are mostly maternity colonies of Pipistrelles) tend to be larger, about 260 strong, in Scotland, but only 50–70 strong in southern England. There are not many counts of bats from the beginning of the century with which to compare modern knowledge, so we cannot extend the trends further back, for most species. A simple count of the number of records for the rarer species cannot help, because there is so much more activity by bat recorders. However, specific surveys of the two most conspicuous cave-hibernating bats, the Greater and Lesser Horseshoe Bats, do give us reference points. Whereas most (vespertilionid) bats tend to tuck themselves away in crevices during hibernation, and are therefore very difficult to count accurately, the two horseshoe bats hang from cave roofs in a very exposed and obvious manner. Thus their presence is readily detected, and they are relatively easy to count. Both are largely confined to limestone areas in the south-west and west of Britain, so their caves are generally known at least to the pot-holing fraternity ('caves' here includes artificial substitutes such as mine tunnels and cellars). Their breeding roosts are also mostly known, adding a second level of accounting. There has been a substantial reduction in the range of the Greater Horseshoe Bat this century. There were several big colonies in Kent and on the Isle of Wight known for instance to Barrett-Hamilton and Hinton (1911); indeed it was first discovered in Britain at the powder mills at Dartford before 1776. There were

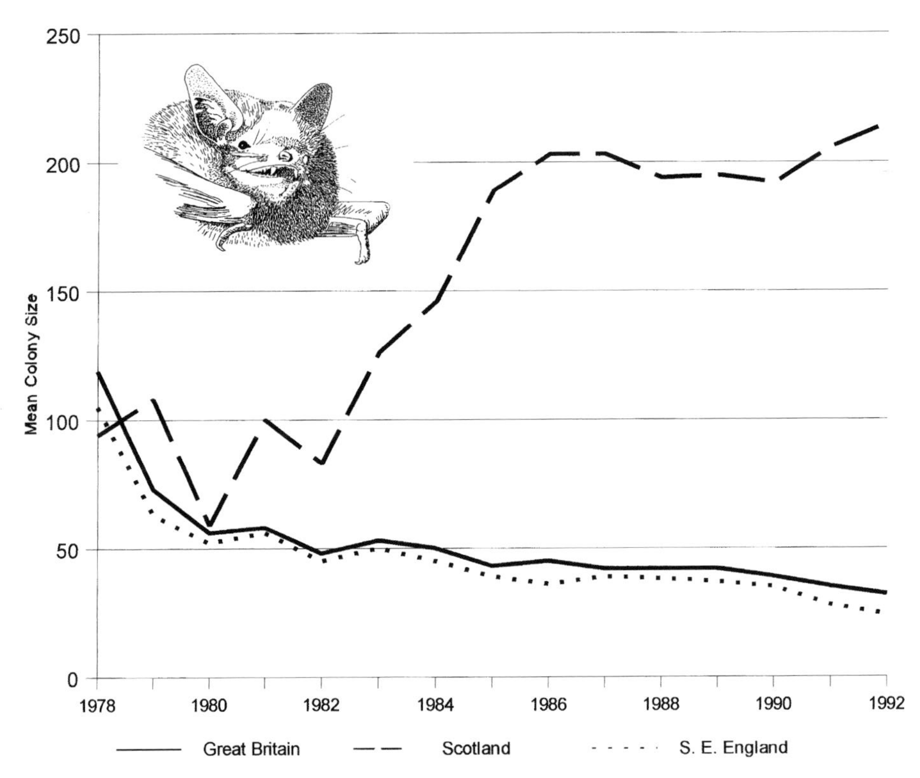

Figure 9.1 Decline in the numbers of bats exiting from a sample of roosts monitored from year to year (data from R.E. Stebbings, in Harris *et al.* 1995).

also records from such places as Henley, Oxfordshire and on the Gower Peninsula, and two individuals taken in Merioneth (Fig. 9.2). Now, the species seems to occur in Wales only in Pembrokeshire, and in England occurs regularly no further east than Dorset, though isolated individuals have been found four times in Surrey. Stebbings (1988) argues that the population declined especially in the period 1950 to 1980, from at least 58 nursery colonies and perhaps 330 000 bats to only 12 sizeable colonies and about 4000 bats, a reduction of around 98%. More detailed knowledge confirms some of these losses. In Dorset, where there used to be three breeding colonies, and an estimated 15 000 bats, there is now only one breeding roost and about 90 bats (Stebbings and Arnold 1987). In Gloucestershire, now near the edge of the species' range, a well-studied population of around 300 bats in 1962 declined over the next 4 years to about 100, apparently as a delayed response to (poor recruitment following) the severe winter of 1962–63, increased to a lower plateau of about 130 bats in 1980–85, but then declined again to below 90 bats following the very late and cold spring of 1986. In Devon, on the other hand, a colony that was also studied quite intensively in the period 1948–55 then averaged 390 bats; a resurvey in 1979 found 304 bats, implying a fairly stable population over this time (Ransome 1989), certainly not the

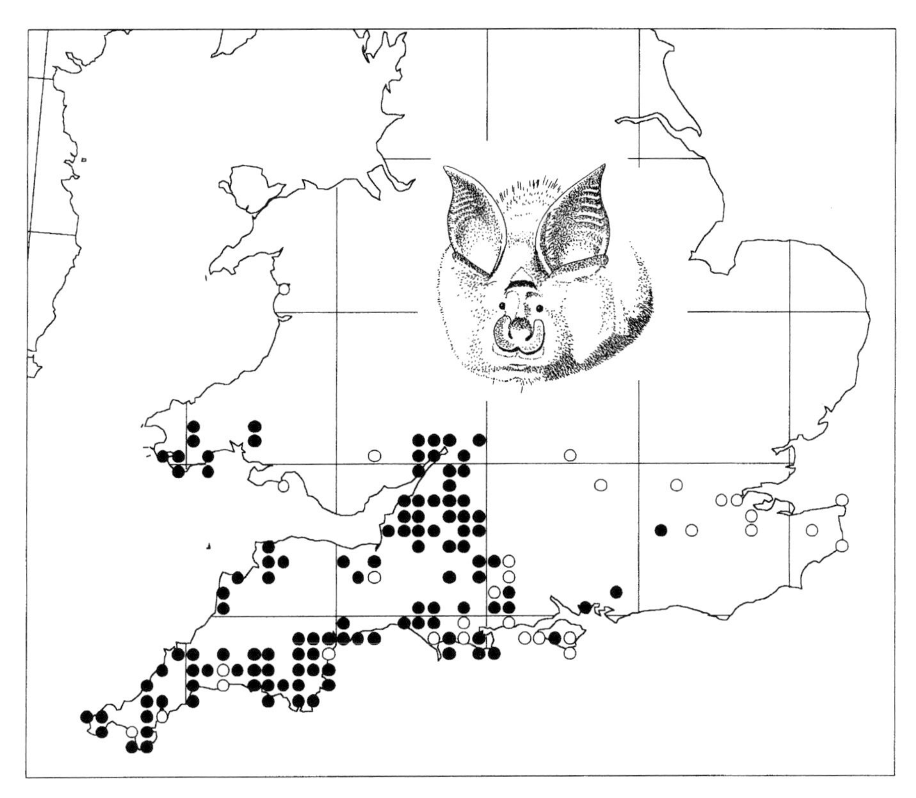

Figure 9.2 The distribution of the Greater Horseshoe Bat (*Rhinolophus ferrumequinum*) in the British Isles this century (after Arnold 1978, 1993). It has withdrawn from the east of its range (○, pre-1950 records).

catastrophic decline suggested by Stebbings (1988). Ransome argued that the population nationally was not so uniformly dense across the whole of the species' earlier range, because of the lack of suitable hibernation sites in many places, and that Stebbings (1988) therefore exaggerated the scale of the decline, but decline there has certainly been. The situation for the Lesser Horseshoe Bat is similar, for around 1900 it was recorded from Ripon and Pateley Bridge in North Yorkshire and from the Peak District (Barrett-Hamilton and Hinton 1911), and there were records from Surrey and Kent as recently as the 1950s. Now it occurs no further north-eastwards than the caves of Denbighshire. Thus its range in Britain has certainly declined this century. However, while in more recent times the Greater Horseshoe Bat seems to have been in population decline as well, the Lesser Horseshoe has been holding its own, even possibly spreading a little further eastwards in Dorset and becoming more common in Wiltshire and south-west England (Harris *et al.* 1995).

For the other bats, there is insufficient earlier information to speculate about the changing status of most of them. The evidence that they depend on hedgerows, for hunting and moving across the landscape (Limpens *et al.* 1989), suggests that hedgerow-loss will have affected them badly. The national distribution survey of feeding bats highlighted the importance of streams, woodland edges and hedgerows,

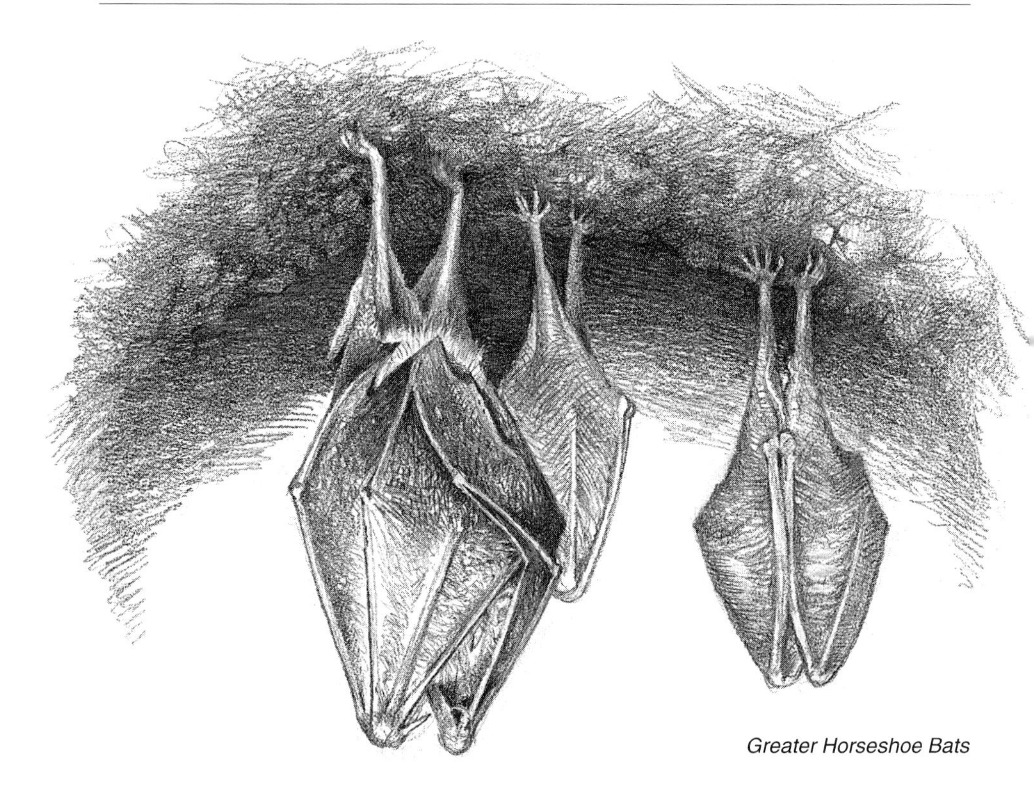

Greater Horseshoe Bats

and demonstrated that bats were still present everywhere, but also highlighted the poverty of agricultural landscapes for them (Walsh and Harris 1996a, 1996b). The fact that they feed entirely on insects, the direct target of all the agricultural insecticides that are now sprayed on farmland (and gardens), suggests that their food supply has diminished. The insecticides have also been used to proof timbers in roof spaces, just where some bats roost, and in the past there have been major kills. It is known now that bats are much more sensitive than most mammals to such insecticides as lindane (Racey and Swift 1986, Shore *et al.* 1990, 1991), partly because of their enormous surface area, and partly because they depend so heavily on fat stores for hibernation (organochlorine pesticides are particularly fat-soluble); fortunately there are now synthetic pyrethroids (permethrin, etc.) which are just as effective at killing woodworm, but not at killing bats (or humans). All these negative factors, and others, are believed to have reduced the numbers of bats, both nationally and internationally, but precise impacts are unquantifiable.

The changed status of two rare species certainly merits notice. The Greater Mouse-eared Bat *Myotis myotis* had a rather dubious status as a British species, though it used to be one of the commonest and best studied species elsewhere in Europe. One was found alive in the grounds of Girton College, Cambridge, in 1888 and taken to Professor Gadow (but might have been a Lesser Mouse-eared Bat *Myotis blythi*), and another was found in the grounds of the British Museum before 1830. There is a suspicion that these may have been 'planted'. However, in 1956, one was found hibernating in a mine in Dorset, and this led to the discovery of a small colony numbering about 12 in 1960. Unfortunately the site became well known, and well disturbed; the colony had died out by 1980. A single, probably vagrant, bat was found in Kent in 1985. Another small hibernating colony had meanwhile been found in Sussex in 1969, and may have numbered around 50 bats. However, none of the females returned to the hibernating site in 1974, and it is believed that the (unknown) nursery roost was destroyed that year. The males lingered on, returning to the hibernaculum in decreasing numbers, and only one male survived from 1985 to 1990. None has been recorded in Britain since then, and the species is now much rarer than it was a century ago in Europe. It may yet return, but is likely never to have more than a tenuous presence in Britain. Meanwhile, it is the only native mammal to have become extinct this century. The converse case is offered by Nathusius' Pipistrelle *Pipistrellus nathusii*, a species that has been expanding in range certainly, possibly also in numbers, in Europe. There has been an increasing number of records in Britain since the first in 1969, 13 up to 1989 and another 20 by 1994. Most have been of single bats, often in September or May, suggesting immigration and departure of migrant bats hibernating in Britain. However, two young animals were found in Cambridgeshire in 1992, suggesting that they had been born locally, and then a Ph.D. student surveying Pipistrelles in Somerset by sound detected the rather different 'song' of this species (Barlow and Jones 1996). Nathusius' Pipistrelle mate, like Common Pipistrelles, in the autumn, hibernate, and then give birth in a nursery colony somewhere else next June. Thus the offspring of any mating in Britain might still be born in Europe: when can we claim a new breeding species? When mating is detected here, or only when young are born here? In fact, in 1997, the species was recorded breeding in Northern Ireland, and suspected in Lincolnshire as well.

Actually, this story hides the identification of another species in Britain. The reason Kate Barlow was recording Pipistrelles in the first place was the discovery that what we have always supposed to be 'the Common Pipistrelle' turns out to be two very similar species, which use rather different frequencies for the main energy in their echolocation pulses, but are otherwise hard to distinguish. Their molecular genetics also turn out to be very different, indeed within the genus *Pipistrellus* they appear not to be very closely related (Barratt *et al.* 1997). Which is the correct one to retain the name *Pipistrellus pipistrellus,* which needs the new name, and what that should be, have not yet been settled. At the moment, everyone is calling them '45 kHz' and '55 kHz' Pipistrelles.

Lagomorphs, hares and rabbits, have had a very chequered career this century. The Rabbit was originally confined to warrens, often on offshore islands (see Chapter 5), and even in the early 19th century it was largely confined to such places in Scotland and Wales. Even in England, Gilbert White barely mentions them, and the suspicion is that they were still uncommon; he would hardly have ignored a common agricultural pest. He makes one reference (Letter VII) to the country people being allowed to destroy one colony because of the inconvenience they caused to deer hunters, something they would struggle to achieve today. The changing agricultural pattern of the 18th century provided them with a better food supply, and the creation of game coverts and suppression of predators in the 19th century allowed them to increase unchecked. By 1900, they were abundant virtually everywhere below 600 m. This remained true throughout the first half of this century, and the Rabbit was by common consent the most serious agricultural pest. It was later estimated that 1950s agriculture was losing £40–50 million annually, despite an annual harvest of about 40 million Rabbits for the fur and meat trade. Indeed, the economic value of the latter, £15 million, was perceived to be as great as the cost to agriculture at the time, though later figures showed how wrong this was (Thompson and Worden 1956).

If the damage done by Rabbits in this country was uncertain, their impact in Australia was well appreciated, and the Commonwealth government had offered a prize for a means of successful control. In Montevideo, Uruguay, all the laboratory Rabbits had been killed in an epidemic of a viral disease in 1896, and the culprit, Myxoma virus, proved to be very specific. The native cotton-tail *Sylvilagus brasiliensis* was a carrier, but not affected, and even hares seemed unaffected. An attempt to control the Rabbits in isolation on Skokholm off Pembrokeshire failed. After similar trials in isolation in Australia, attempts to release the virus in Australia in 1950 also seemed to fail, but then in the wet-season the disease 'caught', and reduced the Rabbit population in Australia to manageable numbers (though it is still common there, still an agricultural pest, and now somewhat resistant to Myxoma). Meanwhile, an experimental laboratory in Switzerland was also investigating the virus, and a colleague of one of the Swiss scientists obtained a sample to try to control the Rabbits within his emparked estate near Paris in June 1952. At the time it was not realized that both mosquitoes and Rabbit Fleas *Spilopsyllus cuniculi* could act as vectors, and the mosquitoes at least were unlikely to respect a park wall. It is arguable that they would not respect the English Channel either, though possible that human intervention resulted in the first report from England, at Edenbridge in Kent, in October 1953. Despite the passing of legislation making it illegal to spread the disease, or diseased Rabbits, in 1954, the disease spread rapidly through England that year, across most of Wales, and into Scotland by July. Parts of northern England and the rest of Scotland were infected during 1955. It became, and remains, endemic everywhere. Its effect was dramatic. Not only did myxomatosis kill about 99% of the Rabbits, and reduce dramatically their impact on farming, it also resulted in a wonderful flowering of all the downland orchids that had been suppressed, encouraged the Field Vole population, and therefore the Weasel population, but reduced the Stoat population, and caused a widespread breeding failure of Buzzards that year. There were fears, too, that Red Foxes would turn more to sheep and poultry, but in the event they were more adaptable, and took more Field Voles (Thompson and Worden 1956, Sumption and Flowerdew 1985).

Rabbits took several years to start a recovery, and they remain both less common and less widely distributed than before 1953. Whereas 94% of cultivable holdings had Rabbits then, by 1970 the figure was only 20%, and in 1986 there were still only about 20% of the pre-myxomatosis Rabbits. Retrospectively, it is believed that there had been about 100 million Rabbits in Britain, before the start of the breeding season, and about 20 million in 1986. The current estimate is 37.5 million (Harris *et al.* 1995; Table 9.1), but there has been a major change in their distribution across the country as well. Where they used to be very abundant in Wales, they are now rather scarce. There is some evidence that where predators are more common (as in Wales), they can hold Rabbits in check, particularly when assisted by regular recurrences of, now weaker, myxomatosis (Trout and Tittensor 1989).

The Brown Hare has also had a somewhat chequered history this century, but a steady, less dramatic, decline. Its distribution seems to have shrunk somewhat, for instance it was described as common in Cornwall at the turn of the century but is now virtually absent (Hutchings and Harris 1996). The best long-term record comes from the game-bags, which show high levels up to the 1920s, a steady long-term decline through to the 1940s, a rise in the 1950s, perhaps in response to the removal of Rabbits by myxomatosis, and then a further decline during the 1960s, 1970s and 1980s (Tapper 1992; Fig. 9.3). This is, of course, a record of what has happened to Brown Hares in the parts of the country best for them, the hunting estates on which they are conserved for hunting. The national survey suggested strongly that they had faired far worse in the country as a whole, and that numbers were lower than had been estimated by extrapolating game-bag densities to the whole country (Hutchings and Harris 1996). Experiments carried out by the Game Conservancy into the effects of gamekeeping, particularly predator control (essentially, killing of foxes, so far as hares are concerned) demonstrate vividly the benefits to Brown Hare populations of this protection (Tapper *et al.* 1996; Fig. 9.4), and the high numbers at the beginning of the century must reflect the high levels, then, of fox control. The declines since are not fully explained, but increasing numbers of predators and the much poorer habitat now presented by our intensively agricultural landscape are major factors. In non-hunting country, it is not worthwhile to retain hedges, or the wide strips round fields that give hares food and cover, while the polarization of agriculture between the

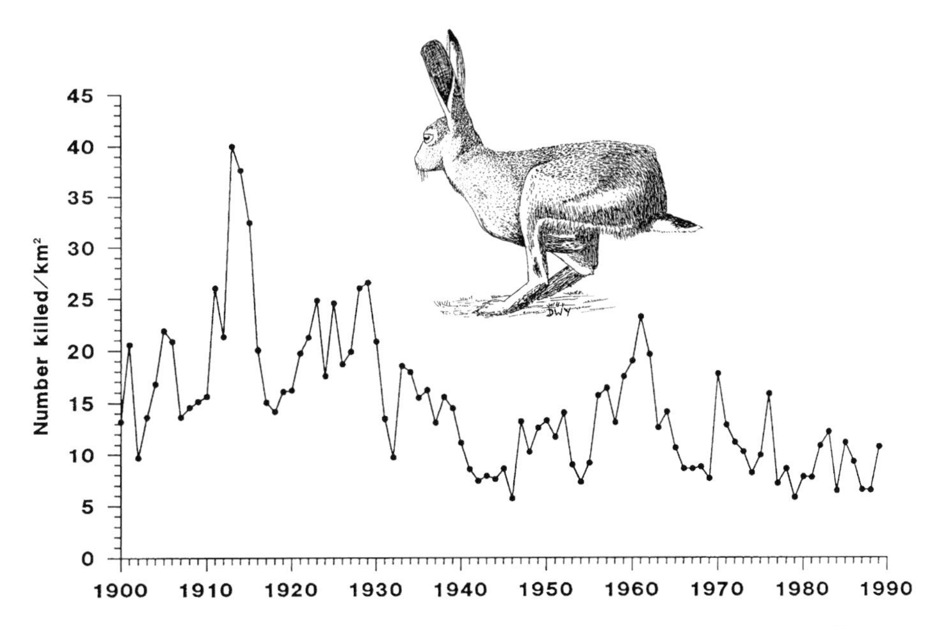

Figure 9.3 The declining abundance of the Brown Hare this century, as documented in the game bags. Even on these game estates, where it is likely to be conserved, its numbers have fallen to about half its numbers in the 1920s. In the wider countryside, as a result of agricultural intensification, it has declined even more steeply. (Reproduced from Tapper 1992 *Game Heritage*, with permission from Dr S. Tapper and the Game Conservancy.)

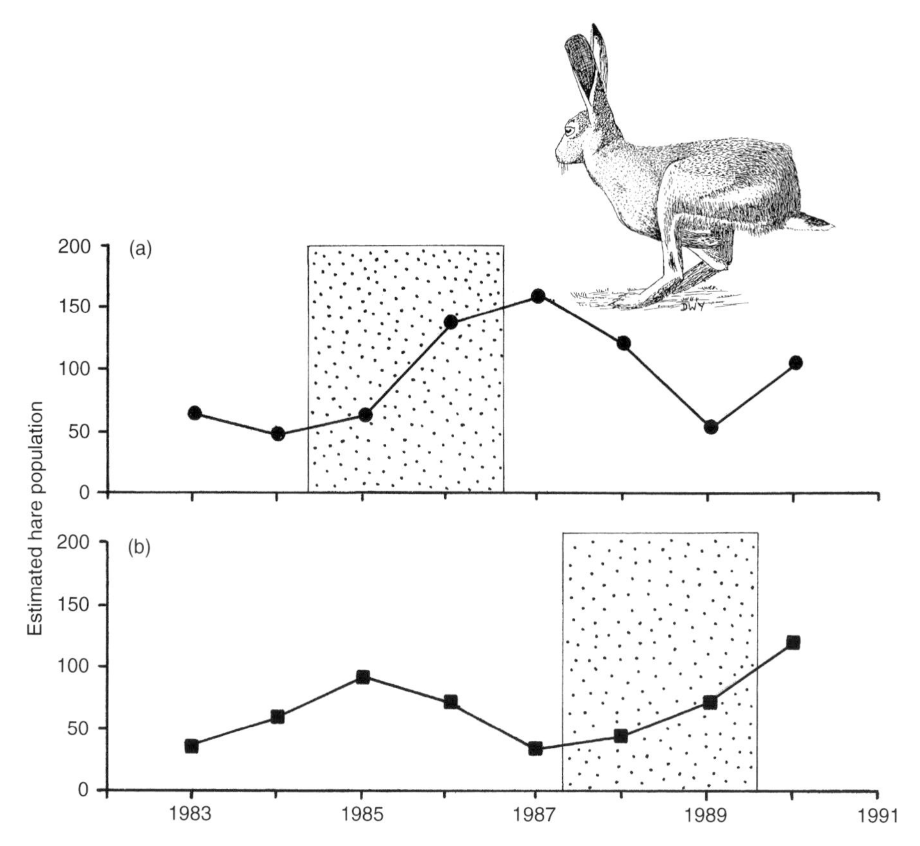

Figure 9.4 The response of the Brown Hare populations on two farms on Salisbury Plain to the activity of a gamekeeper, active from 1985–87 on one farm (a), and then from 1988–90 on the other (b). The control of the Fox population was the only activity that should have affected the hares; protected from predation, they increased first on one farm, and when protection was switched, declined there but increased on the other (from Tapper *et al.* 1991, 1996).

arable east and pastoral west (Fig. 9.5) has removed the mixed farmland that best suited hares, and much more of our farmland wildlife.

At least these changes in arable farmland have had little effect on Mountain Hares, but they too have had an interesting history. They appear to have been native, in historical times, only to Highland Scotland and Ireland. This has tempted many sportsmen in the past to try to diversify their shooting opportunities. The earliest recorded attempts were to Ayrshire, Lanarkshire and Peebleshire in the 1830s and 1840s; these successfully populated the Southern Uplands, and spilt over into England, on the Gilsland Fells in Cumberland. An earlier but unrecorded attempt to introduce them to Wales must have been responsible for their presence before 1830 near Yspytty, Caernarvonshire (Fitter 1959) and releases to the Vaynol Estate near Bangor in about 1885 resulted in populations that lingered in Snowdonia and on the moors of Denbighshire into the 1970s. However, the species seems now to be extinct in Wales,

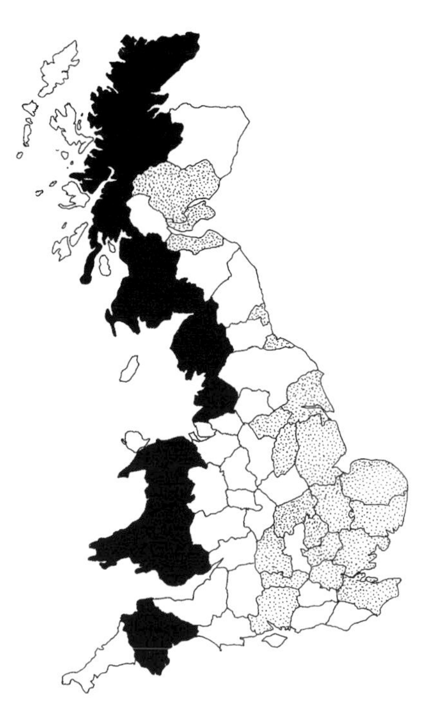

Figure 9.5 The polarization of British agriculture in the 1990s, between the arable east and the pastoral west (after Tapper 1992). In the eastern counties, more than 60% of farmland is ploughed arable land (stipple), while in the west, more than 80% is pasture (black). Even in the remaining counties, there is little truly mixed farming. This has had major impacts on the habitat for mammals.

and in most of the English areas (Lake District, Northumberland) to which it had been introduced. The one successful English population survives in the Peak District, the result of at least three introductions of Perthshire hares in 1870, 1876 and 1880 (Hewson 1956) to the northern moors of the Peak District, around Saddleworth and Penistone. Numbers are now fewer than in the 1930s, and the range of the species a little reduced, but it still occurs in most of the places that Hewson's correspondents mentioned 40 years ago (Yalden 1984b). In Scotland, there have also been numerous introductions, for sporting purposes, to various islands, mostly of Scottish hares, but at least one introduction, to Mull, of both Irish Hares, in 1863, and Scottish ones, the next year. They apparently remained separate for 30 years or more, but now only Scottish hares are present. Other islands with introduced Mountain Hares include Gairsay in Orkney, Islay and Eigg (where they are now extinct), Hoy (Orkney), Jura, Lewis, Raasay, Scalpay, Skye and Vaila and Mainland Shetland (Fitter 1959, Harris *et al.* 1995). There is an intriguing sidelight on their presence on Hoy, for several early accounts of the Orkneys said that Mountain Hares were present there, but died out during the 17th century (Barrett-Hamilton and Hinton 1912). They must have been reintroduced, and thrive still (Berry 1985, Hewson 1995).

The fortunes of the Mountain Hare are closely tied to those of grouse moors, for it prefers short, recently burnt and more nutritious, heather for its diet, and longer heather nearby for cover (Savory 1986). This is precisely what gamekeepers have to supply for Red Grouse, by burning heather in small patches, though grouse prefer to eat slightly older, 4-year-old, heather whereas hares prefer it 2 years old. The decline of grouse moors, both in extent and quality of management, is reflected in the declining bag of Mountain Hares this century to a very low level. From a maximum bag of 4.3/km^2 in 1932, numbers slumped to a minimum of 0.2/km^2 in 1980. However, they have recovered during the 1980s to somewhere near the long-term average at 2.6/km^2 in 1989 (Tapper 1992). Grouse moors are expensive to maintain, and taxed, whereas the alternative land-uses in the uplands, raising sheep and conifers, have been heavily subsidized. Not surprisingly, much moorland has been lost this century (about a third in the Peak District, for example), and what remains is often poor. Mountain Hares themselves are not highly rated as game, indeed the game-bag figures may be suspect for this reason, but the animal is an important prey for predators, notably Golden Eagles, in the uplands. It is important that they remain a common species in the Highlands. The very poor breeding success of the only English pair of Golden Eagles, in the Lake District, may well relate to the scarcity of suitable prey, particularly hares (of either species).

Rodents have already been discussed in previous chapters, so far as the most striking changes are concerned; the decline of the Red Squirrel as the Grey Squirrel advances, the replacement of Black by Brown Rats, the rise and fall of Muskrat and Coypu (Chapter 7). Mention has also been made of the decline of the Water Vole, though not, in detail, how that has been assessed. In a substantial, country-wide survey in 1986–88, Rob Strachan examined 600 m-long stretches of river and canal bank at five sites within a stratified sample of preselected 10-km squares, looking for signs, notably droppings (which Water Voles use to mark territory boundaries), footprints, burrows and feeding signs. These are all quite distinctive, and easily observed under most conditions. As well as checking a sample of 1926 sites selected to cover the country, he visited a further 1044 sites where old naturalists' reports said that Water Voles had been present at various times, up to 80 years ago. This survey not only provided a statement of the present range and, to some extent, abundance (more signs presumably means more voles, even if the relationship cannot be precisely quantified), but also indicated, for the first time, the scale of the decline that has taken place this century. Only 32% of the sites occupied in the period 1900–40 still had Water Voles. Loss of bankside habitat and canalization of rivers, both by overgrazing and dredging for flood control, have had their impact. Mink seem to be the last straw (see Chapter 7); even since the 1989–90 survey, there has been a further decline in the range of the Water Vole, particularly in the Thames basin, from 74% to 24% of positive survey sites. This decline has been correlated with a further increase in the range of the Mink (Strachan *et al.* 1998).

Orkney Voles too have declined by a measurable amount, as their moorland has been 'improved' for agriculture, and this has repercussions for the important raptors (especially Hen Harriers *Circus cyaneus*) that breed on Orkney. Gorman and Reynolds (1993) calculate that non-agricultural habitat in Orkney has declined since 1936 from 63% to 19%, and the number of voles reduced accordingly. Field Voles in Mainland Britain have surely suffered similarly from loss of habitat, without their loss being noticed. Indeed, any losses to agriculture were masked by the changes to forestry in the uplands, for the first few years of exclusion of sheep provide Field Voles with ideal grassy habitat, until the tree canopies close. In some cases, Field Vole plagues were reported in the 1950s like those that occurred in the Southern Uplands in the 1890s (Harris *et al.* 1995).

The other common rodents have had similarly obscure recent histories to the Field Vole. There is little that can be said objectively about the changing status of Bank Vole or Wood Mouse. Indeed, at the turn of the century, Bank Voles were thought to be rather scarce, and it required the advent of the Longworth trap from the 1940s and the notion of searching discarded milk bottles from the 1960s to show that they were in fact quite common in the right habitats. Because most of the predators hunt over grassland, Field Voles are their common prey, and Bank Voles scarce, but the analysis of owl and kestrel pellets had provided earlier naturalists with much of their knowledge. Numbers of both Bank Voles and Wood Mice fluctuate with the level of the autumn seed crop in deciduous woodlands, and this certainly produces dramatic nation-wide slumps in bad years. In turn, Tawny Owls, which do rely on woodland rodents, suffer mixed breeding success, failing completely in years such as 1955 (Southern 1970).

The Yellow-necked Mouse was not recognized until 1894, when it was described as very local. That description would still serve. Its distribution seems in general to be related to that of ancient deciduous woodland, and in detail it seems to decline when that woodland is disturbed, as by coppicing or replanting (Montgomery 1978, Yalden and Shore 1991). (By contrast, most rodents respond to coppicing by becoming much more abundant.) Since much old woodland has been lost this century, it is a fair bet that Yellow-necked Mice are now less abundant than they were in 1900, but there is no direct evidence. However, the Dormouse, which shows a somewhat similar distribution (see Chapter 4), has certainly become less widespread. There have been two recent surveys, in 1973–75 by Hurrell and McIntosh (1984) and the 'Great Nut Hunt' in 1993, a public-participation survey looking for the Hazel nuts opened in a characteristic manner by the species (Bright *et al.* 1996). These demonstrated that the species had gone from seven counties wherein it was reported by Rope (1885), including Yorkshire, Derbyshire, Cheshire, Nottinghamshire, Cambridgeshire, Warwickshire and Northamptonshire, but they also showed that it was still present in several Welsh counties where it had not been reported for decades, and confirmed the persistence of small outlying colonies in Cumbria and Northumberland. This is a species where the old techniques of managing the land by hand, particularly hedge-laying and coppicing, brought it to attention. Mechanization must mean that many records of its presence are lost. Thus it is harder to claim that the abundance of the species has declined than to demonstrate that its range has shrunk. This is one case where steps are being taken to reverse the decline. Dormice, mostly captive-bred, have already been released in one wood in Cambridgeshire and another in Cheshire, thus restoring it to some of its former range. The Dormice are held in a series of cages at the release site while they reorientate themselves, and food is provided at the cages for several weeks after the doors have been opened. Early results are encouraging; released Dormice have bred within their first season in the wild, and successfully hibernated through their first winters (Bright and Morris 1994 and pers. comm.).

Harvest Mice also have a somewhat restricted range, and are similarly easily overlooked in this mechanized age. Their nests are very characteristic, woven from the shredded but still-living leaves of the grasses that support them, and a national survey in 1973–75 (Harris 1979) rediscovered them in several areas where they had been thought extinct. Harvest Mice are very patchy in their distribution, and colonies are often ephemeral. Now, they are rarely found in cereal fields, more likely in hedgerows and ditches, reed-beds and waste ground, anywhere with long grasses. It is supposed that the change to winter-sown wheat (harvested before the Harvest Mice have finished breeding) and shorter stemmed varieties mean that the countryside is less suitable for them, and a resurvey in 1996–97 suggests that they have declined further since the 1970s. This is another case where inadequate data from the turn of the century will preclude us from ever providing a firm index of just how great the decline has been.

Carnivores offer some of the most dramatic and well-documented changes this century. Because of their interest to gamekeepers and others, there are good series of vermin-bag records for many of them, and they were rare enough at the turn of the century to have attracted naturalists' attention. Though generally common, the Red Fox was relatively scarce in parts of Britain, especially in East Anglia and eastern Scotland, 100 years ago, and only since the 1960s has it recolonized those areas. This has accompanied a general increase in abundance, with vermin-bags increasing from $0.6/km^2$ to $2.1/km^2$ over 30 years (Tapper 1992). In part this reflects the general decline in the number of gamekeepers, which has helped other species as well.

Though the Pine Marten was feared to be on the edge of extinction in 1915, and early pleas for its safeguard were made by Lt Cdr E.J. Fergusson, it responded very slowly to the reduced level of persecution. However, by 1926 there were signs that it was becoming more numerous in the north-west Highlands and by 1946 it had spread to the north side of the Great Glen. This seems to have been a barrier for about 10 years, but by 1960 there were 15 records south of the Caledonian Canal (Lockie 1964). A further survey in 1980–82 found them quite widespread in the Highlands, and an introduction attempt had been made with 12 released in Galloway Forest, Dumfriesshire (Velander 1983). In Ireland, the Pine Marten was never so heavily persecuted, and it remained moderately common in the wilder parts of the island, though a survey in 1978–80 found that it had disappeared from many areas, and remained common only in the west (O'Sullivan 1983). Meanwhile, the Pine Marten was also supposed in 1915 to have survived in England, in the Lake District, and in Wales, in Snowdonia. A trickle of records right through to the 1980s suggested that they had indeed survived here and perhaps in a few other sites too, but a sign survey of the most likely areas in 1987–88, calibrated against similar surveys in their known range in Scotland, suggested a population only a third the density of the Scottish ones. Later surveys suggest that the Welsh and English populations might have declined to extinction, and proposals for deliberate reintroduction have been discussed. This is a double mystery; why have these populations not increased this century, as has the Scottish one? how does one explain the definite but irregular sightings that have been made in such areas as the Peak District and North York Moors, not to mention Lake District and Snowdonia? An undoubted recent skull was found in the North York Moors in 1993, for example (Jefferies and Critchley 1994). Given that it is now illegal to kill Pine Martens, but that they are not supposed to be present in North Yorkshire, one could not be surprised that a gamekeeper killed one accidentally and buried the evidence. Given the sporadic record of presence there, I doubt an undocumented population has been living there unnoticed all this time. I suspect that there have been

Field Vole

surreptitious releases, or perhaps escapes, of captive Pine Martens over many years. The animal has always had a particular fascination for those who keep wild mammals in private zoos, for it is usually very active in captivity by daytime, and an attractive as well as rare animal. In the past, before it was protected, Scottish gamekeepers were always willing to dispose of problem animals by sending them south rather than killing them.

The other Highland refugee, the Wild Cat, responded more quickly, or perhaps more overtly, to the reduction in keepering after 1919. As early as 1946, Taylor found them restored to most of their current range, Scotland north of the Central Lowlands. More recent surveys (Jenkins 1962, Easterbee *et al.* 1991) have found them in essentially the same range. However, they have encountered a major problem as they have spread out of their Highland refuge, hybridizing with feral cats in much of the periphery of their range (French *et al.* 1988, Hubbard *et al.* 1992). It seems to have been particularly a problem of the 1940–65 era, when their range and abundance was increasing, but it makes it doubtful whether they could cross the Central Lowlands unaided. There have been records from Galloway recently, but whether this indicates misidentified feral cats, natural spread or, more likely, clandestine releases, remains unknown (Harris *et al.* 1995).

Of the three 'refugees', the Polecat responded most quickly to the relaxation of persecution following the 1914–18 war. As early as the 1920s, there were increasing reports of them in the English–Welsh border counties, and surveys in the 1960s showed that they had recolonized virtually all of Wales, plus the border counties of Shropshire, Herefordshire and Monmouthshire (Walton 1964, 1968). An exception is that supposed remnants in Cumberland and Sutherland in 1915 did not show any recovery, and must have died out. During the 1980s and 1990s, Polecats have continued to spread eastwards into England, reaching across to counties such as Derbyshire and Oxfordshire (Birks 1993). Someone has also released Polecats in Cumbria recently, and they appear to be thriving.

The Otter has had the most chequered career of any of the carnivores this century. Although a familiar presence to anglers and naturalists, only the otter hunts had any quantified data for the first half of the century. These showed a fairly stable bag, of around 400 Otters killed per annum during the 1930s, but during the 1950s Otters became decidedly scarce, at least in England, southern Scotland and Wales. Retrospective analysis, much later, showed that the bag halved, and that the decline began rather sharply in 1957–58 (Chanin and Jefferies 1978). At that time, it was not realized what had happened. Looking back, 20 years later, it was realized that this coincided with the catastrophic declines of Sparrowhawks and Peregrine Falcons, and with the peculiar 'epidemic' of Foxes (Taylor and Blackmore 1961) that turned out to be secondary poisoning by organochlorine pesticides. Eels are a major food of Otters, and very fatty, therefore nutritious, fish. This same fat also accumulates organochlorine pesticides, and even as late as 1988, Londoners were being warned by the Ministry of Agriculture, Fisheries and Food to eat Eels no more than once a week because of their pesticide load (Lean 1988, Brown 1988).

This misfortune for the Otters spawned four fortunate results. The hunts, realizing their prey was threatened with extinction, gave up hunting Otters, and either disbanded, or moved to hunting American Mink. Because of its rarity, the Otter was given legal protection, initially (1978) only in England and Wales but later (1982) in Scotland too. Captive breeding programmes, aimed at bolstering the remnant local populations, particularly in East Anglia, were started by the Otter Trust and others. Lastly, a series of professional surveys, looking for signs of Otters (particularly spraint – droppings – but also footprints, dens and feeding signs) was undertaken. The earliest, in 1977–79, found signs of Otters at only 6% of the sites surveyed in England, 20% of those in Wales and 73% of those in Scotland. Only in Highland Scotland was there a well distributed, and by implication reasonably abundant, population. A survey of Ireland, carried out in 1980–81, showed fortunately that Otters were still widespread there, and present at 92% of sites. This is correlated with the much lower (and later) levels of pesticide use (Jefferies 1989). Repeat surveys in Wales and England in 1984–86 found some recovery, to 39% and 10% positive sites, except that in East Anglia Otters had declined still further (prompting the reintroduction programmes). A third survey of Wales, in 1991, found a further recovery, to 52% of positive sites. The most recent resurvey of England, carried out in 1991–94, has also shown a further significant recovery, to 22% of sites positive, and has also demonstrated the success of the reintroductions to East Anglia, as well as to Yorkshire (Strachan and Jefferies 1996). Even so, extrapolating from these rates of recovery, Strachan and Jefferies (1996) calculate that it will take another 100 years before the population of English Otters recovers to the full its former widespread status (Fig. 9.6).

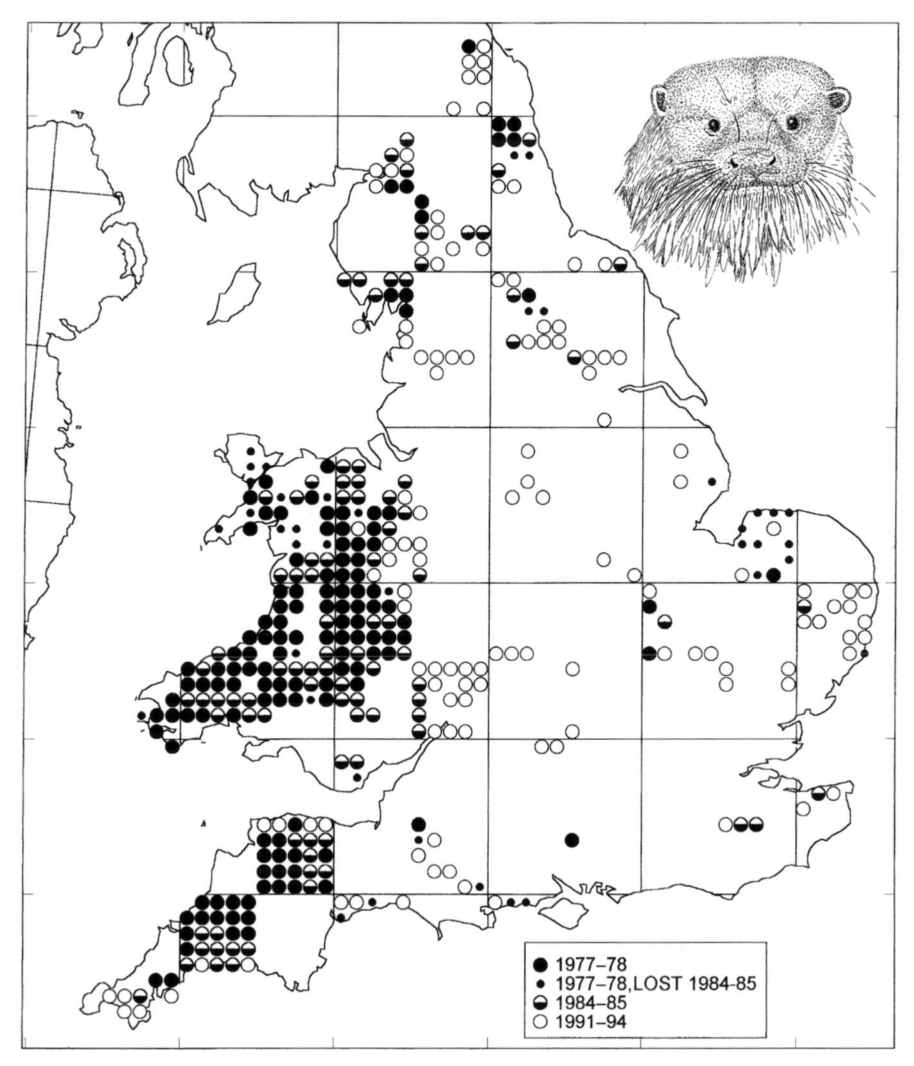

Figure 9.6 The recovery of the Otter population in England and Wales, as shown by the repeated national surveys (after Strachan and Jefferies 1996). In England, only half the 50 × 50 km squares were surveyed.

The other mustelids have had mixed fortunes this century. The Badger was generally regarded as scarce around 1900, a casualty of the general gamekeeping pressure. This may have been true, and certainly was for some parts of England, notably East Anglia where it remains scarce. There is however a strong impression that this largely nocturnal and secretive mammal was greatly under-recorded, in the days before road casualties became so obvious, and before the distinctiveness of badger setts (as contrasted with fox earths or rabbit warrens) was fully appreciated. It was the series of books by Ernest Neal (1948, 1977, 1986), particularly the first, New Naturalist, volume, that drew the attention of naturalists to the possibility of mapping the distribution of Badgers from their setts. The additional demonstration by Kruuk (1978) that each social group was actually defending a territory surrounding its main sett resulted in the further realization that, by concentrating on main setts (and ignoring the various annexes, outliers and supplementary setts that tended to be associated), one could get some idea of the density or abundance of Badgers in different landscapes. The National Badger Survey (Cresswell *et al.* 1990) exploited this fact, and the dedication of some 385 mostly amateur surveyors willing to visit 2455 one-kilometre squares, to arrive at an estimate of 43 000 social groups for the total population of Great Britain. On the further assumption, based on a limited number of cage-trapping studies, that there are about six adult Badgers in a social group, there must be around 250 000 adult Badgers in Britain, and this is one of the better-founded population estimates for a British mammal. A repeat survey in 1995–96 has confirmed the general pattern of abundance across the country, and shown noticeable increases in some areas, particularly the West Midlands, apparently as a consequence of a reduction in the illegal persecution by badger diggers (Wilson *et al.* 1997). Similar surveys have been undertaken in Ireland, where the Badger is similarly distributed very widely, and in similar-sized social groups (about six adults per group). The terrain is ideal Badger country, lush pasture for dairy cattle and therefore good for earthworms, the major food of Badgers in the British Isles. However, there is much less woodland in Ireland, so setts tend to be smaller, and associated with hedgerows and small copses. Overall, the population in Ireland is as large as that in Great Britain, about 200 000 in Eire and 50 000 in Northern Ireland (Smal 1995).

Weasels and Stoats are less well documented, though the vermin bags do give a good idea of their fluctuations. Both survived the depredations of gamekeepers well, and were generally regarded as common at the beginning of the century. Myxomatosis had major but contrasting effects on both their absolute abundance, and their numbers relative to each other. Stoats declined sharply as their main prey disappeared, whereas Weasels became much more abundant. As Rabbits recovered during the 1970s and 1980s, so did Stoats, and Weasels have continued to decline steadily since their peak (Tapper 1992; Fig. 9.7). It is assumed that this reflects the effect of agricultural pressures in removing rough grassland and with it the Field Voles.

Seals have had mixed fortunes. Under legal protection, Grey Seals have increased steadily throughout the century, founding new colonies (St Kilda after 1930, Monachs 1945, Gunna 1955, Scroby Sands 1958, Nave Island 1970, Donna Nook 1970, Isle of May 1977) and increasing sharply at most of the others. From the (dubious) estimate of 500 for the total population in 1914, there were thought to be 4000–5000 by 1928 and 8000 by 1932, when a new seal protection act extended the close season for hunting. Recent estimates are more soundly based, for the white-coated pups are on land for 2–3 weeks before they moult into a 'blue' coat, and another 2 weeks before they swim away, giving a good chance of counting them, often using aerial survey techniques. It is a reasonably secure datum that there would be 3.5 other Grey Seals in a colony for every pup born (its mother, of course, but also several immature seals, and a father perhaps shared with other pups), if the population were stable. Most British Grey Seal colonies are increasing at about 7% per annum, and a multiplier as high as 4.4 might be more appropriate. Estimates of 34 200 in the mid 1960s, 69 000 in the mid 1970s, and 93 500 in 1991 indicate the scope of the increase (Fig. 9.8). This has led to calls for culls, mostly from the fishing industry, and

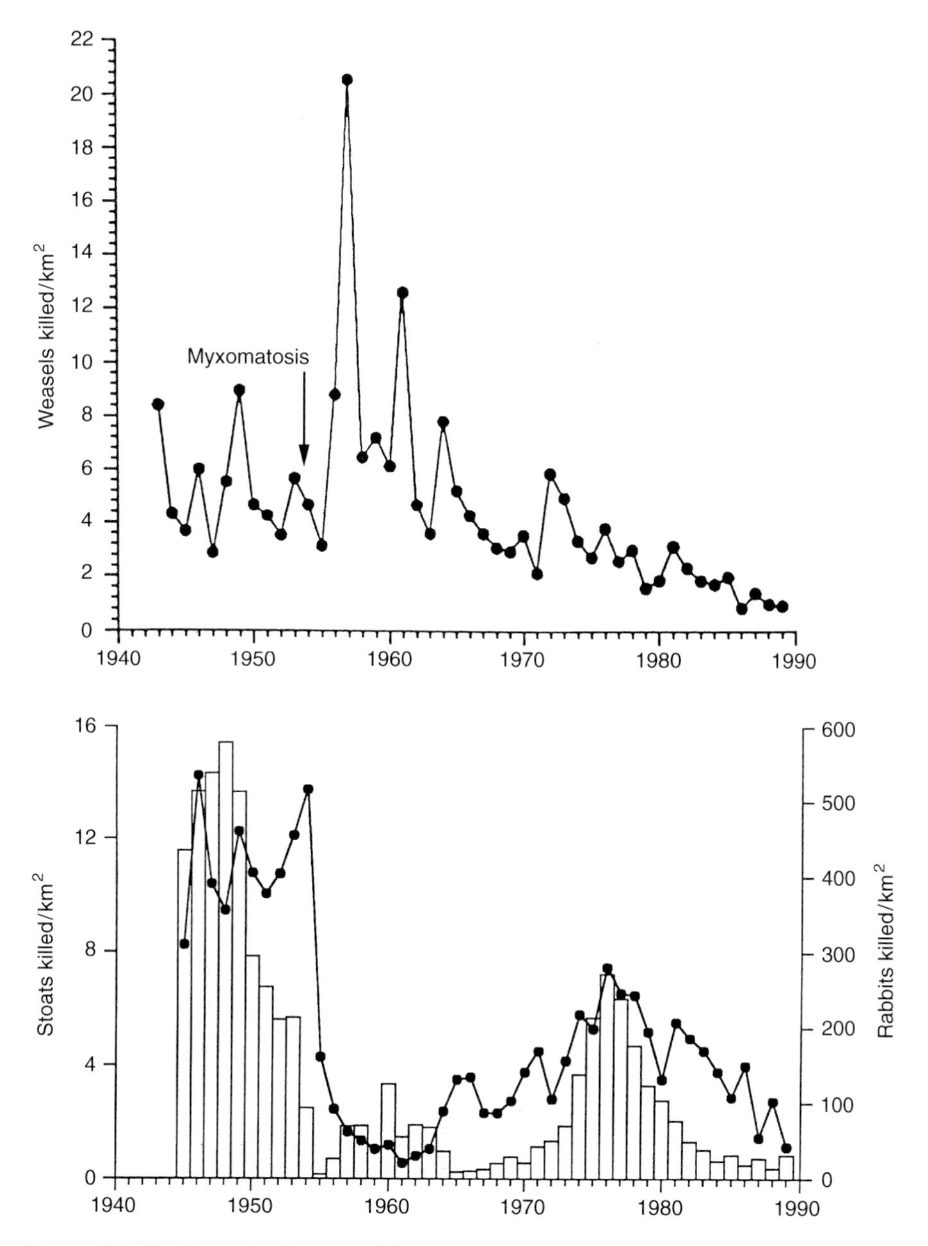

Figure 9.7 The fluctuating fortunes of Stoats and Weasels, as recorded in the vermin bags from keepered estates. In the lower panel, the bag of Rabbits (bars) shows the fate of the Stoats' main food supply, and surely explains much of the fluctuation in its numbers (dots). The Weasel increased on many estates as Rabbits disappeared (upper panel), apparently because longer grass meant more Field Voles, its main prey. (Reproduced from Tapper 1992 *Game Heritage*, by permission of Dr S. Tapper and the Game Conservancy.)

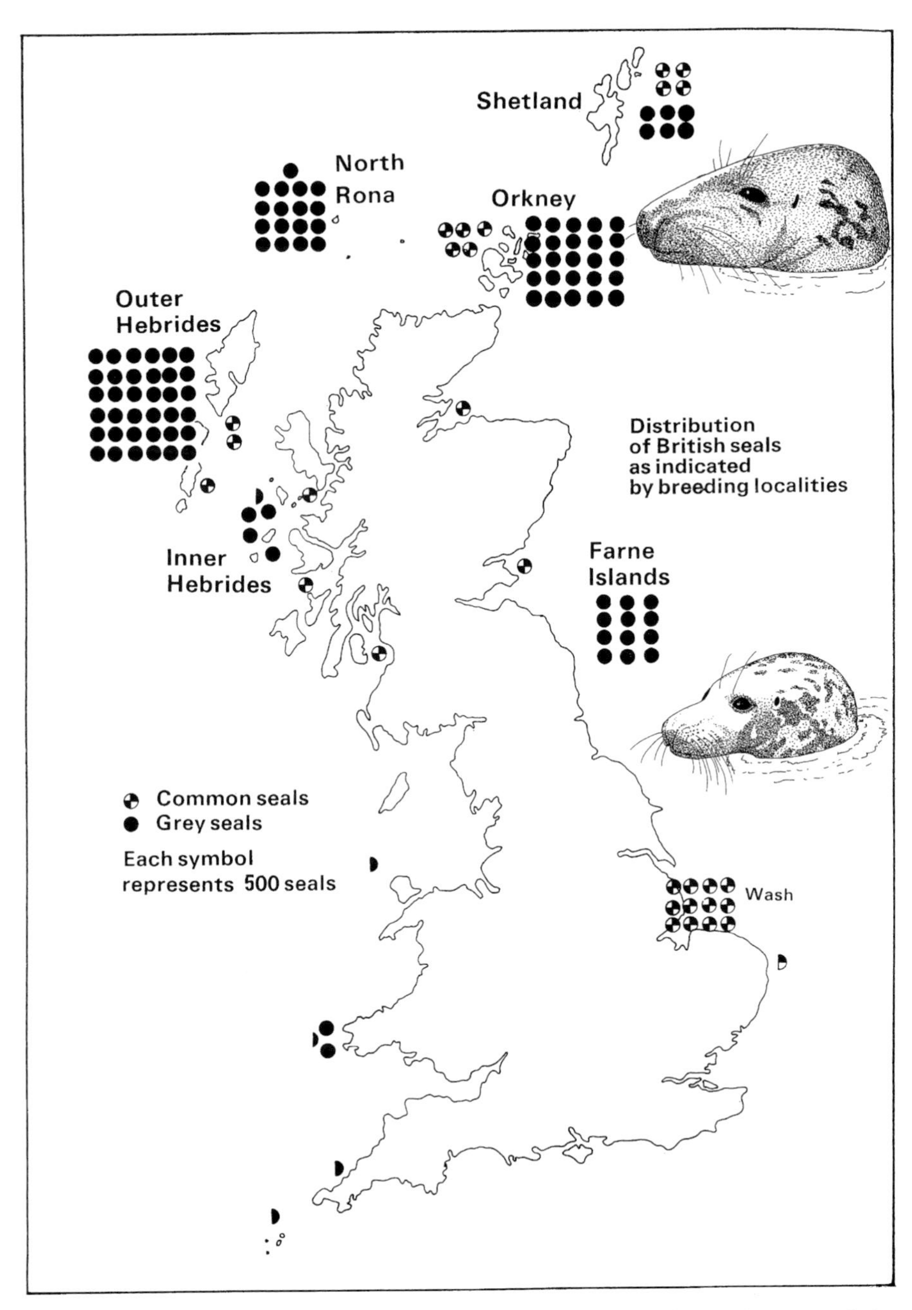

Figure 9.8 The distribution of breeding Grey and Common Seals around the British coasts, with the main colonies indicated (from Bonner 1976 *Stocks of Grey Seals and Common Seals of Great Britain*, by permission of N.E.R.C.).

small-scale commercial culls of 950 pups, for fur, were licensed in the Orkneys from 1962. A similar cull, aimed at reducing the population on the Farnes by 25%, saw a cull of 1001 pups taken in 1963–65, before the National Trust (who own the Farnes) decided that culling was inappropriate in a nature reserve. Larger, fishery protection, culls were authorized for Orkney and the Hebrides in 1977. It was intended that 4000 moulted pups and 900 breeding cows should be culled annually, but such was the public outcry that it was reduced to a pup-only cull in 1978. Neither the effects on the seal population, nor the supposedly beneficial effects on the fisheries, were adequately demonstrated. More limited culls were also reinstated on the Farne Islands in 1972 and 1975. Intended to protect the soils and the seabird colonies (notably Puffins, which nest in burrows in the soil that was being eroded), they removed 132 bulls, 603 cows and 575 pups in 1972, and the follow-up cull three years later removed another 158 bulls, 486 cows and 804 pups (Summers 1978). These culled animals were used to investigate the age structure of the population (the oldest cow was 36 and the oldest bull 27 years; Harwood and Prime 1978), knowledge which still underpins the reconstruction of population dynamics in Britain. However, the National Trust, owners of the Farne Islands, later affirmed that the culls were not consonant with the notion of these islands as a nature reserve, and culls have been discontinued. The number of Puffins breeding there has increased steadily, from 6800 pairs in 1969 to 20 700 pairs in 1984 (Lloyd *et al.* 1991) and 7000 Grey Seals as well. Management now attempts to restrict the seals to rocky islands, where they can do less damage, by patrolling the seabird islands during the Grey Seal breeding season.

The commercial culling of Common Seals continued in a small way in the Shetland Isles, where around 300–400 were killed per year in the 1960s. With the extension of commercial hunting to Grey Seals in the 1960s, extra pressure spread to the Common Seal as well, since their very different breeding seasons allowed complementary culls. The population in the Wash was thought to be yielding 600 per year, the Shetland take went up to 900, and a further 400–600 were taken from the west of Scotland. Later research suggested that these represented only 38% of the annual pup production, but in Shetland were certainly enough to cause a decline. Public antipathy to these culls led to the Conservation of Seals Act 1970, extending breeding season protection to the Common Seal for the first time. Consequently, numbers increased slowly in the Wash, at 3.5% per year from 1969 to 1988, but their status elsewhere is less certain; because they haul out only at low tide, even giving birth between high tides, Common Seals are much harder to count than Grey Seals. However, numbers certainly seemed to have recovered well everywhere from the earlier hunting, when a new disaster hit them, in the shape of a virus infection. This began in the spring of 1988 in the waters of the Kattegat and Skaggerak, moved into the Wadden Sea, and appeared on the Norfolk coast in late July. By the end of September, around 2000 dead Common Seals had been recorded there. This outbreak attracted enormous media attention, as usual with more noise than sense. Initial attempts to claim that it had been passed from dogs turned out to be totally wrong. The disease, subsequently named Phocine Distemper Virus, is certainly related to Canine Distemper, but quite distinct. It is possible that it was passed to Common Seals from arctic species; the failure of northern fish stocks seems to have triggered an exceptional southward migration of Harp Seals that winter. The disease caused a 60% reduction of the population in Danish waters, and 50% of those in the Wash, but it seems to have died out as it spread northwards. Only 10–20% of those on the east coast of Scotland died, and there were no detectable changes in the Common Seal populations of Orkney, Shetland and western Scotland. It seems likely that the passing of the breeding season meant that seals were more dispersed, less likely to pass on the highly infectious disease (Thompson and Hall 1993). At the time, Grey Seals seemed largely unaffected by the virus, though there is some retrospective evidence that there was a 12% mortality, and consequent reduction in breeding output, from some of the eastern colonies (Isle of May, Farnes, Orkney; Harris *et al.* 1995). Grey Seals now seem to be carrying natural immunity, and no resurgence of the disease is expected.

Deer have shown this century an almost uninhibited spread in range and increase in numbers. In Scotland, the counts from the Deer Commission are supposed to allow estate owners to regulate their herds appropriately, but the evidence indicates that most fail to achieve the desired cull. The Forestry Commission has a more definite requirement to manage its deer populations within levels that do not cause damage, but it too has difficulty achieving appropriate control. There is an acceptance that the costs of management, £2.8 million for fencing and culling, are less than the income, £3.6 million from sales of venison and £1.7 million from stalking (cf. Ratcliffe 1989 and Cobham Resource Consultants 1983). In other words, it is not sensible to attempt the impossible task of eliminating deer from woodland, but better to regard the deer as an additional harvest. In lowland Britain, such an approach is more difficult, because of the greater density of people and the risk of accidents, and because of the unfortunate 'bambi effect' of sentimental public perception. Yet there is no doubt that deer, collectively, pose a risk to traffic, a threat to agriculture and horticulture, and a threat to some conservation objectives. How serious any of these threats may be is currently uncertain (Putman and Moore 1998). In the absence of any of their natural predators, there is no doubt that all of the deer will continue to spread geographically, and increase in abundance as well. Thus Roe Deer have spread southwards down either side of the Pennines, across to the east in Kent and up the Cotswolds towards the Midlands (Fig. 9.9). The Muntjac has had its range expansion well mapped, as has the Sika (Chapter 7). The Fallow Deer, less spectacularly, has established itself around many of the parks from which it escaped in two world wars, and is now the most numerous and widespread deer in England. In Scotland, perhaps because the Red and Roe Deer are more numerous there, it is less numerous. There are claims that the three deer are in competition, though it is hard to disentangle the possible effects of competition from the seral changes that occur as a commercial forest matures; the young thicket stage favours Roe Deer, more mature woodland is better for Red Deer, and open woodland with glades favours Fallow (Batcheler 1960). Thus the apparent displacement of Roe by Red and then Fallow might be competition, but more probably is simply a reflection of different habitat preferences and feeding opportunities.

The Present Balance

The conventional species-by-species account of the mammal fauna allows the progress this century of each to be reviewed in turn. A different way of looking at the present mammal fauna of the British Isles is to ask how the overall balance of the species works out. Thanks to the attempt (Harris *et al.* 1995; Table 9.1) to produce

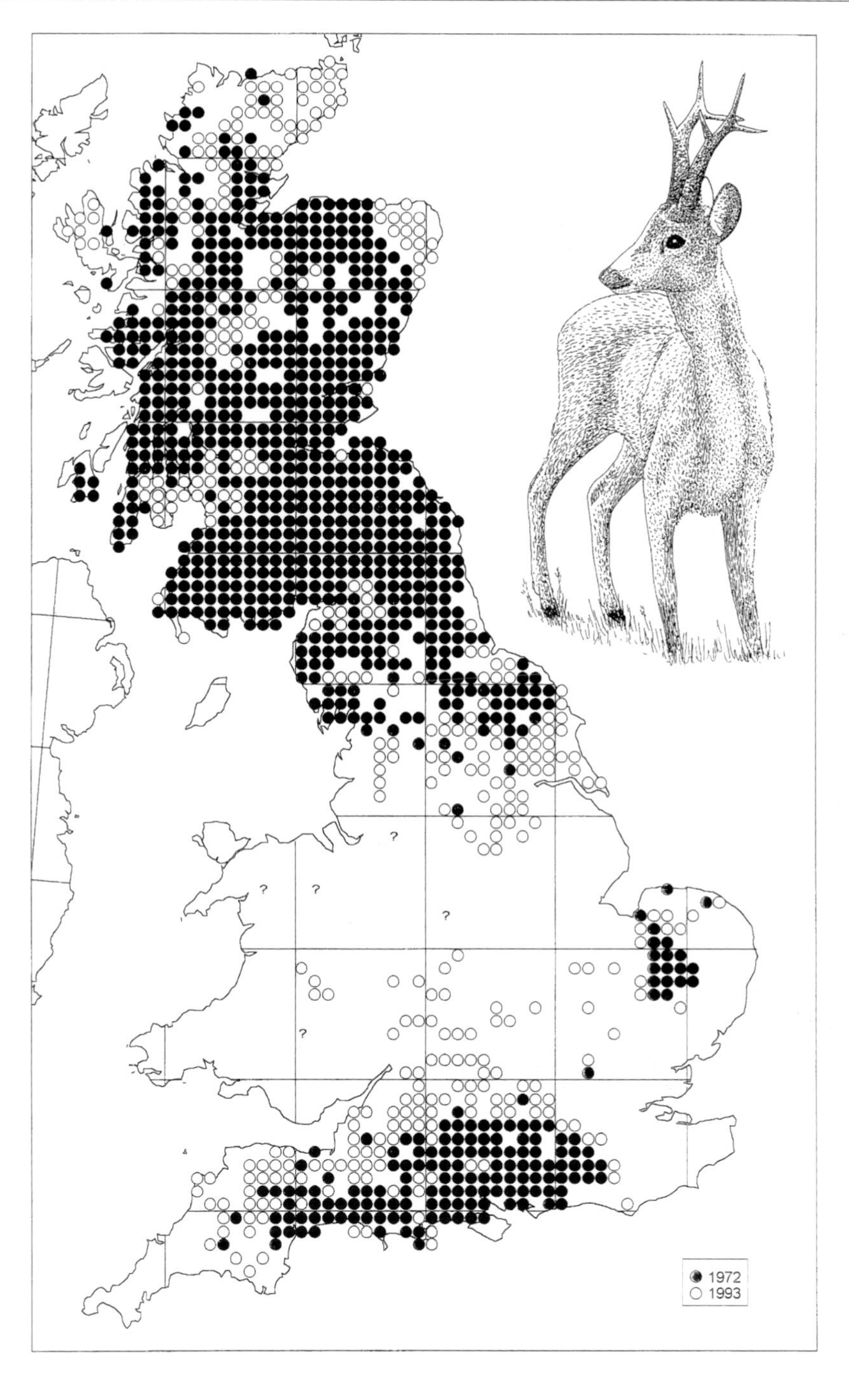

Figure 9.9 The spread of the Roe Deer *Capreolus capreolus* this century (from Clarke 1974, Arnold 1978, 1993).

Table 9.1 The numbers and biomass of mammals in Great Britain. The numbers of wild mammals are as given by Harris *et al.* 1995, and are not modified to allow for more recent information, mentioned in the text. They estimate the spring, pre-breeding season, population. Numbers for domestic mammals come from June agricultural censuses, but are similarly estimates of the breeding stock. For humans, the population over 18 is taken to be the equivalent 'breeding population'. The individual masses (largely from Corbet and Harris 1991) and biomasses for these and domestic mammals come from collaborative work with Stephen Harris, Mary Morris and Pat Morris that is not yet published. There are too few density estimates for Irish mammals to encompass them in this table.

Species	*Population*	*Mass (kg)*	*Biomass (t)*
Greater Mouse-eared Bat	Extinct	0.032	0
Coypu	Extinct	6.25	0
Nathusius' Pipistrelle	?	0.006	?
Red-necked Wallaby	29	12	0.372
Park Cattle	45	150	6.75
Reindeer	80	100	7
Chinese Water Deer	650	15	9.75
Grey Long-eared Bat	1000	0.009	0.009
Black Rat	1300	0.175	0.228
Bechstein's Bat	1500	0.01	0.015
Feral Sheep	2100	20	42
Feral Ferret	2500	0.9	2.25
Wildcat	3500	4.5	15.75
Feral Goat	>3565	45	160.425
Pine Marten	3650	1.5	5.475
Greater Horseshoe Bat	>4000	0.02	0.08
Barbastelle	5000	0.008	0.04
Skomer Vole	7000	0.025	0.175
Otter	>7350	8	58.8
Leisler's Bat	10 000	0.025	0.25
Edible Dormouse	10 000	0.14	1.4
Sika	11 500	50	575
Lesser White-toothed Shrew	14 000	0.005	0.07
Lesser Horseshoe Bat	14 000	0.005	0.07
Serotine	15 000	0.032	0.48
Polecat	15 000	0.9	13.5
Brandt's Bat	30 000	0.006	0.18
Common Seal	35 000	75	2625
Whiskered Bat	40 000	0.006	0.24
Chinese Muntjac	40 000	13	520
Noctule	50 000	0.032	1.6
Grey Seal	93 500	200	18 700
Natterer's Bat	100 000	0.008	0.8
Fallow Deer	100 000	50	5000
American Mink	>110 000	0.8	88
Daubenton's Bat	150 000	0.08	1.2
Red Squirrel	160 000	0.28	44.8
Brown Long-eared Bat	200 000	0.008	1.6
Red Fox	240 000	6	1440
Badger	250 000	10.5	2625
Mountain Hare	350 000	2.8	980
Red Deer	360 000	75	27 000
Weasel	450 000	0.075	33.75

Species	*Population*	*Mass (kg)*	*Biomass (t)*
Stoat	462 000	0.25	115.5
Common Dormouse	500 000	0.018	9
Roe Deer	500 000	20	10 000
Yellow-necked Mouse	750 000	0.03	22.5
Feral Cat	813 000	3.5	2845.5
Brown Hare	817 500	3.3	2697.75
Orkney Vole	1 000 000	0.025	40
Water Vole	1 169 000	0.3	350.55
Harvest Mouse	1 425 000	0.006	8.55
Hedgehog	1 555 000	0.5	777.5
Water Shrew	1 900 000	0.015	28.5
Pipistrelle	2 000 000	0.006	12
Grey Squirrel	2 520 000	0.5	1260
House Mouse	>5 192 000	0.015	77.88
Brown Rat	>6 790 000	0.25	1679.5
Pigmy Shrew	8 600 000	0.004	34.4
Bank Vole	23 000 000	0.02	460
Mole	31 000 000	0.1	3100
Rabbit	37 500 000	1.5	56 250
Wood Mouse	38 000 000	0.02	760
Common Shrew	41 700 000	0.01	417
Field Vole	75 000 000	0.02	2250
Domestic Species			
Horse	750 000	500	375 000
Pig	853 000	150	127 950
Cattle	3 908 900	550	2 149 895
Sheep	20 364 600	45	913 707
Humans	37 866 500	80	3 029 321

estimates for the numbers of each species, we can now attempt such a survey. This suggests that there are about 250 million wild mammals in spring on the main island (Great Britain). Most of them are rodents and insectivores; lagomorphs also are noticeable, numerically, but the other orders are insignificant (Table 9.2). The three common and widespread small mammals, Field Vole, Common Shrew and Wood Mouse, top the list, but the Field Vole is apparently the only wild mammal to exceed the human population numerically. Of the 16 mammals we believe to exceed 1 million, five certainly are introduced (Rabbit, Brown Rat, House Mouse, Grey Squirrel, Orkney Vole), perhaps six (Harvest Mouse). Only one bat figures in this top group, Common Pipistrelle: if this is actually two species (see earlier), they might both just get in the 'top twenty' at 1 million each, but we do not know whether they are equally abundant. As a rough check that the numbers estimated for the commonest mammal are at least sensible and that numbers in Table 9.1 are internally consistent, Jerzy Dyczkowski calculated for me the likely number of young voles born each year, and the likely take of all the predators eating those voles (mammalian as well as avian) to see if the numbers match. Roughly, they do. About 677 million young are born (the 37.5 million female voles produce about five litters each of five young, and some of the early young also breed). About 744 million are eaten by wild predators, most of them by Weasels (about 214 million), Foxes (132 million) and Kestrels (239

Table 9.2 The composition of the wild mammal fauna of Great Britain, by numbers of species in each order, numbers of individuals of each order, and biomass attributable to each order. Data from Table 9.1 (seals omitted).

Order	*Species*	*Individuals (thousands)*		*Biomass (tonnes)*	
Insectivora	6	84769.0	(29.7%)	4357.5	(4%)
Chiroptera	14	2620.5	(0.9%)	18.7	(<1%)
Lagomorpha	3	38667.5	(13.6%)	59927.7	(49%)
Rodentia	15	155523.3	(54.6%)	6217.5	(5%)
Carnivora	11	2357.0	(0.8%)	7234.5	(6%)
Artiodactyla	10	1017.9	(0.36%)	43320.9	(35%)
Marsupialia	1	<0.1	(<0.1%)	0.4	(<1%)
Total	60	284955.2	(100%)	121077.2	(100%)

million). There is great uncertainty about how many Feral Cats there are, and how many voles are eaten by them, easily accommodating any mismatch in the numbers (Dyczkowski and Yalden 1998).

At the opposite end of the list, the rarest species include some fairly dubious contributors. It is a moot point whether Park Cattle, Reindeer, Feral Sheep and Feral Goat deserve a place on a list of British wild mammals. Their 'wildness' varies; the Cattle are totally enclosed but self-regulating populations, the Reindeer wander freely, but are culled to maintain their numbers, some at least of the populations of Wild Sheep and Wild Goats are indeed totally wild and unmanaged, others are culled much as deer to keep numbers limited. Certainly none of these exists in anything like a natural equilibrium across the country as a whole, and their scarcity is an artifact of their restricted distribution. The same is true for Red-necked Wallaby, Chinese Water Deer and Feral Ferret. Of the species which might be 'naturally' rare, the two bats, Grey Long-eared and Bechstein's, have some claim to be regarded as the rarest British mammals. It is unfortunate that these are among the least certain of our population estimates, but there is little doubt that they really are very rare, both in relation to other bats, and in relation to the other mammals as a whole.

There is a problem with these comparisons, that one should expect small mammals to be more numerous than big ones. It is therefore no great surprise to find the small mammals at the top of the list, and some, at least, of the larger ones near the bottom. One might also expect herbivores, at the bottom of the food chain, to be more common than the carnivores that prey on them. One way of investigating these points further is to plot numbers against mass. Actually, it is usual to use the logarithms of the numbers in such comparisons (Greenwood *et al.* 1996; Fig. 9.10). This shows the expected relationship between numbers and body size, allowing the relative numbers of the species to be judged against the 'expected' numbers. In fact, the carnivores do not seem to be scarce, relative to their prey, but one very obvious anomaly is highlighted by this plot. All the bats are very scarce indeed, for mammals of their (very small) size, about two orders of magnitude scarcer than they should be. As it happens, there is one other group of British animals for which a comparably comprehensive set of population estimates is available, the British breeding bird fauna. Rather unexpectedly, the birds are also much scarcer than the (terrestrial) mammals, but the bats fall in line well with their compatriot flyers. Why should bats and birds be so relatively

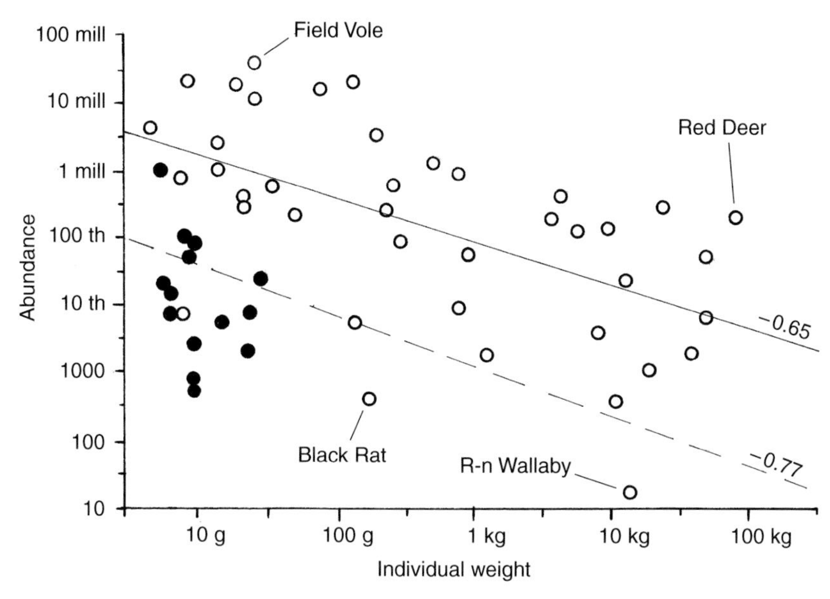

Figure 9.10 The relationship between numbers and individual size (mass) of British mammals (after Greenwood *et al.* 1996). While the smaller mammals are generally more abundant than the larger ones, the bats (filled circles) lie well below the regression line for terrestrial mammals (solid line). They fit well with the relationship for British breeding birds (dashed line), which are also much rarer than terrestrial mammals. Note that these are log–log plots, so that birds (and bats) are at least an order of magnitude rarer than terrestrial mammals.

scarce? Obvious explanations include the notion that flying is energetically expensive, so that a given amount of food or space will support fewer of them, or that they are dependent on scarce resources (cave sites, nest sites, rare habitats) that limit their numbers. In fact, although flying certainly is extravagant, bats at least avoid excessive energy expenditure by hibernating and going into torpor, and their overall energy expenditure is much less than that of a similar-sized shrew. Although scarce habitats or resources might limit some species, it certainly does not explain the relative scarcity of the commonest birds when compared with the commonest mammals. Perhaps birds and bats are adapted to exploit resources which are very patchy, abundant where they occur but scarce generally. That would certainly limit their numbers overall.

This suggests another way of overcoming the problem, in evaluating the fauna, that small mammals ought to be more common by virtue of their size: convert numbers to biomass, that is, multiply numbers by the average mass of each mammal. Since we are considering breeding adults in the spring, changing size with growth is not a concern, but some species show acute sexual dimorphism; average weights of the two sexes, and an assumption of an equal sex ratio, overcome this problem. Of course, this approach merely multiplies any errors in our estimates of numbers by any errors in masses, but the approach provides a valuable extra insight. Instead of suggesting a fauna dominated by small mammals, as it is numerically, it implies a fauna dominated by deer and lagomorphs (Table 9.2), which is surely a more realistic indication of the economic and ecological importance of our mammals. We can make a further useful comparison, by including the domestic ungulates and humans. This makes it clear just how totally the countryside is dominated by these species, and how insignificant is the wild fauna. Only 1.8% of the total mammal biomass in Great Britain is contributed by the wild species; domestic cattle and sheep, particularly, but also pigs and horses provide some 54.3% of the mammal fauna, and humans a further 44% (Table 9.1). Moreover, this calculation is an attempt to score the mammals living in the countryside, and having a direct ecological impact on it; some 7 million domestic cats and 6.9 million domestic dogs, other pets such as rats and rabbits, and laboratory rats and mice, are all additional to this total. Interestingly, the calculated biomass of all the (non-human) mammals now in Great Britain is about 10 times what was estimated to be here in Mesolithic times (Chapter 3), when the more wooded nature of the countryside would have limited the productivity of the vegetation, and therefore the weight of ungulates it could support. This is a measure of the success of the agricultural industry in converting our countryside to productive farmland to support our own burgeoning population; unfortunately, this is exactly the change that has reduced our wild mammal fauna to the small proportion (about 12% by biomass) of what it once was.

The Future

There are three ways in which the future mammal fauna of this country might be predicted. The most obvious is to extrapolate the trends which have already been noted as operating in recent years. A second is to try to forecast likely changes, in the light of various political and other activities. The third is to consider a pro-active conservation campaign to improve the mammal fauna. These can be usefully discussed in turn.

Extending Current Trends

The most obvious trends at present are those which see Chinese Muntjac and Grey Squirrel continuing to spread throughout England, Wales and at least Lowland Scotland, and the Roe Deer spreading back into central England, and into Wales. Wales is at present remarkably free of deer, and that is clearly something which will not last many more years, though the high density of sheep must be more of a deterrent to deer there than elsewhere in Britain. The Otter will surely continue to recover the distribution it lost in central England during the 1960s. Polecats will also continue to spread back into England, though the greater traffic mortality, and perhaps pressure from gamekeepers and poultry farmers, may slow down its progress. At present, though it seems that the Wild Cat is stuck in Highland Scotland, the Pine Martens reintroduced to Galloway may continue to spread, and might eventually colonize northern England.

Not all changes will be positive. Red Squirrels are bound to decline further, in the northern Lake District and in southern Scotland. It remains likely, given the slow rate at which the Grey Squirrel has spread into Highland Scotland, that the Red Squirrel will remain the normal squirrel in the Caledonian Pine Forests, and perhaps also in the planted conifers. In Ireland, too, it looks increasingly as though the Grey may slowly displace the Red Squirrel. The future of the Water Vole looks uncertain, at present, as the Mink continues towards ubiquity. If it is true that Otters and Polecats tend to displace Mink, they might between them also help the Water Vole indirectly. The saddest change seems likely to be the total loss of the Red Deer through introgression with Sika. It seems that the change is already too far advanced to be halted, let alone reversed. It has taken around 30 years to deteriorate to the present state, and it will take at least another century for the whole mainland population to be lost. Conserving at least some native genotypes on the Scottish islands, safe from Sika, seems essential.

Pest control will determine the status of at least two of our introduced mammals, Brown Rat and Rabbit. The Brown Rat is no-one's conservation favourite, and talk of rat-free cities suggests an urban planning ideal that could be realized. This would have repercussions on the predators of urban fringe land, including Barn Owls. The Game Conservancy's records already indicate a long-term decline in the abundance of rural Brown Rats (Tapper 1992). The Rabbit remains, or perhaps has recovered its position as, a major agricultural pest, and could easily be further reduced by novel control techniques, including new diseases. Rabbit Haemorrhagic Disease, another virus first tried in Australia, was reported in Britain in 1994; its effects remain to be assessed. The antipathy to the Badger, as a consequence of its role in harbouring Bovine Tuberculosis, and to the seals from commercial fishermen, could easily see both the subject of legal or illegal culling. At present, the evidence is that Badgers play little part in spreading TB to cattle, but that weather and farming practices play the largest part in determining the variation in cases from year to year (Krebs 1997). Similarly, present evidence is that seals prey very largely on Sand Eels *Ammodytes* spp., of modest concern to commercial fishing, and that overfishing by a poorly regulated industry is the main problem (Anderson *et al.* 1989). In other words, the mammals have been scapegoats. This will not stop the continued clamour for 'action', however ill directed, and this may determine their future.

Given that so many of our mammals are introduced species, one more extrapolation would be the addition of extra aliens. Which? This is a topic discussed by Baker (1990) in a review of the mammals which have already been reported as escapees in the British countryside. Among the most common are further Red-necked Wallabies, so the possibility of an extra colony or two cannot be discounted. The passing of the Dangerous Wild Animals Act 1976 made it necessary for the keepers of potentially dangerous pets to obtain licences from their local authorities, and have their safety measures inspected annually. Rather than face the expense, some pet keepers seem to have turned their pets loose. There was flurry of reports of Racoons *Procyon lotor*, 22 from 1975 to 1988, mostly of isolated individuals. One concerned a pair of cubs, apparently born to a pregnant female that had been reported as an escapee. This is a species which has already shown its adaptability to European conditions by establishing a strong wild colony in Germany which has spread into The Netherlands and France. It is possible that some irresponsible owners also released larger cats, and that the regular summer silly-season reports of black Leopards or Pumas in the wild have some foundation in fact. A pair of Clouded Leopards *Neofelis nebulosa* escaped from a zoo, and one was quickly recaptured: the other survived almost unnoticed for 5 months before an irate farmer trapped the 'fox' that had been killing his sheep (Baker 1990). Asian Clawless Otters (*Aonyx cinerea*) seem to have been breeding in the

Muntjac

Thames basin during the 1980s, in the temporary (we hope) absence of the native Otter; this is a species that does well in captivity, and can obviously tolerate the climate (Strachan and Jefferies 1996). It feeds more on crustaceans than fish, and it is doubtful whether it would survive the ecological competition from a restored population of the natives. Gerbils *Meriones unguiculatus* are very popular pets, and seem to occur in the wild (for instance in owl pellets) with some regularity. Perhaps neither climate nor food supply suit them, but if climate change becomes a reality, or diversification of agriculture produces an appropriate food supply, they might thrive. Golden Hamsters *Mesocricetus auratus* have established themselves for short periods, breeding well and thwarting pest-control actions by going into hibernation over winter, thereby giving the impression that they had been wiped out (Hills 1991). However, they seem to be confined to buildings, and unable to thrive in the wild; an assumption that could yet prove wrong. Black-tailed Prairie Dogs *Cynomys ludovicianus* do very well at semi-liberty in the grounds of several zoos, and they would surely thrive in the wild given half a chance. It must be 12 000 years since we had a ground squirrel wild in Britain, but this is not one we should strive to obtain. Both Indian and African Porcupines (*Hystrix brachyura, H. cristata*) have established themselves in the wild for limited periods, the former breeding successfully near Okehampton during 1969–80. A single pair escaped, but six were eventually killed or captured. Introduced deer have done well, and extra species could easily be added to the fauna. One of the most spectacular escapes recorded by Baker (1990) was a herd of 12 of the rare Chinese Père David's Deer *Elaphurus davidianus* that got out of a farm near Swindon, Wiltshire in 1981. Fortunately, they all returned about 2 years later, but such a mass escape of a herd-living species would obviously be a good way to ensure a successful introduction. The North American White-tailed Deer *Odocoileus virginianus* has established a feral population in Finland, so could surely survive in Britain, but it is not a species often kept here in captivity. Various Asian deer, such as Axis *Cervus axis* and Sambar *C. unicolor* are more likely escapees, though they may be less well suited to the climate, and more likely to be in ecological competition with the existing deer. A single Nilgai *Boselephas tragocamelus*, the largest of the Indian antelopes, survived for 3 years, indicating the potential for other ungulates to be added to the fauna. It should of course be noted that it is now illegal, under Section 14 of the Wildlife and Countryside Act 1981, to release deliberately or allow to escape any non-native species; this includes a number of the established aliens which are specifically listed in Schedule 9 of that Act.

Political and Economic Changes

A greater but unpredictable impact on our mammal fauna will come from economically driven political changes. The reform of the Common Agricultural Policy has been talked about for several years, and given the extent to which the domestic ungulates dominate our mammal fauna (Table 9.1), such changes will undoubtedly be the most potent. On the one hand, support for the production of cereals in the arable half of Britain has powered the destruction of hedges, the 'reclamation' of much marginal land, and thus the marginalization of the wild mammals. In the uplands, similarly, the support for hill sheep has resulted in a doubling of numbers between 1955 and 1991 (from 15 to 31 million in England and Wales). Forestry is also in the balance; support for the further coniferization of the uplands is doubtful, given its dubious economics, but the creation of community forests, primarily deciduous and near to urban areas, has much greater general support. Both reductions in sheep and increasing deciduous forest cover will favour deer and Grey Squirrels. Small mammals and their predators might also benefit, depending on the levels of grazing from larger mammals (in the New Forest, with heavy pressure from deer, cattle and ponies, the ground flora is totally suppressed, and with it the small mammals; Putman 1986). What we have now is a countryside mostly of agricultural land, with small patches of woodland, and a thinning network of hedges, roadsides and riversides connecting them. In the uplands, there are rather more substantial areas of more wild habitat, but even these are broken into 'islands' of suitable habitat by agricultural land in the valleys. Anything that breaks up the habitats of mammals further (more roads, more agriculture) will be harmful, and, conversely, anything that reinstates hedgerows and marginal strips will be helpful. Bright (1993) has assessed the British mammals for their vulnerability to these changes. Those most vulnerable to fragmentation of habitat are the less mobile and slowly reproducing species which are also habitat specialists. Woodland and riparian species, especially those which breed slowly, like Otters and several bats, are most vulnerable. The Dormouse is an extreme example, and one which hedgerow connections would certainly help. A gap only a few feet wide might be enough to break the connection. In Ireland, it has been noted already that the spread of the Bank Vole has been quickest in areas with a good network of hedgerows, and slowed by the agricultural land east of Cork (Smal and Fairley 1984). Even such large and mobile mammals as deer can be inhibited by roads and railways; the failure of Sika Deer in the New

Forest to spread north of the railway line into Fallow Deer range has been mentioned. Better bridges or tunnels over such infrastructure, or underpasses wide enough for wildlife, would also help maintain the connections between bits of habitat. However, road mortality is itself a major problem. It is estimated that some 47 500 Badgers are killed on the roads of Britain each year, about half the total mortality. The British Badger population is strong enough to withstand this attrition, but the much smaller Dutch population faces extinction from the same cause; 220 of an estimated population of 2200 were killed on Dutch roads in 1990 (Wiertz 1993). The Otter, a much rarer animal, also suffers severely from road deaths. The small, largely reintroduced, population in East Anglia has suffered seven known losses to this cause, and road mortality is serious, though difficult to evaluate, in other parts of its range (Strachan and Jefferies 1996). The role of road mortality in limiting Hedgehog populations remains one of the most serious unknowns in the population biology of this species.

Conservation and Reintroductions

A more active conservation programme for British mammals will certainly appeal to anyone likely to have read this far. What would be the most effective targets? In the recent past, much conservation activity has been devoted to 'holding the line'; trying to protect the best bits of habitat and the rarest species. This approach, epitomized by the establishment of nature reserves, is nearly useless for mammals; they are too wide ranging, their populations too thin, to be usefully protected that way. An obvious exception to this dismissal is the protection offered to bat roosts, both breeding sites and hibernacula. Bats are vulnerable enough by virtue of their rarity and their low reproductive rate (one young per year per female), but they make themselves even more so by gathering together in maternity roosts. It is particularly important that the rarer bats are so protected, and therefore encouraging that most of the breeding roosts of Greater and Lesser Horseshoe Bats are known and protected. These two are also vulnerable in hibernation, because they hang from cave roofs in a conspicuous manner, but again many are protected. Of 426 known hibernacula of the Greater Horseshoe, 97, accommodating around 72% of the hibernating population, are at least SSSIs (Sites of Special Scientific Interest), meaning that their importance is known to the site owners, naturalists and planners. For Lesser Horseshoes, which hibernate in smaller groups, the equivalent figures are 909 hibernacula, of which 161 hosting 53% of the bats are SSSIs (Mitchell-Jones *et al.* 1993). Fitting grills on cave entrances, to prevent casual disturbance, has been very successful, as has the negotiation of time-limited (e.g. summer only) access agreements with cavers.

If encouraging other mammals by direct protection of their habitat is difficult, there are certainly ways that their habitat can be improved. Paradoxically, to some, hunting is a good way of promoting habitat improvement. Whether for Foxes, Pheasants or Grey Partridges, hunting promotes the retention or addition of copses and hedgerows, and wider strips of marginal grassland around field boundaries, as nesting and feeding areas for the birds. In the uplands, Red Grouse can only be conserved by promoting good habitat, and grouse shooting has been the only economic counter pressure to conversion of moorlands to either sheep walk or forestry. As habitat for Mountain Hares and their predators, grouse moors are essential. The anti blood-sports lobby seems unable to appreciate just how much money it requires to maintain such habitats, even without the enormous agricultural subsidies acting against their retention; Cobham Resource Consultants (1983) calculated that some £54 million was spent annually by the organizers of hunting activities, mostly those providing shooting, to create and manage the correct habitat. This compares favourably with the total budget for the then Nature Conservancy Council of £32 million in 1986–87 (Nature Conservancy Council, 13th Annual Report).

Another group of mammals could be promoted if riparian habitats were better managed. Several studies of Otters have drawn attention to the importance of riverine tree cover (providing sites for holts), and there is no doubt that Water Voles would do better if bankside vegetation were left rank for a metre or two back from the bank, rather than either mown or grazed short. The most important feature for Otters is undoubtedly their food supply, and particularly Eels. Since these are migratory, the building of dams and weirs has certainly inhibited their numbers, but the biggest problem is undoubtedly still the pollution of the great industrial rivers; the Mersey, Trent, Rother and Don are still too polluted to allow Eel migration, and the Thames recovered only during the 1970s (Harrison and Grant 1976). Recovery of the Otter population in these river basins will follow that of their food supply.

For many, a more active programme of recovery will appeal. Reintroduction of species to areas where they used to occur has already been attempted very successfully with Dormice to Cambridgeshire and Cheshire, and with Otters to East Anglia and North Yorkshire. Such programmes can, in theory, be undertaken by anyone with the appropriate time, expertise and enthusiasm; they do not require a licence from English Nature or the equivalent bodies (though keeping captive rarer mammals to breed animals for release does require a licence). However, it is certainly the expectation now that such releases should follow an approved protocol. The potential new home should be surveyed for adequate food supplies and other habitat requirements, the animals should be bred in captivity, or taken from wild populations that are strong enough to withstand losses, the released animals should be acclimatized in cages at their new site before being released (a 'soft-release'), continue to be fed while they are adapting to freedom, and monitored, which may mean radio-tracking, to check their progress over the next few years. It is no longer acceptable to tip animals out to take their chance. This means that a properly conducted release is an expensive and time-consuming process, requiring a great deal of professional time as well as dedicated amateur support. The procedure was well described for the early Otter releases by Jefferies *et al.* (1986) and for Dormice by Bright and Morris (1994). For Dormice, it was possible to compare the fates of captive-bred with wild-caught Dormice; the captive-bred animals seemed to adapt more slowly, losing weight more rapidly, and moving less far from the release pens. Wild Dormice that were hard-released (turned out immediately) lost more weight and survived less well than soft-released ones. Not all reintroductions adhere to these recommendations. Polecats appear to have been surreptitiously released in the Lake District, apparently without any monitoring, and although the intention was to 'soft-release' the Pine Marten in Galloway, the animals themselves had other ideas. Only 12 animals, in two groups, were available. They escaped from their release pens, and one group simply disappeared. Fortunately the second group stayed near the area of their escape, and founded the new population. After their release, in 1980–81, there were signs of breeding in 1981–83, and then nothing for 8 years. Not until 1991 and 1992 were there further breeding records, including one young kit in a Tawny Owl nest box, proof that the reintroduction had succeeded (Shaw and Livingstone 1992). At present, as remarked, the possibility that Pine Marten might be reintroduced to other sites is being discussed, along with the counter argument that there is no need because they are still in fact present in England and Wales. The possibility that the Wild Cat should be helped across the Central Lowlands to the Southern Uplands has not, so far as I know, been seriously discussed, though it remains a musing. If hybridization with Feral Cats is a serious problem, a matter still under investigation, establishing a genetically 'clean' Wild Cat population would have appeal. The Southern Uplands might not be the right place. On a smaller scale, the Water Vole has also been discussed as a subject for reintroduction, to places where good new riparian habitat has been created. Fortunately, the animal is still sufficiently widespread that there may be little need. Some nature reserves might benefit from its presence, if only as a food supply for the predators (Marsh Harrier *Circus aeruginosus* comes to mind) that ought to share its habitat.

By far the most radical improvement to the British mammal fauna would be the reintroduction of some of the historically extinct species. This **would** require licensing, under the Wildlife and Countryside Act 1981, by the national agencies (English Nature, Scottish Natural Heritage, Countryside Commission for Wales). Thus it would ultimately be a political decision. Though there are precedents, we have been very timid about the subject of reintroductions in Britain. The White-tailed Eagle *Haliaeetus albicilla*, which became extinct in 1916, was reintroduced by bringing 82 young birds from Norway over an 11-year period, 1975–85, and releasing them on Rhum, a National Nature Reserve (Love 1983). The first wild-hatched young fledged in 1985, and a total of 39 has fledged between 1985 and 1994 (Ogilvie 1996). The Large Blue Butterfly *Maculinea arion* has also been reintroduced, using Swedish stock in 1995, after it became extinct in 1979. In Europe, a number of large mammals have been reintroduced to several countries, and the rare remnant populations of others have been encouraged to spread. Wolves, for instance, spread back into France naturally from Italy in 1992, and a small pack of about 10 animals lives in the Mercantour National Park in the Maritime Alps (Poulle *et al.* 1995).

The Beaver has been a popular subject. By the turn of this century, it survived in western Europe only in the Lower Rhone in France, in the Elbe in Germany, and in a small area of southern Norway. Even in eastern Europe, it was extinct in the Baltic States, confined to eastern Poland, and widespread only in Russia. It has been put back into at least 50 river basins across the continent, including sites in Lithuania, Latvia, Finland, Sweden, western Poland, various areas of France and Germany, Austria, Switzerland and the Netherlands. When first suggested for this country 20 or 30 years ago, the response was that it would be too damaging to forestry interests. More knowledge and experience from elsewhere in Europe suggests that Beavers feed largely on deciduous shrubs, particularly willows and Aspen, and that commercial forestry would not be at hazard. Killing trees by flooding their roots might be more serious than feeding on the conifers directly, but even that is a minor problem. Orchards sometimes suffer from the Beaver's attention, and any reintroduction programme would need to have control or compensation packages agreed well in advance. The possible reintroduction of Beavers to Britain, along with a review of continental experience, has been discussed by Macdonald *et al.* (1995), and the notion is at least receiving serious consideration in higher quarters. Beavers reproduce rather slowly, usually rearing only two kits a year, and in the Netherlands (perhaps the nearest ecologically as well as geographically to Britain) the introduced population has barely maintained itself, suffering from excessive emigration. The implication is that a reintroduced population would be easy to control, if that became necessary.

A more surprising species to have been reintroduced to several countries in Europe is the Lynx. Very successful reintroductions have taken place in two areas of Switzerland and in Slovenia. Possibly successful attempts have been made also in Austria, France and Bohemia; attempts in Germany seem not to have succeeded (Breitenmoser and Breitenmoser-Würsten 1990). The Swiss population has spread into France and has caused some problems in sheep-rearing areas. This has not happened elsewhere, and it is not clear whether this represents a peculiarity of French farming practices, an aberrant Lynx population, or a hypersensitive set of farmers. There is a compensation scheme in place, but it may not be sufficiently generous to satisfy the shepherds. It does, however, offer a note of caution for this country. Until the recent radio-carbon date suggesting it survived into Roman times, the supposition had been that it disappeared so long ago from this country as to be an unrealistic subject even for idle speculation. In Białowieża, the main prey are Roe Deer (95 of 138 ungulates killed, the rest being Red Deer calves) and Brown Hares (11% of the weight of prey) (Jędrzejewski *et al.* 1993). These are foods that should be plentiful in Britain, and Lynx might offer some way of balancing the deer populations of our woodlands naturally. Lynx live naturally around Stockholm, even hunting across the allotments on the outskirts of the city, without being any danger to people directly. Perhaps they could be as tolerant of Britons. It is doubtful if Britons could be as tolerant as Swedes.

Another species that has done well in Europe, though without needing reintroduction, is the Wild Boar. It is revealing to see young Wild Boar for sale in the Christmas markets in Paris, and find live animals in the Fontainbleau Forest, a popular Sunday tripper venue, only 60 kilometres south of Notre-Dame. To most Europeans, this is a valuable game animal, and a minor agricultural pest. There is no doubt that Wild Boar can be serious pests of root crops, and can also damage fencing. A car might come second best in a road accident, too. However, as an important piece of the deciduous woodland ecosystem, Wild Boar would be a magnificent presence. They would surely have an important impact on Bracken, regarded as a major forestry weed, because they root out its rhizome system in winter, as a major food item, and they would play an interesting role in burying, perhaps therefore enhancing regeneration of, deciduous tree seeds, including acorns and beech mast. However, it may be that the experiment is already underway. A population of 100–200 Wild Boars has apparently established itself in the Weald, along the Kent–Sussex border. It is reported that some escaped from a farm near Tenterden in the storm of October 1987, when trees crashed down through fencing. Another group, being sent to market from a specialist venison farm, escaped near Aldington, south-east of Ashford. Sightings range across an area about 45 km east to west and 25 km north to south (Fig. 9.11). They have certainly been breeding, and seem sufficiently well established that eliminating them might prove very difficult. Already, farmers are polarized between those who find them a serious pest, not only of wheat crops but also killing lambs, and those who, contrarily, think they can let the stalking for a profit (Tyrer 1997). Though attacks on lambs might seem unlikely, Wild

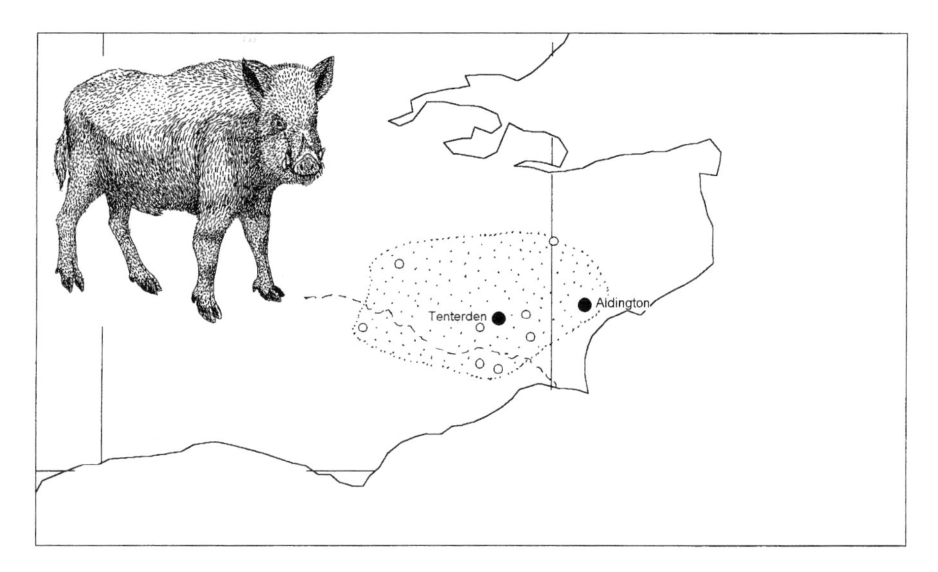

Figure 9.11 The distribution of Wild Boar in the Weald in 1997 (after Tyrer 1997). Escapes from Aldington and Tenterden about 10 years ago have resulted in a population of about 200, spread across 1000 km^2 of Kent and Sussex.

Boar are omnivores, and in Białowieża were the most common scavengers of Lynx kills, even ousting the Lynx themselves.

The most recent of our large mammals to become extinct might be the one best suited to reintroduction, since the habitat should have changed less for it than for the earlier losses. Unfortunately, Wolves have an even worse, and less deserved, reputation than Wild Boar. A seriously reasoned ecological argument was put forward a decade ago (Yalden 1986). There is a large deer population in the Highlands, and the evidence from both Europe and North America is that Wolves prefer to prey on large ungulates. In Białowieża, that means largely Red Deer, though Roe Deer and smaller prey are certainly eaten. Wolves are scared of people, and also of strings of bunting ('fladry'), which means that they have traditionally been very vulnerable to driven hunting (Jędrzejewska *et al.* 1996). The big problem with attempting to reintroduce Wolves would be the political rather than any ecological one, for there is no doubt that sheep would be vulnerable to Wolf predation (Delibes 1990). The suggestion made in 1986 was to introduce Wolves to Rhum, the National Nature Reserve where the Sea Eagles were released. This has a large population of about 1600 Red Deer (but no sheep or shepherds), and the surplus production each year should be enough to sustain a pack of about 20 Wolves. About 250 deer were culled each year to keep the population level, and another 60 died of starvation. At about 80 kg for each hind and 120 kg for each stag, they should supply about 25 000 kg of meat, enough to sustain 19 Wolves; Wolves eat about 3.6 kg as a daily average. The island is only 100 km^2, and it has been argued that this is too small to support a Wolf population. Certainly the much-studied Isle Royale National Park, in Lake Superior, is 5.4 times larger, and seems able to support no more than 20 Wolves comfortably, but then the Moose on which those Wolves prey numbered only about 600 at the time (numbers of both Wolves and Moose have increased and slumped again since Mech (1966) studied them in apparent equilibrium). The Red Deer herd on Rhum is both larger and more productive, because grasses provide more food than the browse on Isle Royale. The evidence from Isle Royale is that Wolves have difficulty preying on Moose, and kill mostly those that are either very young calves, or elderly Moose, often with arthritis, damaged jaws or teeth, or otherwise debilitated. By implication, they would exert a similar selective pressure on Red Deer, which, though less formidable, would rely on fleetness to escape, and would therefore similarly suffer a handicap from diseases like arthritis. Although the Wolves might increase to such an extent as to kill an excessive number of calves during the summer (calves are certainly preyed upon preferentially in Białowieża; Jędrzejewski *et al.* 1992), the Red Deer herd on Rhum has a great deal of spare reproductive capacity; more calves would survive, and more hinds would produce a calf every year, if the population were to be reduced so that there was more food to go round (Clutton-Brock and Albon 1989). Starvation over winter is a major cause of deer mortality on Rhum; those are deer which Wolves could usefully eat.

The danger that Wolves pose to sheep is undoubted. However, the problem is greatest where the natural ungulate prey have been eliminated. In the Appenines in Italy, Wolves have co-existed with sheep rearing throughout this century; the rearing of the sort of sheep-dog that thinks itself a sheep (Maremmas, Pyrenean Mountain Dogs, etc.), and defends the flock against Wolves, is a well proven and effective antidote. Breitenmoser (1998), in reviewing the return of the large predatory mammals to the Alps, points to the necessity of managing both the large mammals and the sheep farming in a way that is sensitive to the different perceptions of rural and urban people.

The danger that Wolves pose to humans is mostly imaginary, the product of fairy tales like Little Red Riding Hood. Allen (1979, p. 402) reports two attempts by newspapers in North America to offer prizes for anyone able to substantiate a case of Wolves attacking humans. The prizes remain unclaimed. Possibly small children have in the past been killed by Wolves in Europe, though the evidence is sparse (Delibes 1990). The major caveat is that Wolves may carry rabies; in France, for instance, between 1851 and 1877, 707 people died of rabies after being bitten by dogs, and another 38 died after being bitten by Wolves (Macdonald 1980). It is thus certain that, in the past, rabid Wolves have attacked people, and justified some of the antipathy that they receive. If Wolves were to be reintroduced, they would certainly be quarantined and vaccinated, probably captive-bred as well. Unfortunately, experience with captive-bred carnivores suggests that they adapt very poorly to the wild; they tend to have lost much of their hunting ability, and sometimes their fear of humans as well (Yalden 1992). While it is acceptable to train herbivores to find food in captivity, carnivores cannot legally be taught to hunt live vertebrate prey. Clearly, any such programme would have to be experimental, carefully monitored throughout, and therefore expensive.

Nevard and Penfold (1978) went rather further, and suggested that Rhum might be a suitable site to attempt a rather larger project to recreate the large mammal fauna of Britain. It would be hard to prevent the creation of a zoo or theme park, rather than a properly functioning ecosystem. At present, there are too few herbivores there to sustain a large guild of carnivores. Either Brown or Mountain Hares might be a useful addition, to support the Golden Eagles. Carrion left by Wolves would benefit both Ravens and Sea Eagles. It would be necessary to introduce Roe Deer if the introduction of Lynx were to be contemplated. There seems insufficient woodland to support them, at present, or Beavers or Wild Boar. There is enough grassland to support another, different, grazer, and the recreated 'Aurochs', bred back from a mixture of different cattle breeds by German zoos, might be more appropriate for a National Nature Reserve than Highland Cattle, currently used to diversify the grazing. On the other hand, there seems little doubt that they would be dangerous to humans (though neither more nor less so than the Buffalo *Synceros caffer* that are a feature of many East African National Parks). The last carnivore to be considered would be the Brown Bear, among the rarest of European large mammals and in principle deserving of more conservation effort. Its food supply may not be present anywhere in Britain, but a tame bear 'Hercules' lived wild for three weeks in August–September 1980 in the Outer Hebrides. After wandering from a photographic shoot for an advertising campaign on Benbecula, he was recaptured 22 days later on North Uist (Henn 1980, Anon 1980). There is no doubt that Brown Bears are also dangerous to humans. Four bilberry-pickers in Slovenia were killed by them in the 1970s, and the large Brown Bear population in Romania is reported to have killed about four people a year during the 1980s (Carpathian Large Carnivore Project 1997); Grizzly Bears (same species, American race) are a known hazard to campers and others in American National Parks, including Yellowstone. Would it be irresponsible to contemplate having such animals in British Nature Reserves? No more irresponsible than taking myself off to the Lion-infested, Elephant-ridden, National Parks in Kenya. Of course wildlife can be dangerous, but it is its wild magnificence that I, and many other tourists, wish to see. So long as it is respected, the wildlife is no more dangerous than the flight there, and much less dangerous than the drive to the airport. To restore something like the full magnificence of the European fauna to at least a small part of these islands would be an appropriate target for the next millenium.

REFERENCES

Aaris-Sorensen, K. (1992) Deglaciation chronology and re-immigration of large mammals. A south Scandinavian example from the Late Weichselian-Early Flandrian. *Courier Forschungs-Institut Senckenberg* **153**: 143–149.

Aguilar, A. (1986) A review of Old Basque Whaling and its Effect on the Right Whales (*Eubalaena glacialis*) of the North Atlantic. *Report of the International Whaling Commission (Special Issue* **10**): 191–199.

Albarella, U. and Davis, S.J. (1996) Mammals and birds from Launceston Castle, Cornwall: decline in status and the rise of agriculture. *Circaea* **12**: 1–156.

Allen, D.L. (1979) *Wolves of Minong*. University of Michigan, Ann Arbor.

Allen, D. and Dalwood, C.H. (1983) Iron Age Occupation, a Middle Saxon Cemetery, and Twelfth to Nineteenth Century Urban Occupation: Excavations in George Street, Aylesbury, 1981. *Records of Buckinghamshire* **25**: 1–60.

Alston, R. (1879) On the specific identity of the British martens. *Proceedings of the Zoological Society of London* **1879**: 468–474.

Altuna, J. (1983) On the relationship between archaeofaunas and parietal art in the caves of the Cantabrian region, pp. 227–238 in Clutton-Brock, J. and Grigson, C. (eds) *Animals and Archaeology. 1. Hunters and their Prey*. BAR International Series **163**, Oxford.

Ambers, J., Matthews, K. and Bowman, S. (1989) British Museum Natural Radiocarbon Measurements 21. *Radiocarbon* **31:** 15–32.

Ammerman, A.J. and Cavalli-Sforza, L.L. (1971) Measuring the rate of spread of early farming in Europe. *Man* **6**: 674–688.

Anderson, M.L. (1967) *A History of Scottish Forestry*. Nelson, London.

Anderson, S.S., Prime, J.H., Harwood, J. and Bonner, N. (1989) British seals – vermin or scapegoats? pp. 251–260 in Putman, R.J. (ed.) *Mammals as Pests*. Chapman and Hall, London.

Andrews, P. (1990) *Owls, Caves and Fossils*. British Museum (Natural History), London.

Anon. (1980) A Grizzly End. *Daily Telegraph* 15 September 1980 (editorial).

Armitage, P.L. (1982) Studies on the remains of domestic livestock from Roman, medieval, and early modern London: objectives and methods, pp. 94–106 in Hall, A.R. and Kenward, H.K. (eds) *Environmental Archaeology in the Urban Context*. CBA Research Report **43**, London.

Armitage, P. and West, B. (1985) Faunal evidence from a late Medieval garden well of the Greyfriars, London. *London and Middlesex Archaeological Society Transactions* **36**: 107–136.

Armitage, P., West, B. and Steedman, K. (1984) New evidence of Black Rat in Roman London. *The London Archaeologist* **4**: 375–383.

Armour-Chelu, M. (1991) The faunal remains, pp. 139–150 in Sharples, N.M. *Maiden Castle. Excavations and Field Survey 1985–6*. English Heritage Archaeological Report **19**, London.

Arnold, H.R. (1978) *Provisional Atlas of the Mammals of the British Isles*. Biological Records Centre, I.T.E., Abbots Ripton.

Arnold, H.R. (1993) *Atlas of Mammals in Britain*. H.M.S.O., London.

Ashbee, P., Bell, M. and Proudfoot, E. (1989) *Wilsford Shaft: Excavations 1960–62*. English Heritage Archaeological Report **11**, London.

Atkinson, R.P.D., Macdonald, D.W. and Johnson, P.J. (1994) The status of the European Mole *Talpa europaea* L. as an agricultural pest and its management. *Mammal Review* **24**: 73–90.

Atkinson,T.C., Briffa, K.R. and Coope, G.R. (1987) Seasonal temperatures in Britain during the past 22,000 years, reconstructed using beetle remains. *Nature, London* **325**: 587–592.

Aybes, C. and Yalden, D.W. (1995) Place-name evidence for the former distribution and status of Wolves and Beavers in Britain. *Mammal Review* **25**: 201–227.

Backhouse, J. (1989) *The Luttrell Psalter*. British Library, London.

Baker, S.J. (1990) Escaped exotic mammals in Britain. *Mammal Review* **20**: 75–96

Balch, H.E. (1928) Excavations at Wookey Hole and other Mendip Caves, 1926–7. *Antiquaries' Journal* **8**: 197–204.

Balch, H.E. (1948) *Mendip – Its Swallet Caves and Rock Shelters*, 2nd edn. Bristol and London.

Bannikov, A.G., Zhirnov, L.V., Lebedeva, L.S. and Fandeev, A.A. (1961) *Biology of the Saiga*. Astrakhan Reserve Laboratory for the Biology of the Saiga. (Translated from the Russian by the Israel Program for Scientific Studies, 1967).

Bard, E., Hamelin, B., Arnold, G. *et al.* (1996) Deglacial sea-level record from Tahiti corals and the timing of global meltwater discharge. *Nature, London* **382**: 241–244.

Barker, G. (1983) The animal bones, pp. 133–150 in Hedges, J.W. *Isbister: A Chambered Tomb in Orkney*. BAR British Series **115**, Oxford.

Barlow, K.E. and Jones, G. (1996) *Pipistrellus nathusii* (Chiroptera: Vespertilionidae) in Britain in the mating season. *Journal of Zoology, London* **240**: 767–773.

Barnosky, A.D. (1985) Taphonomy and herd structure of the extinct Irish Elk, *Megaloceros giganteus*. *Science* **228**: 340–344.

Barnosky, A.D. (1986) 'Big Game' extinction caused by Late Pleistocene climatic change: Irish Elk (*Megaloceros giganteus*) in Ireland. *Quaternary Research* **25**: 128-135.

Barratt, E., Deaville, R., Burland, T.M. *et al.* (1997) DNA answers the call of pipistrelle bat species. *Nature, London* **387**: 138–139.

Barrett, J.H. (1966) Tom Tivey's Hole Rock Shelter, near Leighton, Somerset. *Proceedings of the University of Bristol Spelaeological Society* **11**: 9–24.

Barrett-Hamilton, G.E.H. and Hinton, M.A.C. (1910–1922) *A History of British Mammals*. Gurney and Jackson, London.

Batcheler, C.L. (1960) A study of the relations between roe, red and fallow deer, with special reference to Drummond Hill Forest, Scotland. *Journal of Animal Ecology* **29**: 375–384.

Batzli, G.O. (1993) Food selection by lemmings, pp. 281–301 in Stenseth, N.C. and Ims, R.A. (eds) *The Biology of Lemmings*. Linnean Society Symposium **15**, Academic Press, London.

Beirne, B.P. (1952) *The Origin and History of the British Fauna*. Methuen, London.

Bekenov, A.B., Grachev, Iu. A. and Milner-Gulland, E.J. (1998) The ecology and management of the Saiga antelope in Kazakhstan. *Mammal Review* **28:** 1–52.

Bellamy, A.G. (1995) Extension of the British landmass: evidence from shelf sediment bodies in the English Channel, pp. 47–62 in Preece, R.C. (ed.) *Island Britain: a Quaternary Perspective*. Geological Society, Spec. Publ. **96**, London.

Bennett, K.D. (1988) A provisional map of forest types for the British Isles 5000 years ago. *Journal of Quaternary Science* **4**: 141–144.

Bentley, E.W. (1959) The distribution and status of *Rattus rattus* L. in the United Kingdom in 1951 and 1956. *Journal of Animal Ecology* **28**: 299–308.

Bentley, E.W. (1964) A further loss of ground by *Rattus rattus* L. in the United Kingdom during 1956–1961. *Journal of Animal Ecology* **33**: 371–373.

Bernaldo de Quiros, F.H. (1991) Reflections on the Art of the Cave of Altamira. *Proceedings of the Prehistoric Society* **57**: 81–90.

Berry, R.J. (1969) History in the evolution of *Apodemus sylvaticus* at one edge of its range. *Journal of Zoology, London* **159**: 311–366.

Berry, R.J. (1970) Covert and overt variation, as exemplified by British mouse populations. *Symposium of the Zoological Society of London* **26**: 3–26.

Berry, R.J. (1973) Chance and change in British Long-tailed field mice (*Apodemus sylvaticus*). *Journal of Zoology, London* **170**: 351–366.

Berry, R.J. (1985) *The Natural History of Orkney*. Collins, New Naturalist **70**, London.

Berry, R.J. and Johnston, J.L. (1980) *The Natural History of Shetland*. Collins, New Naturalist **64**, London.

Berry, R.J. and Jakobson, M. (1975) Ecological genetics of an island population of the house mouse. *Journal of Zoology, London* **175**: 523–540.

Berry, R.J. and Rose, F.E.N. (1975) Islands and the evolution of *Microtus arvalis* (Microtinae). *Journal of Zoology, London* **177**: 395–409.

Biddick, K. (1980) Animal bones from the second millenium ditches, Newark Road subsite, Fengate, pp. 217–232 in Pryor, F. (ed.) *Excavations at Fengate, Peterborough, England: The third Report*. Archaeological Monograph **6**. Royal Ontario Museum.

Bilton, D.T., Mirol, P.M., Mascheretti, S. *et al.* (1998). The Mediterranean region of Europe as an area of endemism for small mammals rather than the major source for the postglacial colonisation of northern Europe. *Proceedings of the Royal Society of London B.* **265**: 1219–1226.

Birks, J. (1993) Return of the Polecat. *British Wildlife* **5**: 16–25.

Birtles, T. (1998) Practical deer park management at Tatton Deer Park, pp. 102–109 in Goldspink, C.R., King, S. and Putman, R.J. (eds) *Population Ecology, Management and Welfare of Deer.* Manchester Metropolitan University, Manchester.

Bishop, M.J. (1982) The mammal fauna of the early middle Pleistocene cavern infill site of Westbury-sub-Mendip, Somerset. *Palaeontological Association, Special Papers* **28**: 1–108.

Bliss, L.C. and Richards, J.H. (1982) Present-day Arctic vegetation and ecosystems as a predictive tool for the Arctic-steppe mammoth biome, pp. 241–258 in Hopkins, D.M., Matthews, J.V., Schweger, C.E. and Young, S.B. (eds) *Paleoecology of Beringia.* Academic Press, New York.

Boessneck, J. (1969) Osteological differences between Sheep (*Ovis aries* Linné) and Goats (*Capra hircus* Linné), pp. 331–358 in Brothwell, D. and Higgs, E. (eds) *Science in Archaeology* (revised edition). Thames and Hudson, London.

Boisseau, S. and Yalden, D.W. (1998) The former status of the Crane *Grus grus* in Britain. *Ibis* **140**: 482–500.

Bond, J.M. (1994) Appendix 1: The cremated animal bone, pp. 121–135 in McKinley, J.I. *The Anglo-Saxon Cemetery at Spong Hill, North Elmham. Part VIII: The Cremations.* East Anglian Archaeology Report **69**, Norwich.

Bonner, W.N. (1976) *Stocks of Grey Seals and Common Seals in Great Britain.* Natural Environment Research Publications Series C. **16**, London.

Bourdillon, J. (1993) Animal bone. In Graham, H. and Davies, S. (eds) *Excavations in Trowbridge, Wiltshire 1977 and 1986–88. The Prehistoric, Saxon and Saxo-Norman Settlements and the Anarchy Period Castle.* Wessex Archaeology Report **2**.

Bourdillon, J. and Coy, J. (1980) The animal bones, pp. 79–121 in Holdsworth, P. *Excavations at Melbourne Street, Southampton, 1971–6.* CBA Research Report **33**. York.

Bowman, S.G.E., Ambers, J.C. and Leese, M.N. (1990) Re-evaluation of British Museum Radiocarbon Dates issued between 1980 and 1984. *Radiocarbon* **32**: 59–79.

Boyd, J.M. and Boyd, I.L. (1990) The Hebrides, *New Naturalist* **76**: Collins, London.

Boylan, P.J. (1981) A new revision of the Pleistocene mammalian fauna of Kirkdale Cave, Yorkshire. *Proceedings of the Yorkshire Geological Society* **43**: 253–280.

Bramwell, D. (1964) Excavations at Elderbush Cave, Wetton, Staffordshire. *North Staffordshire Journal of Field Studies* **4**: 46–59.

Bramwell, D. (1971) Excavations at Fox Hole Cave, High Wheeldon, 1961–1970. *Derbyshire Archaeological Journal* **91**: 1–19.

Bramwell, D. (1976) The vertebrate fauna at Wetton Mill Rock Shelter, pp. 40–51 in Kelly, J.H. *The excavation of Wetton Mill Rock Shelter, Manifold Valley, Staffs.* City Museum and Art Gallery, Stoke-on-Trent, Staffs.

Bramwell, D. (1977a) Archaeology and Palaeontology, Chapter 14 in Ford, T.D. (ed.) *Limestones and Caves of the Peak District.* Geo Abstracts, Norwich.

Bramwell, D. (1977b) Birds and voles from Buckquoy, Orkney, pp. 209–211 in Ritchie, A. Excavation of Pictish and Viking-age farmsteads at Buckquoy, Orkney. *Proceedings of the Society of Antiquaries of Scotland* **108**: 174–227.

Bramwell, D., Yalden, D.W. and Yalden, P.E. (1990) Ossom's Eyrie Cave: an archaeological contribution to the recent history of vertebrates in Britain. *Zoological Journal of the Linnean Society of London*, **98**: 1–25.

Breitenmoser, U. (1998) Large predators in the Alps: the fall and rise of Man's competitors. *Biological Conservation* **83**: 279–289.

Breitenmoser, U. and Breitenmoser-Würsten, C. (1990) *Status, Conservation Needs and Reintroduction of the Lynx* (Lynx lynx) *in Europe.* Council of Europe, Strasbourg.

Bright, P.W. (1993) Habitat fragmentation – problems and predictions for British mammals. *Mammal Review* **23**: 101–111.

Bright, P.W. and Morris, P.A. (1992) Ranging and nesting behaviour of the dormouse *Muscardinus avellanarius,* in coppice-with-standards woodland. *Journal of Zoology, London* **226**: 589–600.

Bright, P.W. and Morris, P.A. (1994) Animal translocation for conservation: performance of dormice in relation to release methods, origin and season. *Journal of Applied Ecology* **31**: 699–708.

Bright, P.W., Morris, P.A. and Mitchell-Jones, A.J. (1996) A new survey of the Dormouse *Muscardinus avellanarius* in Britain, 1993–4. *Mammal Review* **26**: 189–195.

Brothwell, D. (1981) The Pleistocene and Holocene Archaeology of the House Mouse and Related Species. *Symposium of the Zoological Society of London* **47**: 1–13.

Brown, A.P. (1977) Late-Devensian and Flandrian vegetational history of Bodmin Moor, Cornwall. *Philosophical Transactions of the Royal Society of London,* B **276**: 251-320.

Brown, D. (1996) Trigfannau'r Arth a'r Blaidd (The dwellings of Bears and Wolves). *Danhaul* **16**: 22–27.

Brown, G. (1988) Insecticide warning to eel eaters. *Daily Telegraph* 21 Oct. 1988.

Browne, S. (1986) The animal bones. In Drewett, P. The excavation of a Neolithic Oval Barrow at North Marden, West Sussex, 1982. *Proceedings of the Prehistoric Society* **52**: 31–51 (pp. 40–41).

Browne, S. (1995) The small mammals. pp. 234–239 in Cunliffe, B. *Danebury: an Iron Age hillfort in Hampshire. Vol. 6: A hillfort community in perspective.* CBA Research Report **102**, London.

Bruce-Mitford, R. (1975) *The Sutton Hoo Ship Burial, Vol. 1. Excavation, Background, The Ship, Dating and Inventory.* British Museum, London.

Bryant, P.J. (1995) Dating remains of Gray Whales from the eastern North Atlantic. *Journal of Mammalogy* **76**: 857–861.

Bryce, J. (1997) Changes in the distributions of Red and Grey Squirrels in Scotland. *Mammal Review* **27**: 171–176.

Buchalczyk, T. (1980) The Brown Bear in Poland. pp. 229–232 in Martinka, C.J. and McArthur, K.L. (eds) *Bears – Their Biology and Management.* Bear Biology Association Conference series 3, Kalispell.

Burleigh, R. (1986) Complimentarity of conventional and accelerator dating: examples in Pleistocene extinctions, pp. 95–98 in Gowlett, J.A.J. and Hedges, R.E.M. (eds) *Archaeological Results from Accelerator Dating.* Oxford University Committee for Archaeology, Oxford.

Burleigh, R., Ambers, J. and Matthews, K. (1982a) British Museum Natural Radiocarbon Measurements XV. *Radiocarbon* **24**: 262–290.

Burleigh, R., Ambers, J. and Matthews, K. (1983) British Museum Natural Radiocarbon Measurements XVI. *Radiocarbon* **26**: 39–58.

Burleigh, R., Ambers, J. and Matthews, K. (1984) British Museum Natural Radiocarbon Measurements XVII. *Radiocarbon* **26**: 59–74.

Burleigh, R., Ambers, J. and Matthews, K. (1985) British Museum Natural Radiocarbon Measurements XVII. *Radiocarbon* **27**: 508–524.

Burleigh, R., Hewson, A. and Meeks, N. (1976) British Museum Natural Radiocarbon Measurements VIII. *Radiocarbon* **18**: 16–42.

Burleigh, R., Hewson, A., Meeks, N., Sieveking, G. and Longworth, I. (1979) British Museum Natural Radiocarbon Measurements X. *Radiocarbon* **21**: 41–47.

Burleigh, R., Matthews, K. and Ambers, J. (1982b) British Museum Natural Radiocarbon Measurements XIV. *Radiocarbon* **24**: 229–261.

Byrne, J.M., Duke, E.J. and Fairley, J.S. (1990) Some mitochondrial DNA polymorphisms in Irish wood mice (*Apodemus sylvaticus*) and bank voles (*Clethrionomys glareolus*). *Journal of Zoology, London* **221**: 299–302.

Callow, P. and Cornford, J.M. (eds) (1986) *La Cotte de St. Brelade 1961–1978. Excavations by C.B.M. McBurney.* Geo Books, Norwich.

Campbell, J.B. (1977) *The Upper Palaeolithic of Britain.* Clarendon Press, Oxford.

Cantor, L. (1983) *The Medieval Parks of England: a Gazetteer.* University Press, Loughborough.

Capanna, E. (1973) Concluding remarks, pp. 681–695 in Chiarelli, A.B. and Capanna, E. (eds) *Cyto-taxonomy and Vertebrate Evolution.* Academic Press, London.

Carpathian Large Carnivore Project (1997) *Annual Report 1996/97.* Wildlife Research Department, Bucharest/Munich Wildlife Society, Munich.

Carter, H.H. (1975) Fauna of an area of Mesolithic occupation in the Kennet Valley, considered in relation to contemporary eating habits. *Berkshire Archaeological Journal* **68**: 1–3.

Carver, M.O.H. (1994) Environment and commodity in Anglo-Saxon England, pp. 1–6 in Rackham, J. (ed.) *Environment and Economy in Anglo-Saxon England.* CBA Research Report **89**, York.

Caseldine, A. (1990) *Environmental Archaeology in Wales.* Department of Archaeology, St David's University College, Lampeter.

Cassels, R. (1984) The role of prehistoric man in the faunal extinctions of New Zealand and other Pacific islands. In Martin, P.S. and Klein, R.G. (eds) *Quaternary Extinctions.* University of Arizona Press, Tucson, pp. 741–767.

Chaline, J. (1972) *Les Rongeurs du Pléistocène Moyen et Supérieur de France.* Centre National de la Recherche Scientifique, Paris.

Chaline, J. (1987) Arvicolid data (Arvicolidae, Rodentia) and evolutionary concepts. *Evolutionary Biology* **21**: 227–310.

Chaline, J. and Brochet, G. (1986) The rodent fauna, pp. 139–143 in Callow, P. and Cornford, J.M. (eds) *La Cotte de St. Brelade 1961–1978. Excavations by C.B.M. McBurney*. Geo Books, Norwich.

Chanin, P.R.F. and Jefferies, D.J. (1978) The decline of the otter *Lutra lutra* L. in Britain: an analysis of hunting records and discussion of causes. *Biological Journal of the Linnean Society* **10**: 305–328.

Chapman, D. and Chapman, N. (1975) *Fallow Deer. Their History, Distribution and Biology*. Terence Dalton, Lavenham.

Chapman, D.I. and Chapman, N.G. (1982) The taxonomic status of feral muntjac deer (*Muntiacus* sp.) in England. *Journal of Natural History* **16**: 381–387.

Chapman, N. (1995) Our neglected species. *Deer* **9**: 360–362.

Chapman, N., Harris, S. and Stanford, A. (1994) Reeves' Muntjac *Muntiacus reevesi* in Britain: their history, spread, habitat selection, and the role of human intervention in accelerating their dispersal. *Mammal Review* **24**: 113–160.

Charles, R. and Jacobi, R.M. (1994) The Lateglacial fauna from Robin Hood Cave, Cresswell: a re-assessment. *Oxford Journal of Archaeology* **13**: 1–32.

Churchill, D.M. (1965) The Kitchen Midden site at Westward Ho!, Devon, England: ecology, age and relation to changes in land and sea level. *Proceedings of the Prehistoric Society* **31**: 74–84.

Claassens, A.J.M. and O'Gorman, F. (1965) The Bank Vole *Clethrionomys glareolus* Schreber: a mammal new to Ireland. *Nature, London* **205:** 923–924.

Clark, J.G.D. (1947) Whales as an economic factor in Prehistoric Europe. *Antiquity* **21**: 84–104.

Clark, J.G.D. and Fell, C.I. (1953) The early Iron Age site at Micklemoor Hill, West Harling, Norfolk, and its pottery. *Proceedings of the Prehistoric Society* **19**: 1–40.

Clarke, H. and Carter, A. (1977) Excavations in King's Lynn 1963–1970. *Society for Medieval Archaeology, Monograph Series* **7**.

Clarke, M. (1974) Deer distribution survey 1967–72. *Deer* **3**: 232–279.

Clarke, W.G. (1922) The Grimes Graves fauna. *Proceedings of the Prehistoric Society* **3**: 431–433.

Clinging, V. and Whiteley, D. (1980) Mammals of the Sheffield Area. *Sorby Record, special series,* **3**: 1–48.

Cloutman, E.W. and Smith, A.G. (1988) Palaeoenvironments in the Vale of Pickering, Part 3. Environmental history at Star Carr. *Proceedings of the Prehistoric Society* **54**: 37–58.

Clutton-Brock, J. (1969) The origins of the dog, pp. 303–309 in Brothwell, D. and Higgs, E. (eds) *Science in Archaeology* (revised edition). Thames and Hudson, London.

Clutton-Brock, J. (1979) Report of the Mammalian remains other than Rodents from Quanterness, pp. 112–133 in Renfrew, C. (ed.) *Investigations in Orkney*. Reports of the Research Committee of the Society of Antiquaries **38**, London.

Clutton-Brock, J. (1981) *Domesticated Animals from Early Times*. British Museum (Natural History), London.

Clutton-Brock, J. (1984) *Excavations at Grimes Graves, Norfolk, 1972–1976. Fascicule 1. Neolithic Antler Picks from Grimes Graves, Norfolk and Durrington Walls, Wiltshire: a Biometrical Analysis*. British Museum, London.

Clutton-Brock, J. (1986) New dates for old animals: the reindeer, the aurochs, and the wild horse in prehistoric Britain. *Archaeozoologia, Mélanges*: 111–117.

Clutton-Brock, J. (1989) Five thousand years of livestock in Britain. *Biological Journal of the Linnean Society* **38**: 31–37.

Clutton-Brock, J. (1991) Extinct species, pp. 571–575 in Corbet, G.B. and Harris, S. (eds) *The Handbook of British Mammals*, 3rd edn. Blackwell Science, Oxford.

Clutton-Brock, J. and MacGregor, A. (1988) An end to medieval reindeer in Scotland. *Proceedings of the Society of Antiquaries of Scotland* **118**: 23–35.

Clutton-Brock, T.H. and Albon, S.D. (1989) *Red Deer in the Highlands*. Blackwell Scientific, Oxford.

Cobham Resource Consultants (1983) *Countryside Sports. Their Economic Significance*. Standing Conference on Countryside Sports, Reading.

Coles, B. and Coles, J. (1986) *Sweet Tracks to Glastonbury*. Thames and Hudson, London.

Coles, J.M. (1971) The early settlement of Scotland: Excavations at Morton, Fife. *Proceedings of the Prehistoric Society* **37**: 284–366.

Collcutt, S.N., Currant, A.P. and Hawkes, C.J. (1981) A further report on the excavations at Sun Hole, Cheddar. *Proceedings of the University of Bristol Speleological Society* **16**: 21–28.

Collinson, M.E. and Hooker, J.J. (1987) Vegetational and mammalian faunal changes in the Early Tertiary of southern England, pp. 259–304 in Friis, E.M., Chaloner, W.G. and Crane, P.R. (eds) *The Origin of Angiosperms and their Biological Consequences*. Cambridge University Press, Cambridge.

Cook, J. and Jacobi, R. (1994) A reindeer antler or 'Lyngby' axe from Northamptonshire and its context in the British Late Glacial. *Proceedings of the Prehistoric Society* **60**: 75–84.

Cooke, A.S. and Farrell, L. (1995) Establishment and impact of muntjac (*Muntiacus reevesi*) on two National Nature Reserves, pp. 48–62 in Mayle, B. (ed.) *Muntjac Deer: their biology, impact and management in Britain.* Deer Society, Trentham, Staffordshire.

Cooke, A.S. and Lakhani, H. (1996) Damage to coppice regrowth by muntjac deer *Muntiacus reevesi* and protection with electric fencing. *Biological Conservation* **75**: 231–238.

Coope, G.R. (1977) Fossil coleopteran assemblages as sensitive indicators of climatic changes during the Devensian (Last) cold stage. *Philosophical Transactions of the Royal Society of London,* B **280**: 313–340.

Coope, G.R. and Pennington, W. (1977) The Windermere Interstadial of the Late Devensian. *Philosophical Transactions of the Royal Society of London,* B **280**: 337–339.

Coope, G.R. and Lister, A. (1987) Late Glacial mammoth skeletons from Condover, Shropshire, England. *Nature, London* **330**: 472–474.

Copley, G.J. (1971) *English Place-Names and their Origins.* David and Charles, Newton Abbot.

Corbet, G.B. (1961) Origin of the British insular races of small mammals and of the 'Lusitanian' fauna. *Nature, London* **191**: 1037–1040.

Corbet, G.B. (1964) Regional variation in the bank-vole *Clethrionomys glareolus* in the British Isles. *Proceedings of the Zoological Society of London* **143**: 191–219.

Corbet, G.B. (1971) Provisional distribution maps of British mammals. *Mammal Review* **1**: 95–142.

Corbet, G.B. (1974) The distribution of mammals in historic times, pp. 179–202 in Hawksworth, D.L. (ed.) *The Changing Flora and Fauna of Britain.* Academic Press, London and New York.

Corbet, G.B. (1975) Examples of short- and long-term changes of dental pattern in Scottish voles (Rodentia; Microtinae). *Mammal Review* **5**: 17–21.

Corbet, G.B. (1978) *The Mammals of the Palaearctic Region: a taxonomic review.* British Museum (Natural History), London.

Corbet, G.B. (1979) Report on rodent remains, pp. 135–137 in Renfrew, C. *Investigations in Orkney.* Reports of the Research Committee of the Society of Antiquaries **38**, London.

Corbet, G.B. (1986) Temporal and spatial variation of dental pattern in the voles, *Microtus arvalis*, of the Orkney Islands. *Journal of Zoology, London* **208**: 395–402.

Corbet, G.B. and Harris, S. (1991) *The Handbook of British Mammals,* 3rd edn. Blackwell's, Oxford.

Corbet, G.B. and Southern, H.N. (1977) *The Handbook of British Mammals,* 2nd edn. Blackwell's, Oxford.

Cordy, J-M., (1991) Palaeoecology of the Late Glacial and early Postglacial of Belgium and neighbouring areas, pp. 40–47 in Barton, N., Roberts, A.J. and Roe, D.A. (eds) *The Late Glacial in North-west Europe.* CBA Research Report. **77**, London.

Cornwall, I.W. and Coles, B.J. (1987) The bones from Meare Village East 1966, pp. 232–233 in Coles, J.M. Meare Village East: The Excavations of A. Bulleid and H. St. George Gray 1932–1956. *Somerset Levels Papers* **13**, Exeter.

Corp, N., Gorman, M.L. and Speakman, J.R. (1997) Ranging behaviour and time budgets of male wood mice *Apodemus sylvaticus* in different habitats and seasons. *Oecologia* **109**: 242–250.

Coxon, P. and Waldren, S. (1995) The floristic record of Ireland's Pleistocene temperate stages, pp. 243–267 in Preece, R.C. (ed.) *Island Britain: a Quaternary perspective.* Geological Society, Spec. Publ. **96**, London.

Coy, J. (1980) The animal bones, pp. 41–51 in Haslam, J. (ed.) A Middle Saxon iron smelting site at Ramsbury, Wiltshire. *Medieval Archaeology* **24**: 1–68.

Coy, J. (1982) The animal bones, pp. 68–73 in Gingell, C. (ed.) Excavations of an Iron Age Enclosure at Groundwell Farm, Blunsdon St Andrews, 1976–7. *Wiltshire Archaeological and Natural History Society Magazine* **76**: 33–75.

Coy, J. (1984) The small mammals and amphibia. pp. 526–527 in Cunliffe, B. *Danebury: an Iron Age Hillfort in Hampshire.* Vol. 2. CBA Research Report **52**, London.

Coy, J. (1987) Non-domestic faunal resources: South West England, pp. 9–29 in Balaam, N.D., Levitan, B. and Straker, V. *Studies in Palaeoeconomy and Environment in South West England.* BAR British Series **181**.

Coy, J. and Maltby, M. (1987) Archaeozoology in Wessex, pp. 204–251 in Keeley, H.C.M. (ed.) *Environmental Archaeology: A Regional Review. Vol. II.* English Heritage, London.

Crabtree, P.J. (1989a) Sheep, horses, swine, and kine; a zooarchaeological perpective on the Anglo-Saxon settlement of England. *Journal of Field Archaeology* **16**: 205–213.

Crabtree, P.J. (1989b) *West Stow, Suffolk: Anglo-Saxon Animal Husbandry.* East Anglian Archaeology Report **47**, Ipswich.

Crabtree, P.J. (1994) Animal exploitation in East Anglian villages, pp. 40–54 in Rackham, J. (ed.) *Environment and Economy in Anglo-Saxon England.* CBA Research Report **89**, York.

Crabtree, P.J. and Campana, D.V. (1991) The faunal remains from Brandon. Unpublished typescript.

Craik, J.C.A. (1997) Long-term effects of North American Mink *Mustela vison* on seabirds in western Scotland. *Bird Study* **44**: 303–309.

Cresswell, P., Harris, S. and Jefferies, D.J. (1990) *The History, Distribution, Status and Habitat Requirements of the Badger in Britain.* Nature Conservancy Council, Peterborough.

Cunliffe, B. (1978) *Iron Age Communities in Britain* (Rev. edn.). Routledge and Kegan Paul, London.

Cunliffe, B. (1979) *Excavations in Bath, 1950–1975.* Committee for Rescue Archaeology in Avon, Gloucestershire and Somerset, Bristol.

Currant, A. (1986) The Lateglacial mammal fauna of Gough's Cave, Cheddar, Somerset. *Proceedings of the University of Bristol Spelaeological Society* **17**: 286–304.

Currant, A. (1987) Late Pleistocene Saiga antelope *Saiga tartarica* on Mendip. *Proceedings of the University of Bristol Spelaeological Society* **18**: 74–80.

Currant, A. (1989) The Quaternary origins of the modern British mammal fauna. *Biological Journal of the Linnean Society* **38**: 23–30.

Currant, A.P. (1991) A Late Glacial Interstadial mammal fauna from Gough's Cave, Somerset, England, pp. 48–50 in Barton, N., Roberts, A.J. and Roe, D.A. (eds) *The Late Glacial in North-west Europe.* CBA Research Report **77**, London.

Dadd, M.N. (1970) Overlap of variation in British and European mammals. *Symposium of the Zoological Society of London* **26:** 117–125.

Darby, H.C. (1977) *Domesday England.* Clarendon Press, Oxford.

Darvill, T.C. and Coy, J.P. (1985) Report on the Faunal Remains from the Mound, Glastonbury. *Proceedings of the Somerset Archaeology and Natural History Society* **129**: 56–60.

David, A. (1991) Late Glacial archaeological residues from Wales: a selection, pp. 141–159 in Barton, N., Roberts, A.J. and Roe, D.A. (eds) *The Late Glacial in North-west Europe.* CBA Research Report **77**, London.

Davis, S.J.M. (1981) The effects of temperature change and domestication on the body size of Late Pleistocene to Holocene mammals of Israel. *Paleobiology* **7**: 101–114.

Davis, S.J.M. (1987) *The Archaeology of Animals.* Batsford, London.

Davis, S.J.M. and Valla, F.R. (1979) Evidence for domestication of the dog 12 000 years ago in the Natufian of Israel. *Nature, London* **276**: 608–610.

Dawkins, W.B. and Sanford, W.A. (1866) *British Pleistocene Mammalia. Introduction.* Palaeontographical Society, London.

Dawkins, W.B. and Jackson, J.W. (1917) The remains of the Mammalia found in the Lake Village of Glastonbury, pp. 641–672 in Bulleid, A. and Gray, A. St.G. (eds) *The Glastonbury Lake Village.* Glastonbury Antiquarian Society, Taunton.

De Jonge, G. and Dienske, H. (1979) Habitat and interspecific displacement of small mammals in the Netherlands. *Netherlands Journal of Zoology* **29**: 177–214.

Delany, M.J. (1964) Variation in the long-tailed field-mouse (*Apodemus sylvaticus* (L.)) in north-west Scotland I. Comparisons of individual characters. *Proceedings of the Royal Society of London* B, **161**: 191–199.

Delany, M.J. (1965) The application of factor analysis to the study of variation in the long-tailed field-mouse (*Apodemus sylvaticus* (L.)) in north-west Scotland. *Proceedings of the Linnean Society of London* **176**: 103–111.

Delany, M.J. and Healy, M. (1966) Variation in the white-toothed shrews (*Crocidura* spp.) in the British Isles. *Proceedings of the Royal Society of London B* **164:** 63–74.

Delany, M.J. and Healy, M.J.R. (1964) Variation in the long-tailed field-mouse (*Apodemus sylvaticus* (L.)) in south-west England. *Journal of Zoology, London* **152**: 319–332.

Delany, M.J. and Healy, M.J.R. (1967a) Variation in the long-tailed field-mouse (*Apodemus sylvaticus* (L.)) in the Channel Islands. *Proceedings of the Royal Society of London* B, **166**: 408–421.

Delany, M.J. and Healy, M.J.R. (1967b) Variation in the long-tailed field-mouse (*Apodemus sylvaticus*) in south-west England. *Journal of Zoology, London* **152:** 319–332.

Delibes, M. (1990) *Status and conservation needs of the wolf* (Canis lupus) *in the Council of Europe member states.* Council of Europe (Nature and Environment Series **47**), Strasbourg.

Dent, A. (1974) *Lost Beasts of Britain.* Harrap, London.

Devoy, R.J. (1985) The problem of a Late Quaternary landbridge between Britain and Ireland. *Quaternary Science Reviews* **4**: 43–58.

Devoy, R.J. (1986) Possible landbridges between Ireland and Britain: A geological appraisal. *Occasional Publications of the Irish Biogeographical Society* **1**: 15–26.

Devoy, R.J.N. (1995) Deglaciation, Earth crustal behaviour and sea-level changes in the determination of insularity: a perspective from Ireland, pp. 181–208 in Preece, R.C. (ed.) *Island Britain: a Quaternary perspective.* Geological Society, Spec. Publ. **96**, London.

Dewar, R.E. (1984) Extinctions in Madagascar: the loss of the subfossil fauna, pp. 574–593 in Martin, P.S. and Klein, R.G. (eds) *Quaternary Extinctions.* University of Arizona Press, Tucson.

Dobney, K. and Mills, A. (1994) *Material assessment of the animal bone from Flixborough.* Environmental Archaeology Unit, York, Report 94/6.

Doncaster, C.P. (1992) Testing the role of intraguild predation in regulating hedgehog populations. *Proceedings of the Royal Society of London, series B* **249**: 113–117.

Drew, C.D. and Piggott, S. (1936) The excavation of Long Barrow 163a on Thickthorn Down, Dorset. *Proceedings of the Prehistoric Society* **2**: 77–96.

Driver, J.C.D. (1990) Faunal remains, pp. 228–244 in Driver, J.C.D., Rady, J. and Sparks, M. (eds) *Excavations in the Cathedral Precincts, 2 Linacre Garden, 'Meister Omers' and St. Gabriel's Chapel. The Archaeology of Canterbury IV.* Kent Archaeological Society, Maidstone.

Dunstone, N. (1993) *The Mink.* T. and A.D. Poyser, London.

Dyczkowski, J. and Yalden, D.W. (1998) An estimate of the impact of predators on the British Field Vole *Microtus agrestis* population. *Mammal Review* **28**: 141–164.

Easterbee, N., Hepburn, L.V. and Jefferies, D.J. (1991) *Survey of the Status and Distribution of the Wildcat in Scotland, 1983–1987.* Nature Conservancy Council for Scotland, Edinburgh.

Ensom, P.C. (1987) Excavations at Sunnydown Farm, Langton Matravers, Dorset: amphibians discovered in the Purbeck Limestone Formation. *Proceedings of the Dorset Natural History and Archaeological Society* **109**: 148–149.

Erbajeva, M.A. (1988) [*Cenozoic Pikas: Taxonomy, Systematics, Phylogeny*]. Akademia Nauk USSR, Siberian Section, Moscow (in Russian).

Evans, C. and Serjeantson, D. (1988) The backwater economy of a fen-edge community in the Iron Age: the Upper Delphs, Haddenham. *Antiquity* **62**: 360–370.

Evans, J.G. and Rouse, A.J. (1992) Small-vertebrate and molluscan analysis from the same site. *Circaea* **8**: 75–84.

Ewart, J.C. (1911) Animal remains, pp. 362–377 in Curle, J. *A Roman Frontier Post and its People. The Fort of Newstead in the Parish of Melrose.* J. Maclehose and Sons, Glasgow.

Fairbanks, R.G. (1989) A 17 000-year glacio-eustatic sea level record: the influence of glacial melting rates on the Younger Dryas event and deep-ocean circulation. *Nature, London* **342**: 637–642.

Fairley, J.S. (1983) Exports of wild mammal skins from Ireland in the Eighteenth Century. *Irish Naturalists' Journal* **21**: 75–79.

Fairley, J.S. (1984) *An Irish Beast Book*, 2nd edn. Blackstaff Press, Belfast.

Fitter, R.S.R. (1959) *The Ark in our Midst.* Collins, London.

Fraser, F.C. (1968) Animal bones from Hod Hill. I. Sites within the Roman Fort, p. 127 in Richmond, D.I. *Hod Hill.* Vol. 2. British Museum, London.

Fraser, F.C. and King, J.E. (1954) Faunal remains, pp. 70–95 in Clark, J.G.D. (ed.) *Excavations at Star Carr.* Cambridge University Press.

French, D.D., Corbett, L.K. and Easterbee, N. (1988) Morphological discriminants of Scottish wildcats (*Felis silvestris*), domestic cats (*F. catus*) and their hybrids. *Journal of Zoology. London* **214**: 235–259.

Funnell, B.M. (1995) Global sea level and the (pen-)insularity of late Cenozoic Britain, pp. 3–13 in Preece, R.C. (ed.) *Island Britain: a Quaternary perspective.* Geological Society, Spec. Publ. **96**, London.

Garrad, L.S. (1972) *The Naturalist in the Isle of Man.* David and Charles, Newton Abbot.

Gascoyne, M., Currant, A.P. and Lord, T.C. (1981) Ipswichian fauna of Victoria Cave and the marine palaeoclimatic record. *Nature, London* **294**: 652–654.

Gee, H. (1993) The distinction between postcranial bones of *Bos primigenius* Bojanus, 1827 and *Bison priscus* Bojanus, 1827 from the British Pleistocene and the taxonomic status of *Bos* and *Bison*. *Journal of Quaternary Science* **8**: 79–92.

Genoud, M. (1988) Energetic strategies of shrews: ecological constraints and evolutionary implications, *Mammal Review* **18**: 173–193.

Genov, P. (1981) Food composition of wild boar in north-eastern and western Poland. *Acta Theriologica* **26**: 185–206.

Gibbard, P.L. (1995) The formation of the Strait of Dover, pp. 15–26 in Preece, R.C. (ed.) *Island Britain: a Quaternary perspective*. Geological Society, Spec. Publ. **96**, London.

Gilbert, J.M. (1979) *Hunting and Hunting Reserves in Medieval Scotland*. John Donald, Edinburgh.

Girling, M.A. (1988) The bark beetle *Scolytus scolytus* (Fabricius) and the possible role of elm disease in the early Neolithic, pp. 34–38 in Jones, M. (ed.) *Archaeology and the Flora of the British Isles*. Oxford University Committee for Archaeology, Monograph **14**, Oxford.

Glue, D.E. (1974) The food of the Barn Owl in Britain and Ireland. *Bird Study* **21**: 200–210.

Godwin, H. (1975) *The History of the British Flora*. Cambridge University Press.

Goodrich, E.S. (1894) On the fossil Mammalia of the Stonesfield Slate. *Quarterly Journal of Microscopical Science* **35**: 407–431.

Gordon, B.C. (1988) *Of Men and Reindeer Herds in the French Magdalenian Prehistory*. BAR International Series **390**, Oxford.

Gorman, M.L. and Reynolds, P. (1993) The impact of land-use change on voles and raptors. *Mammal Review* **23**: 121–126.

Gosling, L.M. and Baker, S.J. (1988) Planning and monitoring an attempt to eradicate coypus from Britain. *Symposia of the Zoological Society of London* **58**: 99–113.

Gosling, L.M. and Baker, S.J. (1989) The eradication of muskrats and coypus from Britain. *Biological Journal of the Linnean Society* **38**: 39–51.

Gosling, L.M., Watt, A.D. and Baker, S.J. (1981) Continuous retrospective census of the East Anglian Coypu population between 1970 and 1979. *Journal of Animal Ecology* **50**: 885–901.

Gould, S.J. (1974) The origin and function of 'bizarre' structures: antler size and skull size in the 'Irish Elk', *Megaloceros giganteus*. *Evolution* **28**: 191–220.

Gowlett, J.A.J., Hall, E.J., Hedges, R.E.M. and Perry, C. (1986) Radiocarbon dates from the Oxford AMS system: *Archaeometry* datelist 3. *Archaeometry* **28**: 116–125.

Grant, A. (1976) Faunal remains, pp. 262–287 in Cunliffe, B. (ed.) *Excavations at Portchester Castle II. Saxon*. Society of Antiquaries, London.

Grant, A. (1984) Animal husbandry, pp. 496–548 in Cunliffe, B. (ed.) *Danebury, an Iron Age Hillfort in Hampshire*. Vol. 2. CBA, London.

Gray, H. St G. (1966) *The Meare Lake Village*. Vol. III. Taunton Castle, Somerset.

Greenwood, J.J.D., Gregory, R.D., Harris, S., Morris, P.A. and Yalden, D.W. (1996) Relations between abundance, body size and species number in British birds and mammals. *Philosophical Transactions of the Royal Society of London B*, **351**: 265–278.

Grigson, C. (1966) The animal bones from Fussell's Lodge Long Barrow, pp. 63–73 in Ashbee, P. The Fussell's Lodge Long Barrow Excavations. *Archaeologia* **100**: 1–80.

Grigson, C. (1969) The uses and limitations of differences in absolute size in the distinction between the bones of aurochs (*Bos primigenius*) and domestic cattle (*Bos taurus*), pp. 277–294 in Ucko, P.J. and Dimbleby, G.W. (eds) *The Domestication and Exploitation of Plants and Animals*. Duckworth, London.

Grigson, C. (1978) The Late Glacial and early Flandrian ungulates of England and Wales – an interim review, pp. 46–56 in Limbrey, S. and Evans, J.G. (eds) *The Effect of Man on the Landscape: the Lowland Zone*. CBA Research Report **21**. London.

Grigson, C. (1982a) Sexing Neolithic domestic cattle skulls and horncores, pp 25–35 in Wilson, B., Grigson, C. and Payne, S. (eds) *Ageing and Sexing Animal Bones from Archaeological Sites*. BAR British Series **109**, Oxford.

Grigson, C. (1982b) Porridge and pannage: pig husbandry in Neolithic England, pp. 297–314 in Bell, M. and Limbrey, S. *Archaeological Aspects of Woodland Ecology*. BAR International Series **146**, Oxford.

Grigson, C. (1983) Mesolithic and Neolithic Animal Bones, pp. 64–72 in Evans, J.G. and Smith, I.F. (eds) Excavations at Cherhill, North Wiltshire, 1967. *Proceedings of the Prehistoric Society* **49**: 43–117.

Grigson, C. (1989) Size and sex – evidence for the domestication of cattle in the Near East, pp. 77–109 in Milles, A., Williams, D. and Gardner, N. (eds) *The Beginnings of Agriculture*. BAR International Series **496**, Oxford.

Grigson, C. and Mellars, P.A. (1987) The mammalian remains from the middens, pp. 243–289 in Mellars, P.A. (ed.) *Excavations on Oronsay*. Edinburgh University Press.

Groot Bruinderink, G.W.T.A., Hazebroek, E. and van der Voot, H. (1994) Diet and condition of wild boar, *Sus scrofa scrofa*, without supplementary feeding. *Journal of Zoology, London* **233**: 631–648.

Gurnell, J. and Pepper, H. (1993) A critical look at conserving the British Red Squirrel. *Mammal Review* **23**: 127–137.

Hall, J. and Yalden, D.W. (1978) A plea for caution over the identification of late Pleistocene *Microtus* in Britain. *Journal of Zoology, London* **186**: 556–560.

Hall, S.J.G. and Hall, J.G. (1988) Inbreeding and population dynamics of the Chillingham cattle (*Bos taurus*). *Journal of Zoology, London* **216**: 479–493.

Hallam, J.S., Edwards, B.J.N., Barnes, B. and Stuart, A.J. (1973) The remains of a Late Glacial elk associated with barbed points from High Furlong, near Blackpool. *Proceedings of the Prehistoric Society* **39**: 100–128.

Hamilton-Dyer, S. (1993) The animal bones, pp. 132–136 in Zienkiewicz, J.D. Excavations at the *Scamnum Tribunorum* at Caerleon. The Legionnary Museum site 1983–5. *Britannia* **24**: 27–140.

Hamilton-Dyer, S. (1996a) The Animal Bones, in Howell, L. and Durden, T. A Grooved Ware Pit on the Seven Barrows All Weather Gallop, Sparsholt. *Oxoniensia* **61**: 21–25.

Hamilton-Dyer, S. (1996b) Animal bones, in Andrews, P. and Crockett, A. Three excavations along the Thames and its tributaries, 1994: Neolithic to Saxon Settlement and Burial in the Thames, Colne and Kennet Valleys. *Wessex Archaeological Reports* **10**: 42 and 157.

Hamilton-Dyer, S. (1997) Birds, fish, amphibians and small mammals, in Naying, N. and Caseldine, A. Excavations at Caldicot, Gwent: Bronze Age Palaeochannels in the Lower Nedern Valley. *CBA Reports* **108**: 234–235.

Harcourt, R. (1971) The animal bones, pp. 188–191 in Wainwright, G.J. and Longworth, I.H. *Durrington Walls: Excavations 1966–1968.* Society of Antiquaries, London.

Harcourt, R. (1974) The dog in prehistoric and early historic Britain. *Journal of Archaeological Science* **1**: 151–176.

Harcourt, R. (1978) The animal bones, pp. 150–160 in Wainwright, G.J. *Gussage All Saints; an Iron Age Settlement in Dorset.* H.M.S.O., London.

Harcourt, R. (1979) The animal bones, pp. 214–215 in Wainwright, G.J. *Mount Pleasant, Dorset, Excavations 1970–71.* Society of Antiquaries, London.

Harman, M. (1978) The animal bones, pp. 177–187 in Pryor, F. *Excavations at Fengate, Peterborough, England: The Second Report.* Archaeological Monograph **5**, Royal Ontario Museum.

Harman, M. (1979) The mammalian bones, pp. 328–332 in Williams, J.H. *St Peter's Street, Northampton: Excavations 1973–6* Northampton Development Corporation Archaeological Monograph 2, Northampton.

Harman, M. (1989) Cremations, pp. 23–25 in Kinsley, A.G. The Anglo-Saxon cemetery at Millgate, Newark-on-Trent, Nottinghamshire. *Nottinghamshire Archaeological Monographs* **2**.

Harman, M. (1993) The mammalian bones, pp. 24–25 in Simpson, W.G., Gurney, D.A., Neve, J. and Pryor, F.M.M. *The Fenland Project Number 7: Excavations in Peterborough and the Lower Welland Valley 1960–1969.* East Anglian Archaeology Report **61**, Cambridge.

Harrington, R. (1973). Hybridization among deer and its implication for conservation. *Irish Forestry Journal* **30**: 64–78.

Harris, S. (1979) History, distribution, status and habitat requirements of the harvest mouse (*Micromys minutus*) in Britain. *Mammal Review* **9**: 159–171.

Harris, S., Morris, P., Wray, S. and Yalden, D. (1995) *A Review of British Mammals: population estimates and conservation status of British mammals other than cetaceans.* JNCC, Peterborough.

Harrison, J. and Grant, P. (1976) *The Thames Transformed.* André Deutsch, London.

Harting, J.E. (1880) *British Animals Extinct within Historic Times.* Trübner, London.

Harwood, J. and Prime, J.H. (1978) Some factors affecting the size of British grey seal populations. *Journal of Applied Ecology* **15**: 401–411.

Hedges, R.E.M., Housley, R.A., Law, I.A., Perry, C. and Gowlett, J.A.J. (1987) Radiocarbon dates from the Oxford AMS system: *Archaeometry* datelist 6. *Archaeometry* **29**: 289–306

Hedges, R.E.M., Housley, R.A., Law, I.A., Perry, C. and Hendy, E. (1988) Radiocarbon dates from the Oxford AMS system: *Archaeometry* datelist 8. *Archaeometry* **30**: 291–305.

Hedges, R.E.M., Housley, R.A., Bronk Ramsey, C. and van Klinken, G.J. (1994) Radiocarbon dates from the Oxford AMS system: *Archaeometry* datelist 19. *Archaeometry* **36**: 417–430

Hedges, R.E.M., Housley, R.A., Bronk Ramsey, C. and van Klinken, G.J. (1995) Radiocarbon dates from the Oxford AMS system: *Archaeometry* datelist 20. *Archaeometry* **37**: 417–430

Henn, C. (1980) TV Bear Still on Loose. *Daily Telegraph* 23 August 1980.

Hewer, H.R. (1974) *British Seals.* Collins, New Naturalist **57**, London.

Hewison, A.J.M. (1995) Isozyme variation in roe deer in relation to their population history in Britain. *Journal of Zoology, London* **235**: 279–288.

Hewson, R. (1956) The mountain hare in England and Wales. *Naturalist* **858**: 107–109.

Hewson, R. (1984) Scavenging and predation upon sheep and lambs in west Scotland. *Journal of Applied Ecology* **21**: 843–868.

Hewson, R. (1995) Mountain Hares *Lepus timidus* on Hoy, Orkney, and their habitat. *Journal of Zoology, London* **236**: 331–337.

Hickling, G. (1962) *Grey Seals and the Farne Islands*. London, Routledge and Kegan Paul.

Higgs, E., Greenwood, W. and Garrard, A. (1979) Faunal report, pp.353–362 in Rahtz, P. *The Saxon and Medieval Palaces at Cheddar: Excavations 1960–62*. BAR British Series **65**, Oxford.

Hills, D. (1991) Ephemeral introductions and escapes, pp. 576–580 in Corbet, G.B. and Harris, S. (eds) *The Handbook of British Mammals*. 3rd edn. Blackwell Science, Oxford.

Hinton, M.A.C. (1952) Remains of small mammals. In Allison, J., Godwin, H. and Warren, S.H. Late-Glacial deposits at Nazeing in the Lea Valley, North London. Philosophical Transactions of the Royal Society of London, B **236**: 169–240.

Hodgson, G.W.I. (1968) A comparative account of the animal remains from *Cortospitum* and the Iron Age site of Catcote near Hartlepools, County Durham. *Archaeologia Aeliana (4th series)* **46**: 127–162.

Hodgson, G.W.I. (1977) *The Animal Remains from Excavations at Vindolanda 1970–1975*. Vindolanda Trust, Bardon Mill, Hexham.

Hooker, J.J. (1980) The succession of *Hyracotherium* (Perissodactyla, Mammalia) in the English early Eocene. *Bulletin of the British Museum (Natural History), Geology* **33**: 101–114.

Hooker, J.J. (1989) British mammals in the Tertiary Period. *Biological Journal of the Linnean Society* **38**: 9–21.

Hooker, J.J. (1994) Mammalian taphonomy and palaeoecology of the Bembridge Limestone Formation (Late Eocene, S. England). *Historical Biology* **8**: 49–69.

Hooker, J.J., Collinson, M.E., van Bergen, P.F. *et al.* (1995) Reconstruction of land and freshwater palaeoenvironments near the Eocene-Oligocene boundary, southern England. *Journal of the Geological Society, London* **152**: 449–468.

Hopwood, A.T. (1939) Excavations at Brundon, Suffolk (1935–37). Part II. Fossil mammals. *Proceedings of the Prehistoric Society* **1939**: 13–29.

Housley, R.A. (1991) AMS dates from the Late Glacial and early Postglacial in north-west Europe: a review, pp. 25–39 in Barton, N., Roberts, A.J. and Roe, D.A. (eds) *The Late Glacial in north-west Europe*. CBA Research Report **77**, London.

Howes, C.A. (1976) The decline of the Otter in South Yorkshire and adjacent areas. *Naturalist* **1976**: 3–12.

Howes, C.A. (1980) Aspects of the history and distribution of Polecats and Ferrets in Yorkshire and adjacent areas. *Naturalist* **105**: 3–16.

Hubbard, A.L., McOrist, S., Jones, T.W. *et al.* (1992) Is survival of European wildcats *Felis silvestris* in Britain threatened by interbreeding with domestic cats? *Biological Conservation* **61**: 203–208.

Huntley, B. (1990) European vegetation history: palaeovegetation maps from pollen data – 13 000 yr BP to present. *Journal of Quaternary Science* **5**: 103–122.

Huntley, B. and Birks, H.J.B. (1983) *An Atlas of Past and Present Pollen Maps for Europe: 0–13 000 Years Ago*. Cambridge University Press, Cambridge.

Huntley, B., Bartlein, P.J. and Prentice, I.C. (1989) Climatic control of the distribution and abundance of beech (*Fagus* L.) in Europe and North America. *Journal of Biogeography* **16**: 551–560.

Hurrell, E. and McIntosh, G. (1984) Mammal Society dormouse survey, January 1975–April 1979. *Mammal Review* **14**: 1–18.

Hutchings, M.R. and Harris, S. (1996) *The Current Status of the Brown Hare* (Lepus europaeus) *in Britain*. J.N.C.C., Peterborough.

Hutterer, R. (1985) Anatomical adaptations of shrews. *Mammal Review* **15**: 43–55.

Izard, K. (1997) The animal bones, pp. 363–370 in Wilmott, T. *Birdoswald. Excavation of a Roman fort on Hadrian's Wall and its successor settlements: 1987–92*. English Heritage Archaeological Report **14**, London.

Jackson, J.W. (1935) Report on the Animal Remains from Pit 5. In Stone, J.F.S., Some discoveries at Ratfyn, Amesbury and their bearing the dating Woodhenge. *Wiltshire Archaeology and Natural History Magazine* **47**: 55–80.

Jackson, J.W. (1953) Archaeology and palaeontology, pp.170–246 in Cullingford, C.H.D. (ed.) *British Caving: an introduction to speleology*. Routledge and Kegan Paul, London.

Jacobi, R.M., Tallis, J.H. and Mellars, P.A. (1976) The southern Pennine Mesolithic and the ecological record. *Journal of Archaeological Science* **3**: 307–320.

James, T.B. (1990) *The Palaces of Medieval England c. 1050–1550*. Seaby, London.

Jarman, M.R. (1972) European deer economies and the advent of the Neolithic, pp. 125–149 in Higgs E.S. (ed.) *Papers in Economic Prehistory*. Cambridge University Press, Cambridge.

Jarrell, G.H. and Fredga, K. (1993) How many kinds of lemmings? A taxonomic overview. pp. 45–57 in Stenseth, N.C. and Ims, R.A. (eds) *The Biology of Lemmings*. Linnean Society Symposium **15**, Academic Press, London.

Jędrzejewska, B., Okarma, H., Jędrzejewski, W. and Milkowski, L. (1994) The effects of exploitation and protection on forest structure, ungulate density and wolf predation in Białowieża Primeval Forest, Poland. *Journal of Applied Ecology* **31**: 664–676.

Jędrzejewska, B., Jędrzejewski, W., Bunevich, A.N. *et al.* (1996) Population dynamics of Wolves *Canis lupus* in Białowieża Primeval Forest (Poland and Belarus) in relation to hunting by humans, 1847–1993. *Mammal Review* **26**: 103–126.

Jędrzejewski, W., Jędrzejewski, B., Okarma, H. and Ruprecht, A.L. (1992) Wolf predation and snow cover as mortality factors in the ungulate community of the Białowieża National Park, Poland. *Oecologia* **90**: 27–36.

Jędrzejewski, W., Schmidt, K., Milkowski, L. *et al.* (1993) Foraging by lynx and its role in ungulate mortality: the local (Białowieża Forest) and the Palaearctic viewpoints. *Acta Theriologica* **38**: 385–403.

Jefferies, D.J. (1989) The changing otter population of Britain 1700–1989. *Biological Journal of the Linnean Society* **38**: 61–69.

Jefferies, D.J. and Critchley, C.H. (1994) A new pine marten record for the North Yorkshire Moors: skull dimensions and confirmation of species. *Naturalist* **119**: 145–150.

Jefferies, D.J., Wayre, P., Jessop, R.M. and Mitchell-Jones, A.J. (1986) Reinforcing the native otter *Lutra lutra* population of East Anglia: an analysis of the behaviour and range development of the first release group. *Mammal Review* **16**: 65–79.

Jenkins, D. (1962) The present status of the wildcat (*Felis silvestris*) in Scotland. *Scottish Naturalist* **70**: 126–138.

Jenkinson, R.D.S. (1983) The recent history of the Northern Lynx (*Lynx lynx* Linné) in the British Isles. *Quaternary Newsletter* **41**: 1–7.

Jenkinson, R.D.S. (1984) *Creswell Crags. Late Pleistocene Sites in the East Midlands*. BAR British Series **122**, B.A.R. Oxford.

Jewell, P.A. (1959) Small mammals of the Bronze Age. *Bulletin of the Mammal Society* **11**: 9–11.

Johnson, W. (ed.) (1970) *Gilbert White's Journals*. David and Charles, Newton Abbot (reprint of 1931 Routledge and Kegan Paul, London, publication).

Jones, R.L. and Keen, D.H. (1993) *Pleistocene Environments in the British Isles*. Chapman and Hall, London.

Jones, R.T. and Serjeantson, D. (1983) The animal bones from five sites at Ipswich. Unpubl. Rep. 13/83 to Ancient Monuments Laboratory, English Heritage, London.

Jones, R.T., Sly, J. and Hocking, L. (1987) The vertebrate remains, pp. 163–170 in Olivier, A.C.H. Excavation of a Bronze Age Funerary Cairn at Manor Farm, near Borwick, North Lancashire. *Proceedings of the Prehistoric Society* **53**: 129–186.

Jope, M. and Grigson, C. (1965) Faunal remains, pp 141–167 in Keiller, A. *Windmill Hill and Avebury*. Clarendon Press, Oxford.

Kenward, R.E. and Holm, J.L. (1989) What future for British red squirrels? *Biological Journal of the Linnean Society* **38**: 83–89.

Kenward, R.E. and Holm, J.L. (1993) On the replacement of the red squirrel in Britain: a phytotoxic explanation. *Proceedings of the Royal Society of London, series B* **251**: 187–194.

Kermack, K.A., Mussett, F. and Rigney, H.W. (1973) The lower jaw of *Morganucodon*. *Zoological Journal of the Linnean Society* **53**: 87–175.

Kermack, K.A., Mussett, F. and Rigney, H.W. (1981) The skull of *Morganucodon*. *Zoological Journal of the Linnean Society* **71**: 1–158.

Kielan-Jaworowska, Z. and Ensom, P.C. (1992) Multituberculate mammals from the Upper Jurassic Purbeck Limestone Formation of Southern England. *Palaeontology* **35**: 95–126.

King, A. (1978) A comparative survey of Bone Assemblages from Roman Sites in Britain. *Bulletin of the Institute of Archaeology, University of London* **15**: 207–232.

King, A. (1988) Animal bones and shells, pp. 260–265 in James, T.B. and Robinson, A.M. *Clarendon Palace*. Research Rep. **45**, Society of Antiquaries, London.

King, C.M. (1990) *The Handbook of New Zealand Mammals*. Oxford University Press, Aukland.

King, C.M. and Moors, P.J. (1979) On co-existence, foraging strategy and the biogeography of Weasels and Stoats (*Mustela nivalis* and *M. erminea*) in Britain. *Oecologia* **39**: 129–150.

King, C.M., Innes, J.G., Flux, M. *et al.* (1996) Distribution and abundance of small mammals in relation to habitat in Pureora Forest Park. *New Zealand Journal of Ecology* **20**: 215–240.

King, J.E. (1962) Report on animal bones. In Wymer, J. Excavations at the Magelomosian sites at Thatcham, Berkshire, England. *Proceedings of the Prehistoric Society* **28**: 255–361.

Kitchener, A. (1987) Fighting behaviour of the Extinct Irish Elk. *Modern Geology* **11:** 1–28.

Kitchener, A.C. and Bonsall, C. (1997) AMS radio-carbon dates for some extinct Scottish mammals. *Quaternary Newsletter,* **83:** 1–11.

Kitchener, A.C. and Conroy, J. (1997) The history of the Eurasian Beaver, *Castor fiber,* in Scotland. *Mammal Review* **27**: 95–108.

Kolfschoten, T. van and Laban, C. (1995) Pleistocene terrestrial mammal faunas from the North Sea. *Mededelingen Rijks geologische Dienst* **52**: 135–151.

Kolstrup, E. (1991) Palaeoenvironmental developments during the Late Glacial of the Weichselian, pp. 1–6 in Barton, N., Roberts, A.J. and Roe, D.A. (eds) *The Late Glacial in North-west Europe.* CBA Research Report, **77**, London.

Kowalski, K. (1980) Origin of mammals of the Arctic tundra. *Folia Quaternaria, Krakow,* **51**: 3–16.

Krebs, J.R. (1997) *Bovine Tuberculosis in Cattle and Badgers.* Ministry of Agriculture, Fisheries and Food, London.

Kristiansson, H. (1990) Population variables and causes of mortality in a hedgehog (*Erinaceous* (sic) *europaeus*) population in southern Sweden. *Journal of Zoology, London* **220**: 391–404.

Kruuk, H. (1978) Spatial organization and territorial behaviour of the European badger *Meles meles*. *Journal of Zoology, London* **184**: 1–19.

Kurtén, B. (1968) *Pleistocene Mammals of Europe.* Weidenfeld and Nicolson, London.

Kurtén, B. (1973) Fossil Glutton (*Gulo gulo* (L.)) from Tornewton Cave, South Devon. *Commentationes Biologicae* **66**: 1–8.

Kvam, T., Overskaug, K. and Sorensen, O.J. (1988) The wolverine *Gulo gulo* in Norway. *Lutra* **31**: 7–20.

Laidler, L. (1982) *Otters in Britain.* David and Charles, Newton Abbot.

Langley, P.J.W. and Yalden, D.W. (1977) The decline of the rarer carnivores in Great Britain during the nineteenth century. *Mammal Review* **7**: 95–116.

Lawson, T.J. (1981) The 1926–7 excavations of the Creag nan Uamh bone caves, near Inchnadamph, Sutherland. *Proceedings of the Society of Antiquaries of Scotland* **111**: 7–20.

Lean, G. (1988) Ministers brought to eel. *Observer* 16 Oct. 1988.

Legge, A.J. (1991) The animal remains from six sites at Down Farm, Woodcutts, pp. 54–100 in Barrett, J., Bradby, R. and Hall, M. *Papers on the Prehistoric Archaeology of Cranborne Chase.* Oxbow Monograph **11**, Oxford.

Legge, A.J. (1992) *Excavations at Grimes Graves, Norfolk, 1972–1976. Fascicule 4. Animals, Environment and the Bronze Age Economy.* British Museum, London.

Legge, A.J. and Rowley-Conwy, P.A. (1988) *Star Carr Revisited: a re-analysis of the large mammals.* Centre for Extra-Mural Studies, Birkbeck College, University of London.

Legge, A.J., Williams, J. and Williams, P. (1988) Animal remains, pp. 90–95 in Moss-Eccardt, J. Archaeological investigations in the Letchworth area 1958–1974. *Proceedings of the Cambridge Antiquarian Society* **77**: 35–103.

Le Sueur, F. (1976) *A Natural History of Jersey.* Phillimore, Chichester.

Lever, C. (1977) *The Naturalized Animals of the British Isles.* Hutchinson, London.

Levitan, B. (1984) The vertebrate remains, pp. 108–138 in Rahtz, S. and Rowley, T. *Middleton Stoney: Excavation and Survey in a North Oxfordshire Parish 1970–1982.* Oxford University Department of External Studies, Oxford.

Levitan, B. (1990) The vertebrate remains. In Bell, M. *Brean Down Excavations 1983–1987.* English Heritage Archaeological Report **15**, London.

Levitan, B. (1993) The vertebrate remains, pp. 257–303 in Woodward, A. and Leach, P. *The Uley Shrines. Excavation of a ritual complex on West Hill, Uley, Gloucestershire 1977–9.* English Heritage Archaeological Report **17**, London.

Levitan, B. and Locker, A. (1987) The vertebrate remains (from Westward Ho!, Devon), pp. 213–222 in Balaam, N.D., Levitan, B. and Straker, V. *Studies in Palaeoeconomy and Environment in South West England.* BAR British Series **181**, Oxford.

Lillegraven, J.A., Kielan-Jaworowska, Z. and Clemens, W.A. (1979) *Mesozoic mammals: the first two-thirds of mammalian history.* University of California Press, Berkeley.

Limpens, H.J.G.A., Helmer, W., van Winden, A. and Mostert, K. (1989) Vleermuizen (Chiroptera) en lintvormige landschapselementen. *Lutra* **32**: 1–20.

Linnard, W. (1982) *Welsh Woods and Forests: history and utilization*. University Press, Cardiff.

Lister, A.M. (1984a) The fossil record of Elk (*Alces alces* (L.) In Britain. *Quaternary Newsletter* **44**: 1–7.

Lister, A.M. (1984b) Evolutionary and ecological origins of British deer. *Proceedings of the Royal Society of Edinburgh* **82B**: 205–229.

Lister, A.M. (1987) Giant Deer and Giant Red Deer from Kent's Cavern, and the status of *Strongyloceros spelaeus* Owen. *Transactions and Proceedings of the Torquay Natural History Society* **19**: 189–198.

Lister, A.M. (1991) Late Glacial mammoths in Britain, pp. 51–59 in Barton, N., Roberts, A.J. and Roe, D.A. (eds) *The Late Glacial in North-west Europe*. CBA Research Report **77**, London.

Lister, A.M. (1995) Sea-levels and the evolution of island endemics: the dwarf red deer of Jersey, pp.151–172 in Preece, R.C. (ed.) *Island Britain: a Quaternary perspective*. Geological Society, Spec. Publ. **96**, London.

Lloyd, C., Tasker, M.L. and Partridge, K. (1991) *The Status of Seabirds in Britain and Ireland*. T. and A.D. Poyser, London.

Lloyd, H.G. (1983) Past and present distributions of red and grey squirrels in Britain. *Mammal Review* **13**: 69–80.

Locker, A. (1977) Animal bones and shellfish, pp. 160–162 in Neal, D.S. Excavations at the Palace of King's Langley, Hertfordshire, 1974–1976. *Medieval Archaeology* **21**: 124–165.

Locker, A. (1987) (The Vertebrate Remains from the Mesolithic Area 3 at Westward Ho!), p. 214 in Balaam, N.D., Levitan B., Straker V., eds *Studies in palaeoconomy and environment in South West England*. BAR British Series **181**, Oxford.

Locker, A. (1994) Faunal remains, pp. 107–110 in Papworth, M. Lodge Farm, Kingston Lacey Estate, Dorset. *Journal of the British Archaeological Association* **147**: 57–121.

Lockie, J.D. (1964) Distribution and fluctuations of the pine marten, *Martes martes* (L.), in Scotland. *Journal of Animal Ecology* **33**: 349–356.

Lodé, T. (1993) Diet composition and habitat use of sympatric polecat and American mink in western France. *Acta Theriologica* **38**: 161–166.

Loftus, R.T., MacHugh, D.E., Bradley, D.G. *et al.* (1993) Evidence for two independent domestications of cattle. *Proceedings of the National Academy of Sciences of the USA* **91**: 2757–2761.

Long, A., Sher, A. and Vartanyan, S. (1994) Holocene mammoth dates. *Nature, London* **369**: 364.

Love, J.A. (1983) *The Return of the Sea Eagle*. Cambridge University Press, Cambridge.

Lowe, V.P.W. (1968) Population dynamics of the Red Deer (*Cervus elaphus* L.) on Rhum. *Journal of Animal Ecology* **38**: 425–457.

Lowe, V.P.W. and Gardiner, A.S. (1974) A re-examination of the subspecies of Red deer (*Cervus elaphus*) with particular reference to the stocks in Britain. *Journal of Zoology, London* **174**: 185–201.

Lowe, V.P.W. and Gardiner, A.S. (1975) Hybridisation between red deer (*Cervus elaphus*) and sika deer (*Cervus nippon*) with particular reference to stocks in N.W. England. *Journal of Zoology, London* **177**: 553–566.

Luff, R. (1982) *A Zooarchaeological Study of the Roman N.W. Provinces*. BAR International Series **137**, Oxford.

Luff, R. (1985) The fauna, pp. 143–149 in Niblett, R. *Sheepen: an early Roman industrial site at Camulodunum*. CBA Research Report **57**, London.

Luff, R. (1993) Animal bones from excavations in Colchester, 1971–85. *Colchester Archaeological Report* **12**. Colchester Archaeological Trust.

Macdonald, D.W. (1980) *Rabies and Wildlife. A biologist's perspective*. Oxford University Press, Oxford.

Macdonald, D.W., Tattersall, F.H., Brown, E.D. and Balharry, D. (1995) Reintroducing the European Beaver to Britain: nostalgic meddling or restoring biodiversity? *Mammal Review* **25**: 161–200.

MacGregor, A. (1989) Animals and the early Stuarts: hunting and hawking at the court of James I and Charles I. *Archives of Natural History* **16**: 305–318.

Macpherson, H.A. (1892) *A Vertebrate Fauna of Lakeland*. Douglas, Edinburgh.

Maltby, M. (1979) *Faunal Studies on Urban Sites. The Animal Bones from Exeter 1971–1975*. Exeter Archaeological Reports, Vol. 2. Department of Prehistory and Archaeology University of Sheffield.

Maltby, M. (1981) Iron Age, Romano-British and Anglo-Saxon Animal Husbandry: A Review of the Faunal Evidence, pp. 155–204 in Jones, M. and Dimbleby, G. (eds) *The Environment of Man: the Iron Age to the Anglo-Saxon Period*. BAR British Series **87**, Oxford.

Maltby, M. (1982) The animal and bird bones. In Higham, R.A., Allan, J.P. and Blaylock, S.R. Excavations at Okehampton Castle, Devon. Part 2. The Bailey. *Proceedings of the Devon Archaeological Society* **40**: 114–135.

Maltby, M. (1983) The animal bones, pp. 47–51 in Collis, J. *Wigber Low, Derbyshire: a Bronze Age and Anglian burial site in the White Peak*. Department of Prehistory and Archaeology, University of Sheffield.

Maltby, M. (1984) The animal bones, pp. 199–212 in Fulford, M.G. (ed.) *Silchester: Excavations in the Defences 1974–1980*. Britannia Monograph **5**, London.

Maltby. M. (1985) The animal bones, pp. 97–112 in Fasham, P. *The Prehistoric Settlement at Winnal Down, Winchester*. Hampshire Field Club & Archaeological Society Monograph 2.

Mansell-Pleydell, J.C. (1889) Bos primigenius, with special reference to Paleolithic and Neolithic man. *Proceedings of the Dorset Natural History and Archaeological Society* **10**: 81–88.

Mansell-Pleydell, J.C. (1895) On the Castoridae, with special reference to Castor fiber. *Proceedings of the Dorset Natural History and Archaeological Society* **16**: 163–170.

Marquet, J.-C. (1993) *Paléoenvironment et chronologie des sites du domaine Atlantique français d'age Pléistocène Moyen et Supérieur d'après l'études des Rongeurs*. D.Sc. Thesis, Université de Bourgogne; privately published by the author, Tours.

Marett, R.R. (1916) The site, fauna and industry of Cotte de St Brelade, Jersey. *Archaeologia* **67**: 75–118.

Martin, P.S. (1967) Prehistoric overkill, pp. 75–120 in Martin, P.S. and Wright, H.E. (eds) *Pleistocene Extinctions*.Yale University Press, New Haven and London.

Martin, P.S. (1984) Prehistoric overkill: a global model, pp. 354–403 in Martin, P.S. and Klein, R.G. (eds) *Quaternary Extinctions*. University of Arizona Press, Tucson.

Matthews, L.H. (1952) *British Mammals*. Collins, New Naturalist **21**, London.

Matthews, L.H. (1982) *Mammals in the British Isles*. Collins, New Naturalist **68**, London.

May, J. (1996) *Dragonby*. Oxbow Monograph **61**, Oxford.

Mayes, P. and Butler, L. (1983) *Sandal Castle Excavations 1964–1973*. Wakefield Historical Publications, Wakefield.

Mayhew, D.F. (1975) *The Quaternary history of some British rodents and lagomorphs*. Ph.D. Thesis, University of Cambridge.

McDonald, R.A., Hutchings, M.R. and Keeling, J.G.M. (1997) The status of the ship rats *Rattus rattus* on the Shiant Islands, Outer Hebrides, Scotland. *Biological Conservation* **82**: 113–117.

McEvedy, C. and Jones, R. (1978) *Atlas of World Population History*. Penguin Books, Harmondsworth.

Mead, J.G. and Mitchell, E.D. (1984) Atlantic Gray Whales, pp. 33–53 in Jones, M.L., Swartz, S.L. and Leatherwood, S. (eds) *The Gray Whale* Eschrichtius robustus. Academic Press, Orlando, Florida.

Mech, L.D. (1966) *The Wolves of Isle Royale*. Fauna of the National Parks of the United States Fauna Series **7**. US Government Printing Office, Washington.

Meddens, B. (1987) *Assessment of the animal bone work for Wroxeter Roman City, Shropshire: from sites at Wroxeter Barker and Wroxeter Webster*. Unpub. Rep. To Ancient Monuments Laboratory, London.

Meddens, B. (1990) *Animal bones from Catterick Bridge (CEU 240), a Roman town (North Yorkshire) excavated in 1983*. Unpub. Rep. 98/90 to Ancient Monuments Laboratory, London.

Meijer, T. and Preece, R.C. (1995) Malacological evidence relating to the insularity of the British Isles during the Quaternary, pp. 89–110 in Preece, R.C. (ed.) *Island Britain: a Quaternary perspective*. Geological Society, Spec. Publ. **96**, London.

Mellars, P.A. (1987) *Excavations on Oronsay*. Edinburgh University Press.

Meylan, A. and Hausser, J. (1978) Le type chromosomique A des *Sorex* du groupe *araneus*: *Sorex coronatus* Millet, 1828. *Mammalia* **42**: 115–122.

Middleton, A.D. (1931) *The Grey Squirrel*. Sidgwick and Jackson, London.

Millais, J.G. (1904) *Mammals of Great Britain and Ireland*. Longmans Green and Co., London.

Miller, G.R. (1912) *Catalogue of the Mammals of Western Europe*. British Museum (Natural History), London.

Mitchell, G.F. (1958) A Late-Glacial deposit near Ballaugh, Isle of Man. *New Phytologist* **57**: 256–263.

Mitchell, G.F. (1977) Raised beaches and sea-levels, pp. 169–186 in Shotton, F.W. (ed.) *British Quaternary Studies: Recent Advances*. Clarendon Press, Oxford.

Mitchell-Jones, A.J., Jefferies, D.J., Stebbings, R.E. and Arnold, H.R. (1986) Public concern about bats (Chiroptera) in Britain: an analysis of enquiries in 1982–83. *Biological Conservation* **36**: 315–328.

Mitchell-Jones, A.J., Hutson, A.M. and Racey, P.A. (1993) The growth and development of bat conservation in Britain. *Mammal Review* **23**: 139–148.

Monaghan, N.T. (1989) The Elk *Alces alces* L. in Irish Quaternary deposits. *Irish Naturalists' Journal* **23**: 97–101.

Montagu, I. (1924) On the remains of Fen Beaver in the Sedgwick Museum – 1. Skulls and Teeth. *Proceedings of the Zoological Society of London* **71**: 1081–1086.

Montgomery, W.I. (1975) On the relationship between sub-fossil and recent British Water voles. *Mammal Review* **5**: 23–29.

Montgomery, W.I. (1978) Studies on the distribution of *Apodemus sylvaticus* (L.) and *A. flavicollis* (Melchior) in Britain. *Mammal Review* **8**: 177–184.

Morris, J. (1973) *The Age of Arthur.* Weidenfeld and Nicholson, London.

Morris, J. (ed.) (1975) *Domesday Book. Surrey.* Phillimore, Chichester.

Morris, P. (1997) *The Edible Dormouse* (Glis glis). Mammal Society, London.

Morris, P.A. and Hoodless, A. (1992) Movements and hibernaculum site in the fat dormouse (*Glis glis*). *Journal of Zoology, London* **228**: 685–687.

Morris, P.A. and Whitbread, S. (1986) A method for trapping the dormouse (*Muscardinus avellanarius*). *Journal of Zoology, London* **210**: 642–644.

Musil, R. (1984) The first known domestication of wolves in central Europe, pp. 23–25 in Grigson, C. and Clutton-Brock, J. (eds) *Animals and Archaeology: 4. Husbandry in Europe.* BAR International Series **227**, Oxford.

Myrberget, S. (1990) Wildlife management in Europe outside the Soviet Union. *NINA Utreding* **018**: 1–47.

Nadachowski, A. (1989) Origin and history of the present rodent fauna in Poland based on fossil evidence. *Acta Theriologica* **34**: 37–53.

Nadachowski, A., Madeyska, T., Rook, E. *et al.* (1989) Holocene snail and vertebrate fauna from Nad Mosurem Staryn Duza Cave (Grodisko near Crakow): palaeoclimatic and palaeoenvironmental reconstructions. *Acta zoologica cracoviensis* **32**: 487–511.

Napier, A.S. (1900) *Old English Glosses.* Clarendon Press, Oxford.

Neal, E. (1948) *Badgers.* Collins, New Naturalist Monograph, London.

Neal, E.G. (1977) *Badgers.* Blanford Press, Poole.

Neal, E. (1986) *The Natural History of Badgers.* Croom Helm, Beckenham.

Neaverson, E. (1940–43) A summary of the records of Pleistocene and Postglacial Mammalia from North Wales and Merseyside. *Proceedings of the Liverpool Geological Society* **18**: 70–85.

Nevard, T.D. and Penfold, J.B. (1978) Wildlife conservation in Britain: the unsatisfied demand. *Biological Conservation* **14**: 25–44.

Newton, E.T. (1891) *The Vertebrate Fauna of the Pliocene Deposits of Britain.* Memoirs of the Geological Survey, UK, London.

Newton, E.T. (1894) The Vertebrate fauna collected by Mr Lewis Abbott from the Fissure near Ightham, Kent. *Quarterly Journal of the Geological Society* **50**: 189–211.

Newton, E.T. (1899) Additional notes on the vertebrate fauna of the rock-fissure at Ightham (Kent). *Quarterly Journal of the Geological Society* **55**: 419–429.

Niethammer, J. (1982) *Handbuch der Säugethiere Europas. Band 2(1) Nagethiere II.* Akademische Verlagsgesellschafte, Wiesbaden.

Noddle, B. (1975) The animal bones, pp. 332–339 in Platt, C. and Coleman-Smith, R. *Excavations in Medieval Southampton: 1953–69.* 2 vols. Leicester University Press, Leicester.

Noddle, B. (1976) Report on the animal bones from Walton, Aylesbury, pp. 269–287 in Farley, M. Saxon and Medieval Walton, Aylesbury: Excavations 1973–4. *Records of Buckinghamshire* **20**: 153–290.

Noddle, B. (1977) The animal bones from Buckquoy, Orkney, pp. 201–209 in Ritchie, A. Excavation of Pictish and Viking-age farmsteads at Buckquoy, Orkney. *Proceedings of the Society of Antiquaries of Scotland* **108**: 174–227.

Noddle, B. (1978) Mammalian bone report, pp.100–119 in Bateman, J. and Redknap, M. *Coventry: Excavations on the Town Wall.* Coventry Museums Monograph Series **2**, Coventry.

Noddle, B. (1980) Identification and interpretation of the mammal bones, pp. 377–409 in Wade-Martins, P. *Excavations in North Elmham Park.* East Anglian Archaeological Report **9**, Dereham.

Noddle, B. (1981) A comparison of mammalian bones found in the 'midden deposit' with others from the Iron Age site of Dun Bhuirg, pp. 38–44 in Reece, R. *Excavations on Iona 1964 to 1974.* Occasional Publication **5**, Institute of Archaeology, London.

Noddle, B. (1983) The animal bones from Knap of Howar, pp. 92–100 in Ritchie, A. Excavation of a Neolithic farmstead at Knap of Howar, Papa Westray, Orkney. *Proceedings of the Society of Antiquaries of Scotland* **113**: 40–121.

Noddle, B. (1985) The animal bones, pp. 84–94 in Shoesmith, R. *Hereford City Excavations. Vol. 3. The Finds.* CBA Research Reports **56**, York.

Noddle, B. (1987) *Animal Bones from Jarrow.* Unpub. Rep. Ancient Monuments Laboratory, London.

Noddle, B. (1989) Cattle and sheep in Britain and northern Europe up to the Atlantic Period: a personal viewpoint, pp. 179–202 in Milles, A., Williams, D. and Gardner, N. (eds) *The Beginnings of Agriculture.* BAR International Series **496**, Oxford.

Noddle, B. (1993) Bones of larger mammals, pp. 97–118 in Casey, P.J., Davies, J.L. and Evans, J. *Excavations at Segontium (Caernarfon) Roman Fort, 1975–1979.* CBA Research Report **90**, London.

Noe-Nygaard, N. (1983) A new find of Brown Bear (*Ursus arctos*) from Star Carr and other finds in the Late Glacial and Post Glacial of Britain and Denmark. *Journal of Archaeological Science* **10**: 317–325.

Norris, J.D. (1967a) The control of Coypus (*Myocastor coypus* Molina) by cage trapping. *Journal of Applied Ecology* **4**: 167–189.

Norris, J.D. (1967b) A campaign against feral Coypus (*Myocastor coypus* Molina) in Great Britain. *Journal of Applied Ecology* **4**: 191–199.

Oakley, K. (1980) Relative dating of the fossil hominids of Europe. *Bulletin of the British Museum (Natural History), Geology* **34**: 1–63.

O'Connor, T.P. (1977) Animal skeletal material, pp. 229–232 in Drewett, P. The excavation of a Neolithic Causwayed Enclosure on Offham Hill, East Sussex, 1976. *Proceedings of the Prehistoric Society* **43**: 201–241.

O'Connor, T.P. (1986) The Garden dormouse *Eliomys quercinus* from Roman York. *Journal of Zoology, London* **210**: 620–622.

O'Connor, T.P. (1987) Why bother looking at archaeological wild mammal assemblages? *Circaea* **4**: 107–114.

O'Connor, T.P. (1989) *Bones from the Anglo-Scandinavian levels at 16–22 Coppergate.* Archaeology of York **15/3**. CBA, London.

O'Connor, T.P. (1991) *Bones from 46–54 Fishergate.* Archaeology of York **15/4**. CBA, London.

O'Connor, T.P. (1992) Pets and pests in Roman and medieval Britain. *Mammal Review* **22**: 107–113.

Ogilvie, M.A. (1996) Rare breeding birds in the United Kingdom in 1994. *British Birds* **89**: 387–417.

Ognev, S.I. (1964) *Mammals of the U.S.S.R. and adjacent countries. Vol. 7. Rodents (part).* Israel Program for Scientific Translations, Jerusalem. (Russian Original, 1950).

Okarma, H., Jędrzejewski, B., Jędrzejewski, W. *et al.* (1995) The roles of predation, snow cover, acorn crop, and man-related factors on ungulate mortality in Białowieża Primeval Forest, Poland. *Acta Theriologica* **40**: 197–217.

Osborne, P.J. (1980) The late Devensian-Flandrian transition depicted by serial insect faunas from West Bromwich, Staffordshire, England. *Boreas* **9**: 139–147.

O'Sullivan, P.J. (1983) The distribution of the Pine Marten (*Martes martes*) in the Republic of Ireland. *Mammal Review* **13**: 39–44.

O'Sullivan, P. (1994) Bats in Ireland. *Irish Naturalists' Journal, Special Zoological Supplement 1994*: 1–24.

Owen, R. (1838) Description of the remains of Marsupial Mammalia from the Stonesfield Slate. *Proceedings of the Geological Society* **3**: 17–21.

Pemberton, J.M. (1993) The genetics of fallow deer production, pp. 129–135 in Asher, G.W. (ed.) *Proceedings of the 1st World Forum on Fallow Deer Farming.* Mudgee, Australia.

Pemberton, J.M., Swanson, G. and Goodman, S. (1998) Management of Scottish Deer in the face of Red-Sika Hybridisation (abstract), p. 118 in Goldspink, C.R., King, S. and Putman, R.J. (eds) *Population Ecology, Management and Welfare of Deer.* Manchester Metropolitan University, Manchester.

Pemberton, J.M. and Smith, R.H. (1985) Lack of biochemical polymorphism in British fallow deer. *Heredity* **55**: 199–207.

Pennington, W. (1977) The Late Devensian flora and vegetation of Britain. *Philosophical Transactions of the Royal Society of London, B.* **280**: 247–271.

Pernetta, J.C. (1973) The animal bones, pp. 112–114 in Robinson, M. Excavations at Copt Hay, Tetsworth, Oxford. *Oxoniensia* **38**: 41–115.

Pernetta, J.C. and Handford, P.T. (1970) Mammalian and avian remains from possible Bronze Age deposits on Nornour, Isles of Scilly. *Journal of Zoology, London,* **162**: 534–540.

Piggott (1962) *The West Kennet Long Barrow.* H.M.S.O., London.

Platt, M.I. (1956) The animal bones, pp. 212–215 in Hamilton, J.R.C. (ed.) *Excavations at Jarlshof, Shetland.* H.M.S.O., Edinburgh.

Platt, F.B.W. and Rowe, J.J. (1964) Damage by the edible dormouse (*Glis glis*) at Wendover Forest (Chilterns). *Quarterly Journal of Forestry* **58**: 228–233.

Poulle, M.-L., Dahier, T., Houard, T. *et al.* (1995) Wolves in France: a natural return. Unpub. Paper, 2nd European Congress of Mammalogy, Southampton.

Powers, R. and Stringer, C.B. (1975) Palaeolithic cave art fauna. *Studies in Speleology* **2**: 266–298.

Preece, R.C. (1986) Faunal remains from radiocarbon-dated soils within landslip debris from the Undercliff, Isle of Wight, Southern England. *Journal of Archaeological Science* **13**: 189–200.

Preece, R.C., Coxon, P. and Robinson, J.E. (1986) New biostratigraphic evidence of the Post-glacial colonization of Ireland and for Mesolithic forest disturbance. *Journal of Biogeography* **13**: 487–509.

Price, C. (1996) Evidence from Holocene and Late Pleistocene small mammal remains, pp. 188–189 in Charman, D.J., Newnham, R.M. and Croot, D.G. (eds) *Devon and East Cornwall: Field Guide.* Quaternary Association, London.

Proctor, C.J., Colcutt, A.P., Currant, A.P. *et al.* (1996) A report on the excavations at Rhinoceros Hole, Wookey. *Proceedings of the University of Bristol Speleological Society* **20**: 237–262.

Pucek, Z., Jędrzejewski, W., Jędrzejewski, B. and Pucek, M. (1993) Rodent population dynamics in a primeval deciduous forest (Białowieża National Park) in relation to weather, seed crop and predation. *Acta Theriologica* **38**: 199–232.

Putman, R.J. (1986) *Grazing in Temperate Ecosystems – large hebivores and the ecology of the New Forest.* Croom Helm, London.

Putman, R.J., Langbein, J., Hewson, A.J.M. and Sharma, S.K. (1996) Relative roles of density-dependent and density-independent factors in population dynamics of British deer. *Mammal Review* **26**: 81–102.

Putman, R.J. and Moore, N.P. (1998) Impact of deer in lowland Britain on agriculture, forestry and conservation habitats. *Mammal Review* **28**: 165–184.

Pyle, C.M. (1994) Some late sixteenth-century depictions of the aurochs (*Bos primigenius* Bojanus, extinct 1627): new evidence from Vatican MS Urb.lat.276. *Archives of Natural History* **21**: 275–288.

Racey, P.A. and Swift, S.M. (1986) The residual effects of remedial timber treatment on bats. *Biological Conservation* **35**: 205–214.

Rackham, D.J. (1979) *Rattus rattus*: the introduction of the black rat into Britain. *Antiquity* **53**: 112–120.

Rackham, D.J. (1989) Animal bones, pp.197–199 in Daniels, R. (ed.) The Anglo-Saxon monastery at Church Close, Hartlepool, Cleveland. *Archaeological Journal* **145**: 158–210.

Rackham, O. (1976) *Trees and Woodland in the British Landscape.* Dent, London.

Rackham, O. (1980) *Ancient Woodland: its history, vegetation and uses in England.* Edward Arnold, London.

Rackham, O. (1986) *The History of the Countryside.* Dent, London.

Rackham, O. (1990) *Trees and Woodland in the British Landscape*, 2nd edn. Dent, London.

Rackham, O. (1994) Trees and woodlands in Anglo-Saxon England: the documentary evidence. pp. 7–11 in Rackham, J. (ed.) *Environment and economy in Anglo-Saxon England.* CBA Research Report **89**, York.

Rahmstorf, S (1995) Bifurcations of the Atlantic thermohaline circulation in response to changes in the hydrological cycle. *Nature, London,* **378**: 145–149.

Rankine, W.F. (1961) The Mesolithic age in Dorset and adjacent regions. *Proceedings of the Dorset Natural History and Archaeological Society* **83**: 91–99.

Ransome, R.D. (1989) Population changes of Greater Horseshoe bats studied near Bristol over the past twenty-six years. *Biological Journal of the Linnean Society* **38**: 71–82.

Ratcliffe, P.R. (1987) Distribution and status of Sika Deer, *Cervus nippon*, in Great Britain. *Mammal Review* **17**: 39–58.

Ratcliffe, P.R. (1989) The control of red and sika deer populations in commercial forests, pp. 98–115 in Putman, R.J. (ed.) *Mammals as Pests*, Chapman and Hall, London.

Rees, W. (1933) *The Historical Map of South Wales and the Border in the Fourteenth Century.* University Press, Cardiff.

Reynolds, J.C. (1985) Details of the geographic replacement of the Red Squirrel (*Sciurus vulgaris*) by the Grey Squirrel (*Sciurus carolinensis*) in eastern England. *Journal of Animal Ecology* **54**: 149–162.

Reynolds, S.H. (1906) *A monograph of the British Pleistocene Mammalia. Vol. II, Part II. The Bears.* Palaeontographical Society, London.

Reynolds, S.H. (1909) *A monograph of the British Pleistocene Mammalia. Vol. II, Part III. The Canidae.* Palaeontographical Society, London.

Reynolds, S.H. (1912) *A monograph of the British Pleistocene Mammalia. Vol. II, Part IV. The Mustelidae.* Palaeontographical Society, London.

Reynolds, S.H. (1929) *A monograph of the British Pleistocene Mammalia. Vol. III, Part III. The Giant Deer.* Palaeontographical Society, London.

Reynolds, S.H. (1934) *A monograph of the British Pleistocene Mammalia. Vol. III, Part VIIa.* Alces (*supplement*). Palaeontographical Society, London.

Reynolds, S.H. (1939) *A monograph of the British Pleistocene Mammalia. Vol. III, Part VI. The Bovidae.* Palaeontographical Society, London.

Ritchie, J. (1920) *The Influence of Man on Animal Life in Scotland.* Cambridge University Press, London.

Roberts, M.B. (1986) Excavation of the Lower Palaeolithic Site at Amey's Earlham Pit, Boxgrove, West Sussex: a preliminary report. *Proceedings of the Prehistoric Society* **52**: 215–245.

Roberts, M.B., Stringer, C.B. and Parfitt, S.A. (1994) A hominid tibia from the Middle Pleistocene sediments of Boxgrove, U.K. *Nature, London* **369**: 311–312.

Roberts, K.A. and Roberts, P.L.E. (1978) Postglacial deposits at Stanstead Abbots. *London Naturalist* **57**: 6–10.

Robertson-Mackay, R. (1987) The Neolithic causewayed enclosure at Staines, Surrey: excavations 1961–63. *Proceedings of the Prehistoric Society* **53**: 23–128.

Rope, G.T. (1885) On the range of the dormouse in England and Wales. *Zoologist*, 3rd Series, **9**: 202–213.

Rowley, I. (1970) Lamb predation in Australia: incidence, predisposing conditions, and the identification of wounds. *CSIRO Wildlife Research* **15**: 79–123.

Ruddiman, W.F., Sancetta, C.D. and McIntyre, A. (1977) Glacial/interglacial response rate of subpolar North Atlantic waters to climatic change: the record in oceanic sediments. *Philosophical Transactions of the Royal Society of London, B* **280**: 119–141.

Ryan, A., Duke, E. and Fairley, J.S. (1996) Mitochondrial DNA in bank voles *Clethrionomys glareolus* in Ireland: evidence far a small founder population and localized founder effects. *Acta Theriologica* **41**: 45–50.

Ryder, M.L. (1971) The animal remains from Petergate, York, 1957–58. *Yorkshire Archaeological Journal* **46**: 418–428.

Sadler, P. (1990) Faunal remains, pp. 462–506 in Fairbrother, J.R. *Faccombe Netherton. Excavation of a Saxon and Medieval Complex*. British Museum, Occasional Paper **74**, London.

Saez-Royuela, C. and Telleria, J.L. (1986) The increased population of the Wild Boar (*Sus scrofa* L.) in Europe. *Mammal Review* **16**: 97–101.

Sainsbury, A.W., Nettleton, P. and Gurmell, J. (1997) Recent developments in the study of parapoxvirus in red and grey squirrels, pp. 105–108 in Gurnell, J. and Lurz, P. (eds) *The Conservation of Red Squirrels,* Sciurus vulgaris *L.* People's Trust for Endangered Species, London.

Saint-Girons, M-C. (1981) Notes sur les mammifères de France 17. Influence d'un été particulièrement sec et chaud sur la dynamique des populations des micro-insectivores. *Mammalia* **45**: 514–515.

Sargent, G. (1995) *The Bats in Churches Project*. Bat Conservation Trust, London.

Savage, R.J.G. (1966) Irish Pleistocene Mammals. *Irish Naturalists' Journal* **15**: 117–130.

Savage, R.J.G. (1969) Pleistocene Mammal Faunas. *Proceedings of the University of Bristol Speleological Society* **12**: 57–62.

Savage, R.J.G. (1989) British mammals of the Mesozoic era. *Biological Journal of the Linnean Society* **38**: 3–7.

Savory, C.J. (1986) Utilisation of different ages of heather on three Scottish moors by red grouse, mountain hares, sheep and red deer. *Holarctic Ecology* **9**: 65–71.

Scaife, R.G. (1982) Late Devensian and early Flandrian vegetation changes in Southern England, pp. 57–74 in Bell, M. and Limbrey, S. (eds) *Archaeological Aspects of Woodland Ecology*. BAR International Series **146**, Oxford.

Scaife, R.G. (1988) The Elm decline in the pollen record of South East England and its relationship to early agriculture, pp. 21–33 in Jones, M. (ed.) *Archaeology and the Flora of the British Isles*. Oxford University Committee for Archaeology, Monograph **14**, Oxford.

Schadla-Hall, R.J. (1988) The early post-glacial in eastern Yorkshire, pp. 25–34 in Manby, T.G. (ed.) *Archaeology in Eastern Yorkshire. Essays in honour of T.C.M. Brewster*. Department of Archaeology and Prehistory, University of Sheffield.

Schaller, G.B. (1977) *Mountain Monarchs*. University of Chicago Press, Chicago.

Schonfelder, M. (1994) Bear-claws in Germanic graves. *Oxford Journal of Archaeology* **13**: 217–227.

Scott, K. (1986a) Man in Britain in the Late Pleistocene: evidence from Ossom's Cave, pp. 63–87 in Roe, D.A. (ed.) *Studies in the Upper Palaeolithic of Britain and Northwest Europe*. BAR International Series **296**, Oxford.

Scott, K. (1986b) The large mammal fauna, pp. 109–137 in Callow, P. and Cornford, J.M. (eds) *La Cotte de St. Brelade 1961–1978. Excavations by C.B.M. McBurney*. Geo Books, Norwich.

Searle, W.G. (1897) *Onomasticum Anglo-Saxonicum*. Cambridge University Press, Cambridge.

Shackleton, N.J. (1977) The oxygen isotope record of the Late Pleistocene. *Philosophical Transactions of the Royal Society of London, B* **280**: 169–182.

Shackleton, N.J., Berger, A. and Peltier, W.R. (1991) An alternative astronomical calibration of the lower Pleistocene timescale based on ODP site 677. *Transactions of the Royal Society of Edinburgh* **81**: 252–261.

Shaw, R.C. (1956) *The Royal Forest of Lancaster*. Guardian Press, Preston.

Shaw, G. and Livingstone, J. (1992) The pine marten – its reintroduction and subsequent history in the Galloway Forest Park. *Transactions of the Dumfries and Galloway Natural History and Antiquarian Society*, 3rd Series, **67**: 1–7.

Sheail, J. (1971) *Rabbits and their History*. David and Charles, Newton Abbot.

Sheail, J. (1988) The extermination of the muskrat (*Ondatra zibethicus*) in inter-war Britain. *Archives of Natural History* **15**: 155–170.

Sheppard, T. (1903) Beavers in East Yorkshire. *Naturalist* **1903**: 109–110.

Sheppard, T. (1922) Vertebrate remains from the peat of Yorkshire: new remains. *Naturalist* **1922**: 187–188.

Shimwell, DW. (1977) Studies in the history of the Peak District landscape: 1. Pollen analyses of some podzolic soils on the limestone plateau. *University of Manchester School of Geography Research Paper* **3**.

Shore, R.F, Boyd, I.L., Leach, D.V., *et al.* (1990) Organochlorine residues in roof timbers and possible implications for bats. *Environmental Pollution* **64**: 179–188.

Shore, R.F, Myhill, D.G., French, M.C., *et al.* (1991) Toxicity and tissue distribution of pentachlorophenol and permethrin in Pipistrelle bats experimentally exposed to treated timber. *Environmental Pollution* **73**: 101–118.

Shorten, M. (1946) A survey of the distribution of the American grey squirrel (*Sciurus carolinensis*) and the British red squirrel (*Sciurus vulgaris leucourus*) in England and Wales. *Journal of Animal Ecology* **15**: 82–92.

Shorten, M. (1954) *Squirrels*. Collins, New Naturalist Monograph, London.

Shuttleworth, C. (1997) The effect of supplemental feeding on the diet, population density and reproduction of red squirrels (*Sciurus vulgaris*), pp. 13–24 in Gurnell, J. and Lurz, P. (eds) *The Conservation of Red Squirrels,* Sciurus vulgaris *L.* People's Trust for Endangered Species, London.

Sidorovich, V.E., Jędrzejewski, B. and Jędrzejewski, W. (1996) Winter distribution and abundance of mustelids and beavers in the river valleys of Białowieża Primeval Forest. *Acta Theriologica* **41**: 155–170.

Simmons, I. (1995) The history of the human environment, pp. 5–15 in Vyner, B. *Moorland Monuments: studies in the archaeology of north-east Yorkshire in honour of Raymond Hayes and Don Spratt*. CBA Research Report 101, York.

Simmons, I.G and Tooley, M.J. (1981) *The Environment in British Prehistory*. Duckworth, London.

Simms, C. (1972) Some North Yorkshire recent bone sites. *Naturalist* **1972**: 113–114.

Simpson, G.G. (1928) *A Catalogue of the Mesozoic Mammalia in the Geological Department of the British Museum*. British Museum (Natural History), London.

Skelcher, G. (1997) The ecological replacement of red by grey squirrels, pp. 67–78 in Gurnell, J. and Lurz, P. (eds) *The Conservation of Red Squirrels,* Sciurus vulgaris *L.*. People's Trust for Endangered Species, London.

Sleeman, D.P. (1986) Ireland's carnivorous mammals – problems with their arrival and survival. *Occasional Publications of the Irish Biogeographical Society* **1**: 42–48.

Smal, C. (1995) *The Badger and Habitat Survey of Ireland*. Stationery Office, Dublin.

Smal, C.M. and Fairley, J.S. (1984) The spread of the Bank vole *Clethrionomys glareolus* in Ireland. *Mammal Review* **14**: 71–78.

Smit, C.J. and van Wijngaarden, A. (1981) *Threatened Mammals of Europe*. Supplementary Volume to *Handbuch der Säugethiere Europas*. Akademische Verlagsgesellschafte, Wiesbaden.

Smith, J.A. (1872) Notice of the discovery of remains of the elk (*Cervus alces* Linn., *Alces machlis* Gray) in Berwickshire; with notes on its occurrence in the British Islands, more particularly in Scotland. *Proceedings of the Society of Antiquaries of Scotland* **9**: 297–350.

Southern H.N. (1954) Tawny Owls and their prey. *Ibis* **96**: 384–410.

Southern H.N. (1968) *The Handbook of British Mammals*, 1st Edn. Blackwell's, Oxford.

Southern H.N. (1970) The natural control of a population of tawny owls (*Strix aluco*). *Journal of Zoology, London,* **162**: 197–285.

Stallibrass, S.M. (1992) *Animal bones from excavations at Lewthwaites Lane, Crown and Anchor Lane and Old Bush Lane, Carlisle, 1982*. Ancient Monuments Laboratory Report **82/92**, London.

Stallibrass, S.M. (1995) Animal bone, pp. 220–224 in Cardwell, P. Excavation of the hospital of St. Giles by Brompton Bridge, North Yorkshire. *Archaeological Journal* **152**: 109–245.

Stebbings, R.E. (1988) *Conservation of European Bats*. Christopher Helm, London.

Stebbings, R.E. and Arnold, H.R. (1987) Assessment of the trends in size and structure of a colony of the greater horseshoe bat. *Symposia of the Zoological Society of London* **58**: 7–24.

Strachan, C., Jefferies, D.J., Burreto, G.R., Macdonald, D.W. and Strachan, R. (1998) The rapid impact of resident American Mink on water voles: case studies from Lowland England. *Symposia of the Zoological Society of London* **71**: 339–357.

Strachan, R. and Jefferies, D.J. (1993) *The Water Vole* Arvicola terrestris *in Britain 1989–1990: its distribution and changing status*. Vincent Wildlife Trust, London.

Strachan, R. and Jefferies, D.J. (1996) *Otter survey of England 1991–1994).* Vincent Wildlife Trust, London.

Straus, L.G. (1983) Terminal Pleistocene faunal exploitation in Cantabria and Gascony, pp. 209–226 in Clutton-Brock, J. and Grigson, C. (eds) *Animals and Archaeology. 1. Hunters and their Prey.* BAR International Series **163**, Oxford.

Straus, L.G. (1986) The end of the Palaeolithic in Cantabrian Spain and Gascony, pp. 81–116 in Straus, L.G. (ed) *The end of the Palaeolithic in the Old World.* BAR International Series **284**, Oxford.

Stuart, A.J. (1982) *Pleistocene Vertebrates in the British Isles.* Longman. London and New York.

Stuart, A.J. (1983) Pleistocene bone caves in Britain and Ireland. *Studies in Speleology* **4**: 9–36.

Stuart, A.J. (1986) Pleistocene mammals in Ireland (pre-10 000 years B.P.). *Occasional Publications of the Irish Biogeographical Society* **1**: 28–33.

Stuart, A.J. (1991) Mammalian extinctions in the Late Pleistocene of northern Eurasia and North America. *Biological Reviews* **66**: 453–562.

Stuart, A.J. (1995) Insularity and Quaternary vertebrate faunas in Britain and Ireland, pp. 111–125 in Preece, R.C. (ed.) *Island Britain: a Quaternary perspective.* Geological Society, Spec. Publ. **96**, London.

Stuart, A.J. and Wijngaarden-Bakker, L.H. van (1985) Quaternary vertebrates, pp. 221–249 in Edwards, K.J. and Warren, W.P. *The Quaternary History of Ireland.* Academic Press, London.

Summers, C.F. (1978) Trends in the size of British grey seal populations. *Journal of Applied Ecology* **15**: 395–400.

Sumption, K.J. and Flowerdew, J.R. (1985) The ecological effects of the decline in rabbits (*Oryctolagus cuniculus*) due to myxomatosis. *Mammal Review* **15**: 151–186.

Sutcliffe, A.J. (1995) Insularity of the British Isles 250,000–30,000 years ago: the mammalian, including human, evidence, pp. 127–140 in Preece, R.C. (ed.) *Island Britain: a Quaternary perspective.* Geological Society, Spec. Publ. **96**, London.

Sutcliffe, A.J. and Kowalski, K. (1976) Pleistocene rodents of the British Isles. *Bulletin of the British Museum (Natural History), Geology Series* **27**: 31–147.

Sutherland, S. (1983) Microfauna identified in the sieve residues from the floor of ST3, pp. 149–150 in Hedges, J.W. (ed.) *Isbister: A Chambered Tomb in Orkney.* BAR British Series **115**, Oxford.

Tallis, J.H. (1991) Forest and Moorland in the South Pennine Uplands in the Mid-Flandrian Period. III The spread of Moorland – Local, Regional and National. *Journal of Ecology* **79**: 401–415.

Tallis, J.H. and Switsur, V.R. (1990) Forest and moorland in the South Pennine Uplands in the Mid-Flandrian Period. II The Hillslope Forests. *Journal of Ecology* **78**: 857–883.

Tapper, S. (1992) *Game Heritage.* Game Conservancy, Fordingbridge.

Tapper, S., Brockless, M. and Potts, D. (1991) The Salisbury Plain predation experiment: the conclusion. *The Game Conservancy Review of 1990*: 87–91.

Tapper, S., Potts, G.R. and Brockless, M.H. (1996) The effect of an experimental reduction in predation pressure on the breeding success and population density of grey partridges *Perdix perdix*. *Journal of Applied Ecology* **33**: 965–978.

Taylor, I. (1911) *Words and Places.* Dent, London.

Taylor, J.C. and Blackmore, D.K. (1961) A short note on the heavy mortality in foxes during the winter 1959–60. *Veterinary Record* **73**: 232–233.

Taylor Page, F.J. (1962) *Roe Deer.* Sunday Times, London.

Tchernov, E. (1984) Commensal animals and human sedentism in the Middle East. In Clutton-Brock, J. and Grigson, C. (eds) *Animals and Archaeology: 3. Early Herders and their Flocks.* BAR International Series **202**, Oxford, pp. 91–115.

Teer, J.G., Nerenov, V.M., Zhirnov, L.V. and Blizniuk, A.I. (1996) Status and exploitation of the saiga antelope in Kalmykia. In Taylor, V.J. and Dunstone, N. *The Exploitation of Mammal Populations.* Chapman and Hall, London, pp. 75–87.

Temple, R.K. and Morris, P.A. (1998) The Lesser White-toothed Shrew on the Isles of Scilly. *British Wildlife* **9**: 94–99.

Thaler, L, Bonhomme, F. and Britton-Davidian, J. (1981) Processes of speciation and semi-speciation in the House Mouse. *Symposium of the Zoological Society of London* **47**: 27–41.

Thompson, H.V. (1953) The edible dormouse (*Glis glis* L.) in England, 1902–1951. *Proceedings of the Zoological Society of London* **122**: 1017–1024.

Thompson, H.V. and Worden, A.N. (1956) *The Rabbit.* Collins, London.

Thompson, P.M. and Hall, A.J. (1993) Seals and epizootics – what factors might affect the severity of mass mortalities? *Mammal Review* **23**: 149–154.

Tipping, R. (1991) Climatic change in Scotland during the Devensian Late Glacial: the palynological record, pp. 7–21 in Barton, N., Roberts, A.J. and Roe, D.A. (eds) *The Late Glacial in north-west Europe.* CBA Research Report. **77**, London.

Tittensor, A.M. and Tittensor, R.M. (1985) The Rabbit Warren at West Dean near Chichester. *Sussex Archaeological Collections* **123**: 151–185.

Toynbee, J.M.C. (1973) *Animals in Roman Life and Art.* Thames and Hudson, London.

Trout, R.C. and Tittensor, A.M. (1989) Can predators regulate wild Rabbit *Oryctolagus cuniculus* population density in England and Wales? *Mammal Review* **19**: 153–173.

Tubbs, C.R. (1974) *The Buzzard.* David and Charles, Newton Abbot.

Tubbs, C.R. (1986) *The New Forest.* Collins, New Naturalist **73**, London.

Tupling, G.H. (1927) *The Economic History of Rossendale.* University Press, Manchester.

Turk, F.A. (1964) Blue and brown hares associated together in a Bronze Age fissure cave burial. *Proceedings of Zoological Society of London* **142**: 185–188.

Turk, F.A. (1978) The animal remains from Nornour: a synoptic view of the finds, pp. 99–103 in Butcher, S.A. (ed.) Excavations at Nornour, Isles of Scilly, 1969–73: the pre-Roman settlement. *Cornish Archaeology* **17**: 29–112.

Turner, A. (1995) Evidence for contact between the British Isles and the European continent based on the distribution of larger carnivores, pp. 141–149 in Preece, R.C. (ed.) *Island Britain: a Quaternary perspective.* Geological Society, Spec. Publ. **96**, London.

Twigg, G.I. (1978) The role of rodents in plague dissemination: a worldwide review. *Mammal Review* **8**: 77–110.

Twigg, G.I. (1992) The Black Rat *Rattus rattus* in the United Kingdom in 1989. *Mammal Review* **22**: 33–42.

Tyers, I., Hillam, J. and Groves, C. (1994) Trees and woodland in the Saxon period: the dendrochronological evidence, pp. 12–22 in Rackham, J. (ed.) *Environment and economy in Anglo-Saxon England.* CBA Research Report **89**, York.

Tyrer, N. (1997) It's back – and this time it's hungry. *Daily Telegraph* 22 Feb. 1997.

Tzedakis, P.C. (1993) Long-term tree populations in northwest Greece through multiple Quaternary climatic cycles. *Nature, London* **364**: 437–440.

Uerpmann, H.-P. (1987) *The Ancient Distribution of Ungulate Mammals in the Middle East.* Reichert, Wiesbaden.

Vartanyan, S.L., Garutt, V.E. and Sher, A.V. (1993) Holocene dwarf mammoths from Wrangel Island in the Siberian Arctic. *Nature, London* **362**: 337–340.

Vaughan, N. (1997) The diets of British bats (Chiroptera). *Mammal Review* **27**: 77–94.

Velander, K.A. (1983) *Pine Marten Survey of Scotland, England and Wales 1980–1982.* Vincent Wildlife Trust, London.

Vereshchagin, N.K. and Baryshnikov, G.F. (1982) Paleoecology of the Mammoth Fauna in the Eurasian Arctic, pp. 267–280 in Hopkins, D.M., Matthews, J.V., Schweger, C.E. and Young, S.B. (eds) *Paleoecology of Beringia.* Academic Press, New York.

Vigne, J.D. (1992) Zooarchaeology and the biogeographical history of the Mammals of Corsica and Sardinia, since the last Ice Age. *Mammal Review* **22**: 87–96.

Vilà, C., Savolainen, P., Maldonado, J.E., Amorim, I.R., *et al.* (1997) Multiple and ancient origins of the domestic dog. *Science* **276**: 1687–1689.

Vince, A. (1994) Saxon urban economies: an archaeological perspective, pp. 108–119 in Rackham, J. (ed.) *Environment and economy in Anglo-Saxon England.* CBA Research Report **89**, York.

Walker, D. and Godwin, H. (1954) Lake stratigraphy, pollen-analysis and vegetational history, pp. 25–69 in Clark, J.G.D. (ed.) *Excavations at Star Carr.* Cambridge University Press.

Walsh, A. and Harris, S. (1996a). Foraging preferences of vespertilionid bats in Britain. *Journal of Applied Ecology* **33**: 508–518.

Walsh, A. and Harris, S. (1996b) Factors determining the abundance of vespertilionid bats in Britain: geographical, land class and local habitat relationships. *Journal of Applied Ecology* **33**: 519–529.

Walton, K.C. (1964) The distribution of the polecat (*Putorius putorius*) in England, Wales and Scotland, 1959–62. *Proceedings of the Zoological Society of London* **143**: 333–336.

Walton, K.C. (1968) The distribution of the polecat *Putorius putorius* in Great Britain, 1963–67. *Journal of Zoology, London* **155**: 237–240.

Warwick, T. (1934) The distribution of the Muskrat (*Fiber zibethicus*) in the British Isles. *Journal of Animal Ecology* **3**: 250–267.

Warwick, T. (1941) A contribution to the ecology of the muskrat in the British Isles. *Proceedings of the Zoological Society of London* **A110**: 165–201.

Waton, P.V. (1982) Man's impact on the chalklands: some new pollen evidence, pp. 75–92 in Bell, M. and Limbrey, S. *Archaeological Aspects of Woodland Ecology*. BAR International Series **146**, Oxford.

Watson, D.M.S. (1931) The animal bones from Skara Brae, pp. 198–204 in Childe, V.G. *Skara Brae, a Pictish Village in Orkney*. Kegan Paul. Trench, Trubner and Co., London.

Webster, G., Fowler, P., Noddle, B. and Smith, L. (1985) The excavation of the Romano-British rural establishment at Barnsley Park, Gloucestershire, 1961–79: Part 3. *Transactions of the Bristol and Gloucestershire Archaeological Society* **103**: 73–100.

West, B. (1993) Birds and mammals from the Peabody Site and National Gallery, pp. 150–168 in Whytehead, R.L. and Cowie, R. Excavations at the Peabody Site, Chandos Place, and the National Gallery. *Transactions of the London and Middlesex Archaeological Society* **40**: 35–176.

West, R.G. (1969) Pollen analyses from interglacial deposits at Aveley and Grays, Essex. *Proceedings of the Geologists' Association* **80**: 271–282.

West, R.G. (1977) *Pleistocene Geology and Biology, with especial reference to the British Isles*. 2nd edn. Longman, London.

White, G. (1788) *The Natural History of Selborne*. B. White, London. (Penguin Paperback Edition, 1977, ed. R. Mabey).

Whitehead, G.K. (1953) *The Ancient White Cattle of Britain and their Descendants*. Faber and Faber, London.

Whitehead, G.K. (1963) *Ancient White Cattle*. Sunday Times Publications, London.

Whitehead, G.K. (1964) *The Deer of Great Britain and Ireland*. Routledge and Kegan Paul, London.

Whitelock, D. (1961) *The Anglo-Saxon Chronicle*. Eyre and Spottiswoode, London.

Wiertz, J. (1993) Fluctuations in the Dutch Badger *Meles meles* population between 1960 and 1990. *Mammal Review* **23**: 59–64.

Wijngaarden-Bakker, L.H. van (1974) The animal remains from the Beaker Settlement at Newgrange, Co. Meath: a first report. *Proceedings of the Royal Irish Academy* **74C**: 313–383.

Wijngaarden-Bakker, L.H. van (1986) The colonization of islands. The mammalian evidence from Irish archaeological sites. *Occasional Publications of the Irish Biogeographical Society* **1**: 38–41.

Willis, K. (1996) Where did all the flowers go? The fate of temperate European flora during glacial periods. *Endeavour* **20**: 110–114.

Wilson, D. (1992) *Anglo-Saxon Paganism*. Routledge, London.

Wilson, G., Harris, S. and McLaren, G. (1997) *Changes in the British Badger Population, 1988 to 1997*. People's Trust for Endangered Species, London.

Wilson, R. (1975) Excavations in Abingdon, 1972–74. Animal bones from the Broad Street and Old Gaol sites. *Oxoniensia* **40**: 105–121.

Wilson, R., Locker, A. and Marples, B. (1989) Medieval animal bones and marine shells from Church Street and other sites in St. Ebbes, Oxford. *Oxoniensia* **54**: 258–268.

Wilson, R. (1993) Report on the bone and oyster shell, pp. 123–133 in Allen, T.G. and Robinson, M.A. *The Prehistoric Landscape and Iron Age Enclosed Settlement at Mingies Ditch, Hardwick-with Yelford, Oxon*. Oxford University Committee for Archaeology, Oxford.

Wingfield, R.T.R. (1995) A model of sea-level changes in the Irish and Celtic seas during the end-Pleistocene to Holocene transition, pp. 209–242 in Preece, R.C. (ed.) *Island Britain: a Quaternary perspective*. Geological Society, Spec. Publ. **96**, London.

Wolz, I. (1993) Das Beutespektrum der Bechsteinfledermaus *Myotis bechsteini* (Kuhl 1818) ermitteltaus Kotanalysen. *Myotis* **31**: 27–68.

Woodman, P.C. (1986) Man's first appearance in Ireland and his importance in the colonization process. *Occasional Publications of the Irish Biogeographical Society* **1**: 34–37.

Woodman, P.C. and Monaghan, N. (1993) From mice to mammoths: dating Ireland's earliest faunas. *Archaeology Ireland* **7**: 31–33.

Woodman, P., McCarthy, M. and Monaghan, N. (1997) The Irish Quaternary fauna project. *Quaternary Science Reviews* **16**: 129–159.

Woodroffe, G.L., Lawton, J.H. and Davidson, W.L. (1990) The impact of Feral Mink *Mustela vison* on Water Voles *Arvicola terrestris* in the North Yorkshire Moors National Park. *Biological Conservation* **51**: 49–62.

Wright, T. (1884) *Anglo-Saxon – Old English Vocabulary*. Trubner, London.

Wymer, J. (1962) Excavations at the Maglemosian sites at Thatcham, Berkshire, England. *Proceedings of the Prehistoric Society* **28**: 255–361.

Yalden, D.W. (1981) The occurrence of the pygmy shrew *Sorex minutus* on moorland, and the implications for its presence in Ireland. *Journal of Zoology, London*, **195**: 147–156.

Yalden, D.W. (1982) When did the mammal fauna of the British Isles arrive? *Mammal Review* **12**: 1–57.

Yalden, D.W. (1984a) The yellow-necked mouse, *Apodemus flavicollis,* in Roman Manchester. *Journal of Zoology, London* **203**: 285–288.

Yalden, D.W. (1984b) The status of the mountain hare *Lepus timidus* in the Peak District. *Naturalist* **109**: 55–58.

Yalden, D.W. (1985) Dietary separation of owls in the Peak District. *Bird Study* **32**: 122–131.

Yalden, D.W. (1986) Opportunities for reintroducing British mammals. *Mammal Review* **16**: 53–63.

Yalden, D.W. (1987) The natural history of Domesday Cheshire. *Naturalist* **112**: 125–131.

Yalden, D.W. (1988) Feral wallabies in the Peak District, 1971–1985. *Journal of Zoology, London,* **215**: 369–374.

Yalden, D.W. (1991) History of the fauna, ch. 2 in Corbet, G.B. and Harris, S. (eds) *Handbook of British Mammals,* 3rd edn. Blackwell Science, Oxford.

Yalden, D.W. (1992) Changing distribution and status of small mammals in Britain. *Mammal Review* **22**: 97–106.

Yalden, D.W. (1995) Small mammals from Viking-age Repton. *Journal of Zoology, London* **237**: 655–657.

Yalden, D.W. (1996a) Historical dichotomies in the exploitation of mammals, pp. 16–27 in Taylor, V.J. and Dunstone, N. *The Exploitation of Mammal Populations.* Chapman and Hall, London.

Yalden, D.W. (1996b) Small mammals from Viking-age Repton. *Journal of Zoology, London* **237**: 655–657.

Yalden, D.W. and Shore, R.F. (1991) Yellow-necked mice *Apodemus flavicollis* at Woodchester Park, 1968–1989. *Journal of Zoology, London* **224**: 329–332.

Yalden, D.W., Morris, P.A. and Harper, J.H. (1973) Studies on the comparative ecology of some French small mammals. *Mammalia* **37**: 257–276.

Yapp, W.B. (1981) Game-birds in Medieval England. *Ibis* **125**: 218–221.

Younger, D.A. (1994) The small mammals from the forecourt granary and the south west fort ditch, pp. 266–268 in Bidwell, P. and Speak, S. (eds) *Excavations at South Shields Roman Fort.* Vol. 1. Society of Antiquaries of Newcastle-upon-Tyne Monograph **4**.

INDEX